D1657236

Green Energy and Technology

More information about this series at http://www.springer.com/series/8059

Walter Leal Filho · Richard Kotter
Editors

E-Mobility in Europe

Trends and Good Practice

Editors
Walter Leal Filho
Faculty of Life Sciences
Hamburg University of Applied Sciences
Hamburg
Germany

Richard Kotter
Department of Geography
Northumbria University
Newcastle upon Tyne
UK

ISSN 1865-3529 ISSN 1865-3537 (electronic)
Green Energy and Technology
ISBN 978-3-319-13193-1 ISBN 978-3-319-13194-8 (eBook)
DOI 10.1007/978-3-319-13194-8

Library of Congress Control Number: 2015937204

Springer Cham Heidelberg New York Dordrecht London

Printed on acid-free paper

Springer International Publishing AG Switzerland is part of Springer Science+Business Media (www.springer.com)

Preface

As European countries face growing pressure to reduce their CO_2 emissions as well as urban air pollution, and generally to find ways to promote sustainable means of transport, a window of opportunity has been opened for clean and more sustainable forms of mobility. Various stakeholders such as supra-national and inter-governmental organisations, governments and their appointed agencies, regional and municipal authorities, private and public transit operators, vehicle manufacturers and the general public have become increasingly aware of and interested in the many possibilities e-mobility offers, both as a tool for reducing emissions of greenhouse gases and pollution, but also as a new (at least in recent times) and interesting type of mobility.

The field of e-mobility as a whole, and the production and deployment of electrical vehicles (EV) in particular, are now widely seen as viable alternatives to traditional means of automotive production and transport/logistics provision, based on the fact that a range of tailpipe emissions and noise can be reduced. Pure EVs, but also hybrid vehicles, are beginning to sell quite well and are progressively taking their places in the mass market, no longer restricted to a few users. Indeed, in countries such as Denmark, Germany, France, Netherlands, Norway, Portugal, Spain, Sweden, the UK, Japan, South Korea, the US and China and Brazil, electromobility is now an established and growing field.

The advantages of e-mobility are many. First of all, this type of mobility is clean at the tailpipe end, with neither carbon, nor noxious gas emissions. Second, it is quiet, i.e. with no noise, a useful feature in inner cities and residential areas. Third, it is the convenience offered by the possibility of having small, compact cars, which take less space and are easier to handle, especially in large cities, and also city distribution (and potentially over longer distances) logistics solutions that can benefit from the above advantages. The challenge of fully decarbonizing and cleaning of electric vehicles in terms of production input and reuse or recycling of electric vehicles depends largely on the energy mix which provides the electric power source, as well as the components input and the manufacturing, reuse and recycling methods. Electric vehicles are also argued and predicted to be able to make a contribution to the electricity smart grid management (vehicle to grid, and grid to vehicle), as well as interfacing with renewable power sources in a smart grid and vehicles to home solutions.

The growth and expansion of e-mobility has reiterated the need for a wider understanding of its many features. Therefore, a publication which documents and promotes trends in the various areas was deemed as useful, and led to the preparation of this volume.

This book is a contribution to address the many information needs seen in the e-mobility area, and is aimed at showcasing how e-mobility is seen and practiced across the North Sea Region and beyond. It is structured along three parts:

Part I—Policy Frameworks and Decision-Making on EV Adoption and Charging Infrastructure Development

Part II—Regional and City Case Studies on E-Mobility Development

Part III—Technological Advancements and User-Friendly Strategies

This book is also an attempt to offer a platform for a range of actors working in different sectors and areas in the e-mobility sectors, such as in R&D, marketing and in applied projects, to share their knowledge about their experiences and limitations in pursuing e-mobility developments.

Technical, but also organizational, economic, political as well as spatial planning elements are highlighted throughout the book, alongside the need for improving knowledge and management of infrastructure. The case studies and experiences involving countries as varied as Denmark, Germany, Netherlands, the UK and Sweden, as well as California, and the reflection and systematic analyses offered by other authors, offer a concrete view of the issues on the ground.

We would like to thank all authors for sharing their knowledge and experience in this book. We hope that the body of information and knowledge amassed in "E-Mobility in Europe: Trends and Good Practice", produced as a follow-up to the Interreg IVB Project "E-Mobility in the North Sea Region", will serve the purpose of illustrating the various sorts of action which are possible, and needed, to better understand and take advantage of the many opportunity such a rapidly growing field offers.

Walter Leal Filho
Richard Kotter

Contents

Part II Regional and City Case Studies on E-Mobility Development

Part III Technological Advancements and User-Friendly Strategies

Contributors

Sjoerd Bakker Faculty of Architecture and the Built Environment, Delft University of Technology, Delft, The Netherlands

Edward Bentley Faculty of Engineering and Environment, Ellison Building, Northumbria University, Newcastle upon Tyne, UK

Martin Birkmeier FIR e.V. an der RWTH Aachen, Institute for Industrial Management at RWTH Aachen University, Aachen, Germany

Martin Borgqvist SP Technical Research Institute of Sweden AB (External Expert for Lindholmen Science Park), Lund, Sweden

Louise Bunce Department of Psychology, University of Winchester, Winchester, UK

Daniel Busse Institute of Applied Thermodynamics and Combustion Technology, Aachen University of Applied Sciences, Aachen, Germany

Matthias Deindl FIR e.V. an der RWTH Aachen, Institute for Industrial Management at RWTH Aachen University, Aachen, Germany

Peter van Deventer Governor's Office of Planning and Research, State of California & Province of North-Holland, Sacramento, The Netherlands; Province of Noord-Holland/Consulate General of the Netherlands, San Francisco, CA, USA

Eiman Y. ElBanhawy Department of Architecture and the Built Environment, Faculty of Engineering and Environment, Northumbria University, Newcastle upon Tyne, Tyne and Wear, UK

Thomas Esch Institute of Applied Thermodynamics and Combustion Technology, Aachen University of Applied Sciences, Aachen, Germany

Andreas Fink Faculty of Economics and Social Sciences, Institute of Computer Science, Helmut-Schmidt-University, Hamburg, Germany

Dirk Fornahl Centre for Regional and Innovation Economics (CRIE), Bremen University, Bremen, Germany

Simon Árpád Funke Fraunhofer Institute for Systems and Innovation Research ISI, Karlsruhe, Germany

Sidharta Gautama Department of Telecommunications and Information Processing (TELIN), Ghent University, Ghent, Belgium

Dominique Gillis Department of Telecommunications and Information Processing (TELIN), Ghent University, Ghent, Belgium

Till Gnann Fraunhofer Institute for Systems and Innovation Research ISI, Karlsruhe, Germany

Holger Haas Wirtschaftsförderung Region Stuttgart GmbH, Stuttgart, Germany

Tobias Held Department of Mechanical Engineering and Production Management, Hamburg University of Applied Sciences, Hamburg, Germany

Colin Herron Zero Carbon Futures, Future Technology Centre, Sunderland, UK

Michael Hillgärtner Department of Electrical Engineering and Information Technology, University of Applied Sciences Aachen (FH Aachen), Aachen, Germany

Timm Kannstätter Universität Duisburg-Essen, CAR—Center for Automotive Research, Duisburg, Germany

Richard Kotter Department of Geography, Faculty of Engineering and Environment, Northumbria University, Newcastle upon Tyne, UK

Jochen Kreutzfeldt Department of Mechanical Engineering and Production Management, Hamburg University of Applied Sciences, Hamburg, Germany

Robert Kuhfuss Department Electrical Systems, Fraunhofer Institute for Manufacturing Technology and Advanced Materials IFAM, Bremen, Germany

Steffen Kümmell IAV GmbH, Berlin, Germany

Gill Lacey Faculty of Engineering and Environment, Ellison Building, Northumbria University, Newcastle upon Tyne, UK

Walter Leal Filho Hamburg University of Applied Sciences (HAW Hamburg), Research and Transfer Centre, Applications of Life Sciences, Hamburg, Germany

Maarten Linnenkamp MRA-Electric, Amsterdam, The Netherlands

Franziska Mannke Hamburg University of Applied Sciences (HAW Hamburg), Research and Transfer Centre, Applications of Life Sciences, Hamburg, Germany

Sebastian Meerschiff Universität Duisburg-Essen, CAR—Center for Automotive Research, Duisburg, Germany

Roman Muntaniol Institute of Applied Thermodynamics and Combustion Technology, Aachen University of Applied Sciences, Aachen, Germany

Benjamin Myklebust Transport Technology Advisor, Zero Emission Resource Organisation (ZERO), Oslo, Norway

Giuseppe Pace Department of Telecommunications and Information Processing (TELIN), Ghent University, Ghent, Belgium

Patrick Plötz Fraunhofer Institute for Systems and Innovation Research ISI, Karlsruhe, Germany

Ghanim Putrus Faculty of Engineering and Environment, Ellison Building, Northumbria University, Newcastle upon Tyne, UK

Michael Rahier Mobile Media and Communication Lab at the University of Applied Sciences Aachen, Aachen, Germany

Kathrin Rath Hamburg University of Applied Sciences (HAW Hamburg), Research and Transfer Centre, Applications of Life Sciences, Hamburg, Germany

Rolf Reiner Innovationhouse Deutschland GmbH, Stuttgart, Germany

Thomas Ritz Mobile Media and Communication Lab at the University of Applied Sciences Aachen, Aachen, Germany

Marco Roscher FIR e.V. an der RWTH Aachen, Institute for Industrial Management at RWTH Aachen University, Aachen, Germany

Ben Rubin M Public Affairs (Past: California Governor's Office/OPR), New York, USA

R.M. Van Schelven Kwink Groep, The Hague, The Netherlands

Ivana Semanjski Department of Telecommunications and Information Processing (TELIN), Ghent University, Ghent, Belgium

Stephen Shaw Faculty of Business and Law, London Metropolitan University, London, UK

Martijn van der Steen Netherlands School of Public Administration (NSOB), The Hague, The Netherlands

Tessa T. Taefi Department of Mechanical Engineering and Production Management, Hamburg University of Applied Sciences, Hamburg, Germany

Jan Jacob Trip Faculty of Architecture and the Built Environment, Delft University of Technology, Delft, The Netherlands

M.J.W. van Twist Erasmus University Rotterdam (EUR), Rotterdam, The Netherlands

Johanna Vogt Hamburg University of Applied Sciences (HAW Hamburg), Research and Transfer Centre, Applications of Life Sciences, Hamburg, Germany

Ramona Wallenborn Mobile Media and Communication Lab at the University of Applied Sciences Aachen, Aachen, Germany

Josey Wardle Zero Carbon Futures, Future Technology Centre, Sunderland, UK

Noreen Wernern Centre for Regional and Innovation Economics (CRIE), Bremen University, Bremen, Germany

Christine van 't Hull MRA-Electric, Amsterdam, The Netherlands

Part I
Policy Frameworks and Decision-Making on Charging Infrastructure Development

Fostering Sustainable Mobility in Europe: The Contributions of the Project "E-Mobility North Sea Region"

Walter Leal Filho, Kathrin Rath, Franziska Mannke, Johanna Vogt, Richard Kotter, Martin Borgqvist, Benjamin Myklebust and Peter van Deventer

Abstract Efforts towards the promotion of sustainable mobility across Europe need to be supported by technological, political and strategic decisions. In the field of technology, the quest for sustainable mobility can be greatly supported by use of electric vehicles. Apart from the well-known benefits related to reduction of CO_2 emissions, electric mobility may also contribute to reduced air pollution, less noise and thus an increase in the quality of life, especially in urban centres. This paper presents the experiences gathered as part of the project "North Sea Region Electric Mobility Network (E-Mobility NSR)", co-financed by the Interreg IVB North Sea Programme, with the aim of promoting electric mobility in the North Sea Region (NSR). The main objectives of the project are described, along with its structure,

W. Leal Filho (✉) · K. Rath · F. Mannke · J. Vogt
Hamburg University of Applied Sciences (HAW Hamburg),
Research and Transfer Centre, Applications of Life Sciences,
Ulmenliet 20, 21033 Hamburg, Germany
e-mail: walter.leal@haw-hamburg.de

R. Kotter
Department of Geography, Faculty of Engineering and Environment,
Northumbria University, Newcastle upon Tyne NE1 8ST, UK
e-mail: richard.kotter@northumbria.ac.uk

M. Borgqvist
SP Technical Research Institute of Sweden AB (External Expert for Lindholmen Science Park), Scheelevägen 27, 223 70 Lund, Sweden
e-mail: martin.borgqvist@sp.se

B. Myklebust
Transport Technology Advisor, Zero Emission Resource Organisation (ZERO),
Maridalsveien 10, 0178 Oslo, Norway
e-mail: benjamin@zero.no

P. van Deventer
Province of Noord-Holland/Consulate General of the Netherlands,
Houtplein 33, Haarlem, the Netherlands/One Montgomery, San Francisco,
CA 94104, USA
e-mail: petervandeventer@gmail.com

W. Leal Filho and R. Kotter (eds.), *E-Mobility in Europe*,
Green Energy and Technology, DOI 10.1007/978-3-319-13194-8_1

partnership and a set of results reached to date. The paper is complemented by an overview of future needs and opportunities, so as to further support the development of electric mobility policies and practices in the North Sea Region.

Keywords E-mobility · North Sea Region · Policy · Drivers · Interreg

1 Introduction

The term *e-mobility* refers to vehicles that rely on plug-in electricity for their primary energy, regardless of whether they have an auxiliary internal combustion engine for range extension or for keeping the battery charged (battery electric vehicles, plug-in hybrid electric vehicles and extended range electric vehicles) (Fédération Internationale de l'Automobile 2011).

E-mobility technology can be applied to all forms of transportation including rail, shipping and heavy duty trucks (Deason 2010). It also requires a specific infrastructure that includes EV charging stations, connected within a network grid infrastructure based on electric energy and which supplies the energy for recharging (Trip et al. 2012).

E-mobility is associated with the shift to a new network which consists of established members of the automotive industry (Capgemini 2012) but also a large range of start-up companies (Bakker 2012)—some of whom have failed while others (such as Tesla) have become innovation trendsetters, governments, energy providers with an active role in the process by investing in charging infrastructure, and new participants such as automotive suppliers (e.g. battery producers) (Schwedes et al. 2013), IT providers and providers of battery charging/changing services (Capgemini 2012), with Better Place a notable casualty. The energy industry, motivated by the creation of new markets, potentially including consumers in a smart grid context (Kotter 2013; Putrus et al. 2013), is considered the driving force of e-mobility (Schwedes et al. 2013) as well as governments needing to co-invest in networks (Walker 2014). Furthermore, the second life of electric vehicle batteries may become commercially appealing for grid support (Lacey et al. 2013a).

In the 1990s, interest in e-mobility began to resurface (after initial interest lasting from the beginning of the twentieth century to the 1920/1930s in the United States and Europe). It was motivated by an international economic crisis in the automotive industry and a turning point in the debate on climate change (Radkau 2014). In 2007, the issue appeared on the global agenda for the second time. The reasons included an international economic crisis of previously unknown proportions, which put additional pressure on the automobile industry. Today, e-cars are discussed in connection with renewable energy sources (Schwedes et al. 2013).

Today, the energy issue is considered the main driver for the development of the e-mobility sector in Europe. The EC has focused their activity on security of energy supply, climate protection and competitiveness. E-mobility is supported by an innovation budget of approximately five billion euros and the 'Green Cars Initiative'. The latter is focused on sustainable forms of transport and mobility (Praetorius 2011) and designed to support the automotive industry during economic downturns (van Deventer et al. 2011). In addition, the EC is supporting the introduction of alternative forms of transportation technology via public procurement of clean and energy efficient vehicles and establishment of market regulations (e.g. for pollutant emissions) (Deason 2010), as well as funding for a range of research and demonstration projects on electric vehicles (Walker 2012). Among the EU countries that are actively developing their e-mobility sector are Spain, France, the Netherlands, the UK, Denmark, Germany, Sweden and Norway (see Trip et al. 2012 for NSR countries).

Spain: The Electric Mobility Plan (MOVELE) is outlined in the Integral Strategy as a means to promote EVs between 2010 and 2014. The initiatives undertaken include financial aid for purchasing EVs and the introduction of Charging Manager roles (e.g. selling electricity necessary to recharge EVs and installing new charge points in public spaces). The final goal is a total of 250,000 EVs by the end of 2014 (Colmenar-Santos et al. 2014).

France: E-mobility has been supported by the French government via a system of tax rebates and political and financial encouragement for the French automotive industry (Hildermeister and Villareal 2011), and has also been strongly fostered by the mayor of Paris (Dijk et al. 2013), as well as in other NSR areas such as La Rochelle, including city logistics with electric vehicles.

The Netherlands: E-mobility plays a key role in the strategic planning of the Dutch government (van Deventer et al. 2011). Between 2009 and 2011, the Ministry of Economic Affairs carried out its nationwide Action Plan for Electric Driving (Trip et al. 2012). The government also promotes electric vehicles through tax credits and outreach support (van Deventer et al. 2011). Further measures include "Green Deals" with local and regional governments, e.g. MRA Electric (Amsterdam Metropolitan Region), stimulation of public-private partnerships, support of transnational cooperation such as the Coast to Coast e-Mobility programme and signing of international bilateral agreements (Van Deventer 2015; see chapter in this book). In 2011, 2,000 new electric vehicles were registered; in 2015, this number is expected to reach 20,000, with one million EVs and a 20 % market share being projected for 2020 (Trip et al. 2012). As of April 2014, more than 35,000 EVs are on the road in the Netherlands. In December 2013, 25 % of all cars sold were either full electric vehicles or plug-in hybrid electric vehicles.

Denmark: The EDISON (Electric vehicles in a Distributed and Integrated market using Sustainable energy and Open Networks) project is one of the well-known projects promoting electro mobility, partly financed by the Danish government and focused on the cooperation of public and private research organisations, international companies (such as Siemens and IBM) and local authorities

(Trip et al. 2012). The government also supports the sector through purchase subsidies and tax reliefs that amount to more than 30 % of the 2010 EV price (Zsilinszky et al. 2011).

Sweden: The Swedish automotive industry is involved in several e-mobility projects and initiatives concerning both personal and commercial vehicles. This includes research and demonstration activities as well as the development and launch of production models on the market. The Volvo Car Corporation has launched a platform called *Roadmap: Sweden* together with energy companies and other industry companies such as ABB designed to foster e-mobility. The recently-founded company NEVS (National Electric Vehicle Sweden) develops electric cars under the Saab brand name in former Saab facilities.

The Swedish government has implemented an incentive system that gives rebates to low emission vehicles. In addition, funding is provided for automotive research through the FFI-programme, some of which is used for e-mobility-related projects. Regarding further policies and incentives for e-mobility, the government is currently in an investigation and evaluation phase following research on fossil-independent vehicles in Sweden.

Norway: In Norway, incentives for EVs are provided by a package of attractive tax breaks and benefits in use. In practice, many EVs have the same price as comparable fossil fuel-powered vehicles, and savings in the form of toll-free roads and low fuel costs make EVs attractive. As of April 2014, there are more than 27,000 battery electric vehicles in Norway, with EV sales so far at about 13 % of total sales. Thus, of a total of about 2.5 million vehicles in Norway, EVs now make up more than 1 %. An agreement across political party lines is in place designed to maintain the incentives until at least 2017, or until there are 50,000 EVs on the road. In the same agreement, a goal is defined to reach an average of 85 g of CO_2/km in 2020. To achieve the 2020 goals, strong incentives to further reduce average emissions are needed, and EV incentives are at the core of this effort.

Germany: The government has implemented a national e-mobility platform to develop appropriate framework conditions for electric vehicles in a broad stakeholder process. The major national programme of *Modellregionen Elektromobilität* (electric mobility model regions, 2009–2011) incorporated eight German metropolitan regions and was supported financially by the Federal Ministry of Transport, Building and Urban Development (BMVBS) (Zsilinszky et al. 2011). In 2010, the government launched a new initiative designed to put one million battery-powered vehicles and 500,000 fuel cell vehicles on Germany's roads by 2020 (Praetorius 2011; Deason 2010).

United Kingdom: Since 2009, the UK central government has, largely through its inter-ministerial Office for Low Emissions Vehicles (OLEV), invested in public procurement in eight "plugged-in place" regional consortia with different structures. This is in line with the UK's generally preferred approach of encouraging market-based innovation, i.e. divergent routes towards implementation, rather than pre-defined standards (as is the case in countries such as the Republic of Ireland and Spain). In addition, there have been other central government (e.g. Department of Transport) funds for electric, sustainable mobility, notably for buses, as well as research and innovation. Furthermore, large projects to test development of smart grids have been funded by the Department of Climate Change and Energy and other green tax levy sources via the energy sector regulator (Ofgem). At the same time, there has been significant investment in charging infrastructure from the private sector, which is the source of the majority of charging points across the UK.

Such infrastructure is increasingly becoming integrated via mergers and acquisitions of firms as well as agreements to allow access, which to some extent has helped to overcome issues of interoperability, including via pay-as-you-go options in addition to membership subscriptions. Issues of interoperability between Japanese (CHAdeMO) and European (Type 2) charge point infrastructure are only slowly being addressed through retrofitting. The devolved governments of Scotland and Wales have developed their own funding programmes, each with its own particular focus, and the devolved province of Northern Ireland has linked up with the Republic of Ireland's network and centralised approach. In most cases, the lead for the uptake of e-mobility needs to come from local authorities and other public bodies (police, ambulance services, etc.) with some governmental and industry support, although the automotive industry (especially Nissan) and energy production and distribution industry (e.g. British Gas and Northern Powergrid) have made major contributions. Recently, the focus of central government funding has been directed more towards home charging and charging infrastructures such as those for railways (Kotter and Shaw 2013; Kotter 2013). Among the companies working on e-mobility related projects are Enel (2014), Vattenfall (2013), Scania and Siemens (Scania 2014), and BMW (Dempster 2012).

Proponents of the development of the e-mobility sector have discussed a variety of environmental and economic advantages. One of the central arguments for e-mobility is that it will contribute to sustainable transport development (Schwedes et al. 2013), and will allow better usage of energy resources (Döppers and Iwanowski 2012). According to the KPMG analysis, within the next 15 years, e-mobility will achieve sales of two million vehicles in Western Europe (KPMG International 2014). An IDTechEx report, "Electric Boats, Small Submarines and Autonomous Underwater Vehicles (AUV) 2014–2024" reveals that the market for electric watercraft, including vehicles on and under water, will increase from $2.6 billion to $7.3 billion in 2024. There is also a new market for waterborne electric aircraft (Harrop and Das 2014).

However, experts identify an even greater number of challenges and obstacles that need to be addressed and overcome in all related fields.

Technology: E-mobility represents a fundamental technological change for the automotive industry (Capgemini 2012). One of the main problems is battery life. Batteries are still expensive, have low energy intensity and generally take a long time to charge (Skog 2012). Electrically driven vehicles with one battery charge cover smaller distances compared to conventional vehicles (Döppers and Iwanowski 2012).

Standardisation bodies and industries need to agree on common standards and protocols for battery charging systems and their associated information and communication systems (Fédération Internationale de l'Automobile 2011).

There is no voluntary European industrial agreement on the type of connectors (both AC and DC) (Union of the Electricity Industry 2012).

E-mobility requires development and integration of an appropriate infrastructure which includes public charging stations and a smart grid (van Deventer et al. 2011).

The electricity distribution system requires coordination in order to avoid unsustainable peak loads (Union of the Electricity Industry 2012).

There is still no preferred form of technology; therefore, research needs to develop various technology routes in parallel until a preferred form of technology is identified (Taylor Wessing & Technische Universität München 2011).

Economics: The e-mobility sector requires extensive funding (Schwedes et al. 2013). On the other hand, the initial investment is still much higher than for conventional cars (Praetorius 2011) and the sector does not seem to be attractive enough for business to invest in (Skog 2012). There is also uncertainty about the price of electric cars (van Deventer et al. 2011).

Environment: The main concern is the environmental impact of the high volume of electricity that needs to be generated (van Deventer et al. 2011). Doubt exists that electric cars are cleaner than their predecessors, as they consume 'dirty coal-based electricity' (Praetorius 2011).

The development of the sector also requires legislative support, adaptation of the automotive industry to changes in customer expectations and consideration of new ways of distribution (Taylor Wessing & Technische Universität München 2011).

The logistics of energy consumption are expected to change radically. New, mobile consumers will be empowered with the freedom to choose their own supplier, in the same way as traditional electricity consumers. Current expectations are that 80 % of recharging activity will take place at home (Lo Schiavo et al. 2011). In addition, an increased number of various interest groups requires extensive cooperation and efficient, strong coordination (Davies et al. 2012).

Thus, e-mobility is a complex issue that, together with ambitious goals and potential advantages, requires many radical changes in all conventional fields, including the automotive industry areas and beyond to changes in consumer behavior.

Several regions and cities in the North Sea Region are currently seeking to develop strategies and action plans to stimulate e-mobility by encouraging the use of electric vehicles for local transport. However, many activities are not yet fully worked out or interlinked, neither on a transnational nor on an interregional level, meaning that valuable human and financial resources are not being put to full use. This restricts the concept of e-mobility to individual cities or regions and leads to a limited use of e-mobility between these cities and regions, a situation characterised by untapped potential. There needs to be learning across regions and cities within countries and beyond in the NSR and further afield. This suggests that there is therefore an existing need for projects that aim to develop the conditions to steer current and future developments, connect the various networks to form a true transnational "backbone" or "grid" within the NSR (and beyond) for e-mobility and increase accessibility in the region.

It is against this background that the “North Sea Region Electric Mobility Network (E-mobility NSR)” project has been set up. The project contributes to prioritise 3 (Improving the accessibility of places in the NSR) and 4 (Promoting competitive and sustainable communities) of the Interreg IVB North Sea Programme.

2 The Aims of the Project

The main aim of the project E-Mobility NSR is to increase accessibility by fostering e-mobility and stimulating the use of electric vehicles across the NSR, placing additional focus on its links to freight and logistics. It is argued that the launching of the project “North Sea Region Electric Mobility Network (E-mobility NSR)” is well-timed, as it may help the EU meet its climate commitments and address some of the key logistical questions and problems involved in promoting e-mobility. Furthermore, the project may unlock substantial future investments which are needed within the next 10 years, thereby ensuring it offers value for money in the medium and the long term. An expected main result is the use of electric vehicles to travel across the North Sea Region more easily. The area of electric mobility is therefore a fast growing field, and one of great strategic value to the North Sea Region. For this reason, the project not only helps foster the scientific and technological developments in the NSR in an area of strategic interest, but also supports regional economic development and assists towards increasing its accessibility and achieving greater territorial integration. Furthermore, e-mobility contributes to reducing CO_2 emissions, improving air quality as it helps the EU to achieve the goal of reducing its emissions by 20 % by 2020.

Relating to the overall objectives, the intent of the project is to reach the following specific aims:

- To provide state-of-the-art information which may help policy development in e-mobility in the NSR;
- To provide insight into gaps and needs in respect of infrastructure, logistics and preliminary standards for multicharging techniques;
- To develop an NSR smart grid concept with charging points, thereby increasing accessibility in the region;
- To provide a long-term basis upon which regional and local governments as well as other relevant stakeholders in the NSR may dedicate themselves to e-mobility by creating physical or virtual e-mobility information centres in each participating region or city, among other measures.
- To integrate urban freight logistics dimensions into the e-mobility network, promoting better accessibility and cleaner cities by stimulating the use of electric vehicles as a more efficient solution.

3 The Project Partnership

The partnership comprises 11 organisations from within the NSR, comprising all countries in the North Sea Region: Belgium (Flanders Region), Denmark, Germany, the Netherlands, the United Kingdom, Norway and Sweden. The partnership includes universities, economic development agencies, cities, local and provincial governments, NGOs and public enterprises:

Hamburg University of Applied Sciences (DE): Representing the lead partner of E-mobility NSR, the interdisciplinary team of the "Applications of Life Sciences" Research and Transfer Center deals with EU projects in climate change, sustainability and renewable energy (e.g. Leal Filho and Vasoulous 2014).

FDT—Association of Danish Transport and Logistics Centres (DK): FDT disseminates knowledge and promotes the use of open, neutral transport centres in Denmark and other European countries. Within the project, FDT coordinates the theme of electric transport solutions.

Lindholmen Science Park (SE): Lindholmen Science Park is an innovative, collaborative environment connecting research, innovation and education in transport, ICT and media in a distinctive working space. Within the project, the interdisciplinary Swedish team places special focus on fast charging.

Delft University of Technology (NL): The Department of Urban and Regional Development of Delft University of Technology contributes to the project with research on transport and on special development including e-mobility (Bakker 2012; Bakker and van der Vooren 2012; Bakker and Trip 2013a; Sierzchula et al. 2014). The Dutch team deals with state-of-the-art policy analysis, NSR-wide stakeholder surveys and transnational learning.

Høje-Taastrup Kommune (DK): Høje-Taastrup Kommune is located in the Greater Copenhagen region and aims to foster e-mobility within the municipality and beyond. Within the project, the Danish team develops the best practices for the introduction of e-mobility information services.

Northumbria University (UK): Northumbria University is involved in various research activities in e-mobility in the UK. One particular focus of the British team is its focus on a broad range of battery-related issues, V2G and smart grid issues (Putrus et al. 2013; Lacey et al. 2013a, b, c; Jiang et al. 2014), but also on spatial analysis and simulation research based on real data (ElBanhawy et al. 2012a, b; ElBanhawy et al. 2013a, b; ElBanhawy and Nassar 2013; ElBanhawy and Dalton 2013a, b; ElBanhawy and Anumba 2013; ElBanhawy 2014a; ElBanhawy et al. 2014; ElBanhawy 2014b, c), as well as work on business models based on energy optimisation (Conti et al. 2014), and UK policy trends (Kotter 2013; Walker 2014).

Province of Noord-Holland (NL): The Province of Noord-Holland, located in the northwest of the Netherlands, seeks to develop strategies to promote and enhance e-mobility. In the framework of the project, a vast range of charging infrastructure has been set up and an innovative technology pilot, the 'solar road', has been demonstrated. Furthermore, the Dutch School of Public Administration (NSOB, van der Steen) with support from a Dutch consultancy group has been commissioned to undertake comparative policy learning analysis in the NSR, Europe more widely and extending to California in the United States.

Zero Emission Resource Organisation—ZERO (NO): Zero Emission Resource Organisation or ZERO is an environmental organisation in Norway which aims for the reduction of CO_2 emissions and also deals with the increase of electric mobility. The Norwegian NGO widely promotes and communicates sustainable mobility solutions, including an annual international conference.

Faculty of Business and Law, London Metropolitan University (UK): The team at the Faculty of Business and Law (FBL, previously the Cities Institute) of London Metropolitan University delivers evidence-based research for urban policy makers and practitioners. One research focus is on e-mobility, and within the project, FBL contributes to transnational planning issues.

WFB Wirtschaftsförderung Bremen GmbH (DE): WFB is responsible for the development, strengthening and marketing of Bremen as a business location on behalf of the Free Hanseatic City of Bremen with a focus on aerospace, wind energy, maritime economics and logistics. It represents a multifunctional actor landscape, connecting various sectors and agents to promote sustainable mobility.

Flanders Region, represented by Universiteit Gent (BE): The University of Ghent deals with the improvement of battery systems in electric vehicles. The Belgium team utilises laboratory performance and field tests to monitor and improve electric propulsion systems, especially in electric cars and electric buses, but also monitors the (energy) use of these vehicles through data logging, and links this to ICT-based real-time spatial intelligence, connected to a range of other e-mobility smart mobility demonstration projects in Flanders.

The activities undertaken as part of the project

The "North Sea Region Electric Mobility Network (E-mobility NSR)" project is structured in seven work packages (WP) which are briefly described below. In each case, a dedicated partner is responsible for leading and monitoring the individual WP:

WP1: Management and administration (HAW Hamburg, DE):
This WP comprises a sound management and administration scheme to enable the proper implementation and monitoring of the project.

WP2: Information and communication (HAW Hamburg, DE):
The goal of WP2 is to communicate and disseminate information about the project and its results within the North Sea Region and beyond. In order to reach this goal, a multilingual project website has been installed and is continuously updated. Regularly issued news items and newsletters have ensured the dissemination of project activities in the framework of international scientific events and conferences.

WP3: Inventory of state of the art and stakeholder analysis (TU Delft, NL):
WP3 aims to provide a state-of-the-art knowledge base for the project, to make an in-depth analysis of the role of public and private stakeholders and consumers in the further implementation of electric mobility, and to encourage discussion,

knowledge exchange and transnational learning between project partners and relevant stakeholders (Bakker et al. 2014). In particular, this WP has delivered concrete insights into gaps and needs in the field of e-mobility in the NSR and to compile state-of-the-art-information on e-mobility in the NSR. Furthermore, as part of this work package, an extensive policy comparison (in terms of conduct and performance) of all countries in the NSR was performed and benchmarked against the leading region worldwide: California.

WP4: Development of a transnational e-mobility plan (Lindholmen Science Park, SE):
In general, WP4 works to attract the attention of policy makers, industry representatives, academics and the general public for cross-border e-mobility travel and the creation of a virtual e-mobility route in the North Sea Region. In this regard, WP4 focuses on identifying cross-border hubs within the NSR, where electric travel and transport have great potential but need harmonisation of policy and infrastructure in order to be realised on a broader scale. A transnational plan in the sense of an implementation guidance has been developed and a virtual e-mobility route has been developed, connecting circle and hub corridors in the NSR.

WP5: Smart grid solutions—(Flanders Region, represented by Universiteit Gent, BE)
The aim of WP5 is the development of a North Sea Region smart grid concept with charging points. In order to implement this aim, a survey of existing smart grids and expert preferences in this regard has been conducted, supported by a specialist trans-national seminar. To develop a true "smart" grid concept, energy from renewable resources such as solar and wind power has been included in a smart grid designed to integrate electricity generation, storage and distribution using digital technology. In this regard, sustainable energy generation is essential to achieve a system that is truly sustainable from well to wheel.

WP6: Set up of transnational electric mobility information centres (EMIC)—(Høje-Taastrup Kommune, DK)
The project partners of this WP have been working on setting up virtual or physical information centres in their partner cities to inform the interest groups about the overall topic of e-mobility in their region. Available or planned infrastructures have been used to the greatest extent possible in this regard. Strategic recommendations have been issued based on the experiences of electric mobility information centres (EMICs). These EMICs may facilitate transition to electric mobility solutions throughout the North Sea Region by providing tailored information for distinctive uses and forms of technology.

WP7: Promoting efficient and effective urban freight logistics solutions in enhancing regional accessibility—(FDT—Association of Danish Transport and Logistics Centres, DK)
Freight distribution in urban areas causes significant problems relating to air pollution and noise. New, cleaner, efficient and environmental solutions are needed to increase efficiency and create a better urban environment. The objective of WP7 is

to integrate urban freight logistics into the e-mobility network in the NSR by promoting cleaner and more efficient city logistics solutions. The ambition of this WP is better integration of EU policy in the development of concepts for efficient city logistics solutions. Besides analyses of freight transport sector, demands for cleaner transport solutions and user needs for driver-assisting ICT solutions, a comparative analysis of European and international examples of schemes for electrified vehicles in city distribution concepts has been developed.

4 Results Achieved

All achievements are publically accessible via the project website, as well as a repository of other relevant reports and studies on the info-pool. All presentations from the Transnational Conferences (Hamburg, October 2011; Newcastle, March 2012; Gothenburg, October 2012; Copenhagen, March 2013; Haarlem, October 2013; London 2014) can be accessed via the 'events' part of the project website.

Key results and outputs produced within the project up by the time this paper was written (May 2014) are listed below.

4.1 Electric Mobility Policies in the North Sea Region Countries

Input from various project partners has helped OTB/TU Delft (Trip et al. 2012) issue a baseline report on "Electric mobility policies in the North Sea Region countries". This discusses the state of the art and the expected development of electric mobility, policy initiatives from NSR countries and a brief comparative analysis of these policies. This was supported by a supplementary report (Walker 2012) reviewing all relevant EU projects in the field of e-mobility at the time as a baseline for the project.

4.2 European Consultation on E-Mobility

In the framework of European stakeholder consultation, different policy measures were ranked in terms of their effectiveness, efficiency and political feasibility. Organised by E-Mobility NSR partner Delft University of Technology, participants from various cities across the North Sea Region discussed the effectiveness, efficiency and feasibility of a wide range of policy measures. These different policy measures were ranked for their feasibility and importance. As a result of this consultation process, the team from Delft University and involved E-Mobility NSR partners produced a report (Bakker et al. 2012), and OTB/TU Delft published additional information on this subject (Bakker and Trip 2013b).

4.3 Mapping of Public and Private E-Mobility Awareness Needs in the North Sea Region

Høje Taastrup Kommune et al. (2013a) issued a report which provides recommendations gathered to address current e-mobility information gaps and awareness needs. This information serves as valuable input for an appropriate set-up of e-mobility information centres in the North Sea Region and provides an indication of the kind of information in the public and private sphere that an EMIC infrastructure would have to address. This report holds both English and Danish texts, and project beneficiaries have contributed with important and relevant data.

Mapping of current public e-mobility events, showcasing projects, demonstrations and dissemination in *as well as e-mobility activities, campaigns and information channels in the North Sea Region*

In the framework of E-Mobility NSR's work package 6, two reports by Høje-Taastrup Kommune et al. (2013b, c), provide, respectively, a mapping of public e-mobility events, showcasing projects, demonstrations and dissemination in the period of 1 Sept. 2011 to 31 July 2012, as well as a "snapshot" compilation of current e-mobility activities, campaigns and information channels aimed at fostering market penetration of EVs in the North Sea Region. The main objective of these actions was to provide relevant, important background information in order to develop transnational guidelines on how to set up so-called e-mobility information centres.

4.4 Stakeholder Strategies for Realisation of Electric Vehicle Recharging Infrastructure

An international analysis of current stakeholder strategies and agents involved in the setting up of EV recharging infrastructure was conducted by TU Delft (Bakker and Trip 2013a). The report shows how various stakeholders have approached the build-up of a public recharging infrastructure for electric vehicles. In the resulting report, the following dynamics for each of the seven North Sea Region countries are described: Which stakeholders have taken part in the realisation of the infrastructure, what have they done and why have they done so? Differences between countries stem from national and regional ambitions regarding e-mobility and subsequent policy measures, but also from the structure of the energy sector and prevalent electricity production methods.

4.4.1 Electric Vehicle Charge Point Map Websites in the North Sea Region

This interim report (Lilley et al. 2013) is a review of the EV charge point (station) map websites in the North Sea Region (NSR) with the aim of identifying whether there are any patterns or noticeable gaps in the information presented by the interactive EV charge point tools. For each example of the charge point (station) map website, a review has been undertaken by visiting the charge point (station) map website and recording whether the site contains the information which is of key importance from an EV user perspective, such as an interactive map; any information on the charger power of the charge points (stations); the type of connection of the charge points (stations); the addresses of the charge points (stations) and further helpful details.

4.4.2 Micro to Macro Policy Level Investigation

The “Micro to Macro Policy Level Investigation” report issued by Northumbria University and London Metropolitan University (Kotter and Shaw 2013) considers how the step change to mainstream market acceptance of Electric Vehicles (EVs) is being supported by macro-level policy to secure economic as well as environmental benefits. Particular reference is made to the UK, where interventions include grants to purchase new plug-in cars and vans, tax exemptions, and match-funding from the government for “plugged-in places”: pilot schemes designed to stimulate innovation and development of EV infrastructure at the meso-level of areas within the country. There has also been a supplementary report on “Methodologies for Mutual Learning” (Shaw and Evatt 2013). Myklebust (2013) has also looked into a controversial Norwegian EV supporting policy in Oslo concerned with the use of bus lanes.

4.4.3 E-Mobility Helpdesk for Local Governments

The first e-mobility helpdesk in the Netherlands was launched in 2013 during Ecomobiel Rotterdam, the largest annual exhibition and trade fair in the Netherlands for sustainable mobility. There is a growing need for information on electric mobility in the Netherlands, especially on a local level, where there are currently more than 35,000 electric vehicles on the roads, with one million expected by 2025. Consequently, MRA-Electric has created a helpdesk dedicated specifically to local governments which have questions about the infrastructure needed for this increase in the amount of electric vehicles.

4.4.4 Development of an Electric Mobility Information Centre (EMIC)

The e-mobility NSR team in the Høje-Tastrup Kommune (HTK) has decided to take its electric mobility information centre (EMIC) directly to the doorsteps of potential users in the Copenhagen region for optimal outreach. The purpose of the Danish mobile EMIC is to inform public and private companies and users who are actively seeking information about green mobility solutions in the future. The mobile unit offers a wide range of informational material to visitors on local, national and transnational e-mobility matters and solutions. Moreover, companies and other municipalities within the Copenhagen region can book the mobile EMIC centre and expert through HTK. Similarly, an EMIC was set up in Amsterdam, the Netherlands as part of the MRA-electric initiative. The EMIC serves as a centre for information gathering and distribution for local and regional governments, knowledge institutions and private businesses.

4.4.5 Co-housing as a Test Model for Shared Mobility

E-mobility NSR partner Ghent University, Belgium, is running several field tests on electric cars (EVs) and electric buses. With the goal of collecting information on EVs' charging and consumption behaviour in daily transport operations, one field test focuses on EV sharing in co-housing. Co-housing is a special type of collaborative housing in which residents actively participate in the running of their own neighbourhoods. Within this scheme, the sharing of goods (bikes, cars, household machines, common areas) and services is highly accepted. All in all, four co-housing teams represent the test population, with two co-housing teams situated in the city centre of Ghent and the other two teams in semi-urban areas, the suburbs of Ghent and Brussels.

4.4.6 Standardisation of EV Recharging Infrastructures

The report, compiled by TU Delft (Bakker et al. 2014), provides a concise history of the emergence of various EV recharging standards and an overview of the standards used in the North Sea Region (NSR). The global competition between various recharging standards is fuelled by international industrial competition, regional electricity grid conditions, and the diverging interests of car manufacturers and electric utilities. From this, three standards have emerged for regular charging (AC) and two standards have emerged for fast charging (DC). In practice, the existence of various standards, both for plugs and for payment systems, implies that EV drivers cannot simply charge from any charging station, let alone make cross-border trips. This report shows which plug standards are installed in the seven NSR countries and to what extent people can roam between different regions and networks. Despite recent announcements of an EU standard, it is likely that diversity of plugs and payment system will remain a problem for EV drivers in the foreseeable future.

4.4.7 A Modelling Tool to Investigate the Effect of Electric Vehicle Charging on Low Voltage Networks

It is known that charging EVs in the evening can place additional strain on low voltage networks, and if many people charge their EVs at this time, a power outage could result. Charging EVs after midnight is not problematic. A Northumbria University research paper (Lacey et al. 2013c) demonstrates and explains the development of an IT tool which can simulate the power flow, voltage and current over a 24 h profile. The user can input the number of houses, shops, etc., and the number of EVs and when they are charged. Graphs illustrate where and when the stress points are, so work can be undertaken to solve the issues.

4.4.8 Analysis of User Needs for ICT Solutions Assisting the Driver

The corresponding workshop, "ICT solutions for electric urban distribution vehicles", was held on 26 September 2013 at E-Mobility NSR partner Lindholmen Science Park, Sweden, in close cooperation with E-Mobility NSR partner FDT. Focusing on ICT solutions for electrified urban transport, the goal was to provide a state-of-the-art overview of existing technology solutions and standards as well as insights into the potential for certain technology applications. Furthermore, future ideas and related work were elaborated.

4.4.9 The Effect of Cycling on the State of Health of the Electric Vehicle Battery

The E-Mobility NSR team at Northumbria University published another research paper on the effect of cycling on EV batteries (Lacey et al. 2013b). Their paper provides an analysis of the experimental results available for lithium-ion battery degradation which has been used to create a model of the effect of the identified parameters on the ageing of an EV battery.

4.5 Experiences from the Norwegian–Swedish Cooperation on Electric Vehicle Infrastructure

A Lindholmen Science Park (Borgqvist 2012b) report provides a brief summary on experiences and outcomes from a bilateral cooperationbetween Norway and Sweden regarding electric vehicle infrastructure, which could serve as a background document for future similarcollaborations. Through the Swedish Energy Agency coordinating the interaction of a range of agencies with regard to investigating jointcross-border activities, such as: road signs for directions to and indication of charge spots, efforts to establish two demonstration paths forelectric vehicles along highways, common evaluation of charge infrastructure, monitoring of standards and safety issues, availability tocharge infrastructure and debiting systems that functions on both sides of the border, cooperation within the identified areas and effortsto stimulate activities and new projects within each area, as well the possibility to harmonise certain policies were investigated. This bilateralcooperation has stimulated several spin-off activities and new projects, such as an infrastructure project along the E6 between Oslo andGothenburg as well as a project concerning mapping charge spots within the Nordic countries.

4.5.1 Comparative Analysis of European Examples of Schemes for Freight EVs Released

FDT Association of Danish Transport and Logistics Centres, together with TU Delft, HAW Hamburg, Lindholmen Science Park and ZERO have issued a report entitled "Comparative Analysis of European Examples of Schemes for Freight Electric Vehicles", which, in 350 pages and citing more than 60 cases, presents experiences in utilising electric vehicles for goods deliveries in urban areas. Case studies cover initiatives realised in seven North Sea Region countries: Belgium, Denmark, Germany, the Netherlands, Norway, Sweden and the United Kingdom. The report gives a comprehensive overview of initiatives in progress and completed (both publicly supported and entirely privately financed). The main aim of the report is to identify challenges, strengths and opportunities associated with utilisation of electric vehicles for goods distribution, as experienced by users. An important part of the report is also the Appendix, where technical specifications of all 36 identified types of electric freight vehicles used for distribution in urban areas are presented. This has since been presented and published at the 4th LDIC Congress on Logistics (Taefi et al. 2014), as well as other European/EC fora (Laugesen 2014).

4.5.2 Experiences from the Gothenburg fast charging project

Experiences from the Gothenburg fast charging project for electrical vehicles and *Experience Electrical Vehicle*: The case of fast charging

Lindholmen Science Park has issued three study reports on its test and demonstration project in Gothenburg. The results provide interesting insights into how fast charging works, from the user's perspective as well as other perspectives. An international workshop found that, despite the fact that several installations are already in place, there is not yet a functioning business model for fast charging and it will most likely take several years before companies make any direct revenue (depending on the rate of electric car/vehicle rollout) (Borgqvist 2012a). Two stations were installed for testing and demonstration of fast charging, to explore whether the CHAdeMO standard on which they are based is technically mature. A common impression among the project participants was that, despite the fact that both cars and chargers ready for fast charging are available on the market, there are still one or two development steps necessary to determine whether fast charging is a mature and easily implemented form of technology, and one ready for full-time operation (Granström and Gamstedt 2012). The ability to quickly and fully charge the vehicle was emphasised as opposed to the flexibility and ability to use the vehicle more often. This emphasis is contradicted by chargers only being filled to 80 %. Furthermore, the fast charger takes longer in winter (45 min as compared to 20 min). All of this lowers the appeal and usage of fast charging. The participants that did find that there was an added value of the fast charger were those that would use the charger to increase the efficiency of the vehicle, in contexts such as car pools. Improvements by participants focused on the handling of the fast charger, with recommendations including changes to the handle and the weather shield built at the test site (Nilsson 2012).

4.5.3 Transnational Learning

Transnational learning: The Transition to electric mobility: spatial aspects and multi-level policy-making.

In a project report, Bakker et al. (2014) review some lessons learned within WP3, Activity 3.7 specifically focuses on 'transnational' learning, and brings out some questions to address further. This aimed to stimulate discussion and the exchange of knowledge and experiences between the project partners, by organising two expert meetings based on explorative discussion papers:

- 'Spatial aspects of the transition to electric vehicles', Haarlem, 9 October 2013. Hans Nijland from PBL Netherlands Environmental Assessment Agency in The Hague attended as an external expert. During the meeting the participation of project partners was facilitated by the use of electronic voting devices.
- 'A systemic policy mix to support electric mobility development and adoption', London, 11 April 2014. This expert meeting was embedded in the final conference of the E-Mobility NSR project and was attended by a large number of participants from various fields of expertise. Various aspects of the theme were introduced by Dena Kasraian Moghaddam and Sjoerd Bakker from Delft University of Technology and Rogier van Schelven (Kwink Groep) with Dr Martijn van der Steen (Netherlands School of Public Adminstration). These meetings focused on themes that emerged from the project, which were considered as relevant for partners and activities of the E-Mobility NSR project, but also as relatively new and unexplored. The starting point for both meetings was a discussion paper written by Delft University of Technology. The report therefore presents the two discussion papers that preceded the two meetings.

4.5.4 The Road that Converts Sunlight into Electricity

A unique and truly inspiring project began in October 2013 in Krommenie, a small Dutch town, part of the municipality of Zaanstad, belonging to the province of Noord-Holland approximately 15 km north-east of Haarlem. The SolaRoad is a road that can convert sunlight into electricity in the same way as a solar panel. To demonstrate this innovation, politicians from the regional and local governments got on their bikes and cycled down the road, generating enough energy together to release a shower of confetti flowers for a symbolic launch of SolaRoad. It is envisaged that, by next year, cyclists will be able to ride on SolaRoad, which will constitute the world premiere of this technology.

4.5.5 Coast-to-Coast E-Mobility Connection

Western Europe and the west coast of the United States face many similar challenges concerning air quality, oil dependence, job creation, congestion and sustainable urban regions. Transportation in these areas is already mainly sustainable but still offers great opportunities for change and growth. E-mobility is one of these promising opportunities, for which Norway, California and the Netherlands are global leaders. Recently, Matt Rodriquez, Secretary of the California Environmental Protection Agency, and Wilma Mansveld, Secretary of the Dutch Ministry of Infrastructure and the Environment, signed an important Memorandum of Understanding (MoU) on e-mobility cooperation and stimulation. In addition, the California Plug-In Electric Vehicle Collaborative and Coast to Coast e-Mobility partnership signed a multi-year agreement to cooperate on knowledge transfer and business development.

"The Coast to Coast E-Mobility Connection" (C2C) is a public–private partnership (www.Coast2CoastEV.org) which aims to promote knowledge and innovation exchange between the governments, universities and companies of the Netherlands and the United States, strengthen bilateral relations between decision-makers and help position private organisations in relevant e-mobility markets. In April 2013, C2C set up a dedicated Holland E-Mobility House, located within the Netherlands Consulate General in San Francisco, to organise seminars, trade missions, study tours and other programmes to support the partnership's aims. C2C communicates news related to developments in e-mobility in the U.S. and the Netherlands through various channels. Through its role as "liaison" C2C has a unique position for many interest groups. Since its launch, the Holland E-Mobility House has made great strides, achieving several significant results. This includes the signing of a Transnational Agreement between California and the Netherlands, a Governor's Conference on e-mobility defining three specific projects for transnational cooperation, a research paper comparing United States and Dutch policies supporting the introduction of electric mobility (van Deventer et al. 2011), an exchange between California Governor's Office and Dutch government trading policy advisors for 3 months to learn and cooperate on e-mobility adoption, and a study tour hosted by the University of California at Davis (UC Davis) to experience state-of-the-art e-mobility developments in the Netherlands, as well as investments by Spijkstaal, a large Dutch e-mobility company, in California. Investments and market development are provided by e-Traction. Together with the Los Angeles County Economic Development Corporation (LAEDC) and the University of California at Los Angeles (UCLA), grant funding from the California government was secured to set up an e-mobility centre in Southern California involving more than 20 million people. Together with assistance from the Ministry of Economic Affairs, United States investors are being provided with incentives to set up their businesses in the Netherlands, such as Tesla and Zero Motorcycles, and various programmes and pilot opportunities have been created. In 1 year, the transnational initiative C2C has grown from 8 to 20 organisations from the Netherlands and California.

5 Conclusions

The E-Mobility NSR project has provided a concrete contribution towards a better understanding of the need for e-mobility and the requirements to be fulfilled for development of e-mobility in the North Sea Region. Practical activities on the one hand and the research and studies undertaken as part of the project on the other not only serve the purpose of demonstrating the suitability of this sector, but have also served the purpose of fostering more partnerships and unlocking further investments in research and infrastructure.

References

Bakker, S., (2012). The competitive environment of electric vehicles: An analysis of prototype and production models. *Environmental Innovation and Societal Transitions*, *2*, 49–65.

Bakker, S., & Trip, J. J. (2013a). *Stakeholder strategies regarding the realization of an electric vehicle recharging infrastructure*. Project report. Retrieved from http://e-mobility-nsr.eu/fileadmin/user_upload/downloads/info-pool/3.2_E-MobilityNSR_International_stakeholder_analysis.pdf.

Bakker, S., & Trip, J. J. (2013b). Policy options to support the adoption of electric vehicles in the urban environment. *Transportation Research Part D: Transport and Environment*, *25*, 18–23.

Bakker, S., & van der Vooren, A. (2012). Challenging the portfolio of powertrains perspective: Time to choose sides. In *Proceedings of EVS26 Los Angeles*. California. Retrieved May 6–9, 2012, from http://www.academia.edu/2071174/Challenging_the_portfolio_of_powertrains_perspective_time_to_choose_sides.

Bakker, S., Maat, K., & Trip, J. J. (2014). *Transition to Electric Mobility: Spatial Aspects and Multi-level Policy-making*. Project report, NSR E-mobility. TU Delft. Retrieved from http://e-mobility-nsr.eu/info-pool/.

Borgqvist, M. (2012a). *1st International Workshop on Experiences and the future of EV Fast Charging*. Project report. Gothenburg: Lindholmen Science Park. Retrieved from http://e-mobility-nsr.eu/fileadmin/user_upload/downloads/info-pool/Report_WS_fast_charging_120315.pdf.

Borgqvist, M. (2012b). *Experiences from the Norwegian—Swedish Cooperation on Electric Vehicle Infrastructure*. Project report. Gothenburg: Lindholmen Science Park. Available at: http://e-mobility-nsr.eu/fileadmin/user_upload/downloads/info-pool/Experiences_from_NO_SE_cooperation.pdf.

Capgemini. (2012). *Managing the Change to e-Mobility*. Retrieved from http://www.capgemini.com/resource-file-access/resource/pdf/Managing_the_Change_to_e-Mobility___Capgemini_Automotive_Study_2012.pdf.

Colmenar-Santos, A. et al. (2014). Macro economic impact, reduction of fee deficit and profitability of a sustainable transport model based on electric mobility. Case study: City of León (Spain). *Energy*, 65, 303–318. Retrieved 26 March, 2014, from http://linkinghub.elsevier.com/retrieve/pii/S0360544213010475.

Conti, M., Kotter, R., & Putrus, G. (2014). Energy efficiency in electric and plug-in hybrid electric vehicles and its impact on total cost of ownership. In D. Beeton & G. Meyer (Eds.), *New Business Models for Electric and Hybrid Electric Vehicles*. Weinheim: Wiley-VCH.

Davies, H. et al. (2012). ENEVATE Project—Electric vehicle market drivers and e-mobility concepts. In *EEVC European Electric Vehicle Congress* (pp. 1–8). Brussels. Retrieved from http://www.enevate.eu/Workpackage3/wp3_paper.pdf.

Deason, K. (2010). *Governmental Programs on E-Mobility*. Ulm, Germany. Retrieved from http://www.iphe.net/docs/Events/uect/final_docs/Ulm_Workshop_Report_FINAL.pdf.

Dempster, P. (2012). A future vision for sustainable e mobility. Retrieved from http://www.stanford.edu/class/me302/PreviousTerms/2012-01-17%20BMWtech_E-Mobility.pdf.

Dijk, M., Orsato, R. J., & Kemp, R. (2013). The emergence of an electric mobility trajectory, *Energy Policy*, *52*, 135–145.

Döppers, F.-A., & Iwanowski, S. (2012). E-mobility fleet management using ant algorithms. *Procedia—Social and Behavioral Sciences* (Vol. 54, pp. 1058–1067). Retrieved 26 March, 2014, from http://linkinghub.elsevier.com/retrieve/pii/S1877042812042838.

E-Mobility NSR website. (2014). Info Pool. Retrieved from http://e-mobility-nsr.eu/info-pool/.

ElBanhawy, E. Y. (2014a). *Spatiotemporal analysis of e-mobility recharging facilities*. Paper presented at ARCOM Doctoral Workshop, London: University College London.

ElBanhawy, E. Y. (2014b). A hybrid simulation user-equilibrium basis for identifying EV charging hotspots in transportation network. In *Proceedings of IEEE Transportation Electrification Conference and Expo (ITEC 2014)*, Michigan. Inner Urban Core, State College.

ElBanhawy, E. Y. (2014c). Analysis of space-time behaviour of electric vehicle commuters. The experience of the metropolitan and inter-cities scales. In *Proceedings of IEEE Transportation Electrification Conference and Expo (ITEC 2014)*, Michigan.

ElBanhawy, E. Y., & Anumba, C. (2013). *E-mobility System, State College, USA*. Paper presented at Charging Infrastructure Conference, Novi, Michigan.

ElBanhawy, E. Y., & Dalton, R. C. (2013a). *Model of Transports to simulate EV system*. Paper presented at GeoComputation Conference, China.

ElBanhawy, E. Y., & Dalton, R. C. (2013b). *Syntactic approach to electric mobility in metropolitan areas NE 1 district core, segment map*. Paper presented at Space Syntax Sympouim, Korea.

ElBanhawy, E. Y., & Nassar, K. (2013). A movable charging unit for green mobility. *ISPRS—International Archives of the Photogrammetry, Remote Sensing and Spatial Information Sciences XL-4/W1*. (pp. 77–82). doi:10.5194/isprsarchives-XL-4-W1-77-2013. Retrieved from http://www.int-arch-photogramm-remote-sens-spatial-inf-sci.net/XL-4-W1/77/2013/.

ElBanhawy, E. Y., Dalton, R., Thompson, E., & Kotter, R. (2012a). Heuristic Approach for Investigating the Integration of Electric Mobility Charging Infrastructure in Metropolitan Areas: An Agent-based Modelling Simulations. In *2nd International Symposium on Environment Friendly Energies and Applications (EFEA)* (pp. 74–86). Newcastle Upon Tyne, UK. Retrieved from http://e-mobility-nsr.eu/fileadmin/user_upload/downloads/info-pool/EFEEA2012-EE-_FINAL.PDF.

ElBanhawy, E. Y., Dalton, R., Thompson, E., Kotter, R. (2012b). Real-Time E-Mobility Simulation in Metropolitan Area. In A. Henri, P. Jiri; H. Jaroslav, M. Dana (Eds.), *30th eCAADe Conference* (pp. 533–546). Czech Republic, Prague: Technical University in Prague; Faculty of Architecture. Retrieved from http://e-mobility-nsr.eu/fileadmin/user_upload/downloads/info-pool/Real-Time_Electric_Mobility_Simulation_in_Metropolitan_Areas.pdf.

ElBanhawy, E. Y., Dalton, R., Nassar, K. (2013a). Integrating space-syntax and discrete-event simulation for e-mobility analysis. *American Society of Civil Engineers* 1 (91). Retrieved from http://e-mobility-nsr.eu/fileadmin/user_upload/downloads/info-pool/Integrating_space_syntax_and_discrete_event_simulation.pdf.

Elbanhawy, E. Y., Dalton, R., Thompson, E. (2013b). Eras of electric vehicles: Electric mobility on the verge. Focus Attention Scale. *ISPRS—International Archives of the Photogrammetry, Remote Sensing and Spatial Information Sciences XL-4/W1*. Retrieved from http://e-mobility-nsr.eu/fileadmin/user_upload/downloads/info-pool/Eras_of_Electric_Vehicles.pdf.

ElBanhawy, E. Y., Dalton, R., & Anumba, C. (2014). Agent based modeling of Electric Vehicles. In *Proceedings of IEEE Transportation Electrification Conference and Expo (ITEC 2014)*, Michigan.

Enel. (2014). E-mobility. Retrieved March 25, 2014, from http://www.enel.com/en-GB/innovation/project_technology/zero_emission_life/mobile_sustainability/e-mobility.aspx.

FDT Association of Danish Transport and Logistics Centres, TU Delft, HAW Hamburg, Lindholmen Science Park and ZERO. (2013). *Comparative analysis of European examples of schemes for freight electric vehicles*. Compilation Report. Retrieved from http://e-mobility-nsr.eu/fileadmin/user_upload/downloads/info-pool/E-Mobility_-_Final_report_7.3.pdf.

Fédération Internationale de l'Automobile. (2011). *Towards E-Mobility: The Challenges Ahead*. Retrieved from http://www.lowcvp.org.uk/assets/reports/emobility_full_text_fia.pdf.

Gothenburg: Lindholmen Science Park. Retrieved from http://e-mobility-nsr.eu/fileadmin/user_upload/NEWS/Analysis_of_user_needs_for_ICT/7.5_Report_Analysis_of_user_needs_for_ICT.pdf.

Granström, R., & Gamstedt, H. (2012). *Experiences from the Gothenburg fast charging project for electrical vehicles*. Project report. Retrieved from http://e-mobility-nsr.eu/fileadmin/user_upload/downloads/info-pool/FastCharge_Pilot_GOTHENBURG.PDF.

Harrop, P., & Das, R. (2014). *Electric Boats, Small Submarines and Autonomous Underwater Vehicles (AUV) 2014–2024*. Retrieved from http://www.idtechex.com/research/reports/electric-boats-small-submarines-and-autonomous-underwater-vehicles-auv-2014-2024-000371.asp.

Hildermeister, J., & Villareal, A. (2011). Shaping an emerging market for electric cars: How politics in France and Germany transform the European automotive industry. *European Review of Industrial Economics and Politics*, No. 3. Retrieved from http://revel.unice.fr/eriep/index.html?id=3329.

Hoje-Taastrup Kommune et al. (2013a). *Mapping of public & private E-Mobility awareness needs in the North Sea Region*. Project report. Retrieved from http://e-mobility-nsr.eu/fileadmin/user_upload/downloads/info-pool/E-Mobility_NSR_Activity_6.6_Report.pdf.

Hoje-Taastrup Kommune et al. (2013b). *Mapping of public E-Mobility events, showcasing projects, demonstrations and dissemination in the period of 01/09/2011 to 31/07/2012*. Retrieved from http://e-mobility-nsr.eu/fileadmin/user_upload/downloads/info-pool/E-Mobility_NSR_Activity_6.4_Report.pdf.

Hoje-Taastrup Kommune et al. (2013c). *Mapping of current E-Mobility activities, campaigns and information channels aimed at penetrating the market in the North Sea Region*. Compilation report. Retrieved from http://e-mobility-nsr.eu/fileadmin/user_upload/downloads/info-pool/E-Mobility_NSR_Activity_6.2_Report.pdf.

Jacobsson, S. (2013). *Analysis of user needs for ICT solutions assisting the driver*.

Jiang, T., Putrus, P., Gao, Z., Conti, M., McDonald, S., & Lacey, G. (2014). Development of a decentralized smart charge controller for electric vehicles. *International Journal of Electrical Power & Energy Systems, 61*, 355–370.

Kotter, R., & Shaw, S. et al. (2013). *Micro-to-Macro Investigation*. Project report. Retrieved from http://e-mobility-nsr.eu/fileadmin/user_upload/downloads/info-pool/E-mobility_3.6._Main_Report_April_2013.pdf.

Kotter, R. (2013). The developing landscape of electric vehicles and smart grids: a smart future? *International Journal of Environmental Studies, 70*(5). Special Issue: Renewable Energies and Smart Grid—The Solution for Tomorrow's Energy.

KPMG International (2014). *KPMG's Global Automotive Executive Survey 2014*. Retrieved from www.osd.org.tr/yeni/wp-content/uploads/2014/01/KPMG-Global-2014.pdf.

Lacey, G., Putrus, G., & Salim, A. (2013a). The Use of Second Life Electric Vehicle Batteries for Grid Support. In *Proceedings of EuroCon 2013*. July 1–4, 2013, Zagreb, Croatia. Retrieved from http://e-mobility-nsr.eu/fileadmin/user_upload/downloads/info-pool/The_Use_of_Second_Life_Electric_Vehicle_Batteries_for_Grid_Support.pdf.

Lacey, G., Putrus, G. A., Jiang, T., & Kotter, R. (2013b). The effect of cycling on the state of health of the electric vehicle battery. In *Proceedings of UPEC Conference September 2013*. Dublin: Dublin Institute of Technology. Retrieved from http://e-mobility-nsr.eu/fileadmin/user_upload/downloads/info-pool/Northumbria2013_CyclingEffectsOnEVbattery.pdf.

Lacey, G., Putrus, G, Bentley, E., Johnston, D., Walker, S., & Jiang, T. (2013c). A modelling tool to investigate the effect of electric vehicle charging on low voltage networks. In *Proceedings of EVS27 Barcelona*. Spain. Nov 17–20, 2013. Retrieved from http://e-mobility-nsr.eu/fileadmin/user_upload/downloads/info-pool/Barcelona_EVS27_GL_Conference_Proceedings_EVS27.pdf.

Laugesen (2014). Best fact stakeholder external presentation to policy makers in Brussels in January 2014: Clean Urban Freight Logistics Solutions.http://www.bestfact.net/wp-content/uploads/2014//02/BESTFACT_Brussels:30Jan_Day2.SA.1_FTD.pdf.

Leal, W., & Vasoulous, V. (Eds.). (2014). *Global Energy Policy and Security*. Berlin: Springer.

Lilley, S., Kotter, R., & Evatt, N. (2013). *A review of electric vehicle charge pointmap websites in the NSR*. Interim report. Retrieved from http://e-mobility-nsr.eu/fileadmin/user_upload/downloads/info-pool/Review_of_NSR_charge_point_maps_interim_report_June_2013_final.pdf.

Lo Schiavo, L. et al. (2011). *Changing the Regulation for Regulating the Change Innovation-driven regulatory developments in Italy: smart grids, smart metering and e-mobility*, Milan. Retrieved from www.iefe.unibocconi.it, http://e-mobility-nsr.eu/fileadmin/user_upload/downloads/info-pool/FastCharge_Pilot_experiences.pdf.

Myklebust, B. (2013). *EVs in Bus Lanes—A Controversial Incentive in Oslo, Norway*. Paper presented at EVS27, Barcelona, Nov 2013.

Nilsson, M. (2012). *Experience Electrical Vehicle. The case of fast charging*. Gothenburg: Lindholmen Science Park. Retrieved from http://e-mobility-nsr.eu/fileadmin/user_upload/downloads/info-pool/FastCharge_Pilot_experiences.pdf.

Praetorius, B. (2011). E-Mobility in Germany: A research agenda for studying the diffusion of innovative mobility concepts. In *ECEEE 2011 Summer Study: Energy efficiency first: The foundation of a low-carbon society* (pp. 885–890). Retrieved from http://proceedings.eceee.org/papers/proceedings2011/4-174_Praetorius.pdf?returnurl=http%3A%2F%2Fproceedings.eceee.org%2Fvisabstrakt.php%3Fevent%3D1%26doc%3D4-174-11.

Putrus, G. A., Bentley, E., Binns, R., Jiang, T., & Johnston, D. (2013). Smart grids: Energising the future. *International Journal of Environmental Studies, 70*(5). Special Issue: Renewable Energies and Smart Grid—The Solution for Tomorrow's Energy.

Radkau (2014). *The Age of Ecology*. (English version) Cambridge: Polity.

Scania. (2014). Scania tests next-generation electric vehicles. Retrieved March 25, 2014, from http://newsroom.scania.com/en-group/2014/03/13/scania-tests-next-generation-electric-vehicles/.

Schwedes, O., Kettner, S., & Tiedtke, B. (2013). E-mobility in Germany: White hope for a sustainable development or Fig leaf for particular interests? *Environmental Science & Policy, 30*, 72–80. Retrieved March 26, 2014, from http://linkinghub.elsevier.com/retrieve/pii/S1462901112001839.

Shaw, S., & Evatt, N. (2013). *Methodologies for Mutual Learning. Supplementary paper to main report on Micro to Macro Investigation*. London: Cities Institute, London Metropolitan University. Retrieved from http://e-mobility-nsr.eu/fileadmin/user_upload/downloads/info-pool/E-mobility_3.6._Supplementary_Paper_to_Main_Report_April_2013.pdf.

Sierzchula, W., Bakker, S., Maat, K., & van Wee, B. (2014). The influence of financial incentives and other socio-economic factors on electric vehicle adoption. *Energy Policy*, *68*(Issue C), 183–194.

Skog, D. (2012). Conference report: "Enhancing the business case of electric transport." Retrieved March 25, 2014, from http://eharbours.eu/uncategorized/conference-report-electric-transport-is-the-way-of-the-future.

Taefi, T. T., Kreutzfeldt, J., Held, T., Konings, R., Kotter, R., Lilley, S., Baster, H., Green, N., Laugesen, M. S., Jacobsson, S., Borgqvist, M., Nyquist, C. (2014). Comparative Analysis of European examples of Freight Electric Vehicles Schemes. A systematic case study approach with examples from Denmark, Germany. The Netherlands, Sweden and the UK. In K.-D. Thoben, & H. Kotzab (Eds.) *Proceedings of the 4th LDIC Congress*—International Conference on Dynamics in Logistics, Feb 10–14, 2014. Bremen University, Berlin: Springer.

Taylor Wessing & Technische Universität München. (2011). *Future eMobility. The future of electric mobility in Germany*. Retrieved from http://www.taylorwessing.com/uploads/tx_siruplawyermanagement/Taylor_Wessing_TU_M%C3%BCnchen_Future_eMobility_Study_EN.pdf.

Trip, J.J., Lima, J., & Bakker, S. (2012). *Electric mobility policies in the North Sea Region countries*. Retrieved from http://www.northsearegion.eu/files/repository/20130716113551_3.3_-_E-mobility_policies_in_the_NSR_countries.pdf.pdf.

Union of the Electricity Industry. (2012). *Facilitating e-mobility: EURELECTRIC views on charging infrastructure*, Brussels. Retrieved from http://www.eurelectric.org/media/27060/0322_facilitating_emobility_eurelectric_views_-_final-2012-030-0291-01-e.pdf.

van Deventer, P. et al. (2011). *Governing the transition to e-mobility: small steps towards a giant leap*. Retrieved from http://www.nsob.nl/EN/wp-content/uploads/2010/06/e-mobility-webversie.pdf.

Vattenfall. (2013). What we do in E-mobility. *Research and development*. Retrieved March 25, 2014, from http://www.vattenfall.co.uk/en/what-we-do-in-e-mobility.htm.

Walker, S.L. (2012). *A review of European projects in the field of electric vehicles*. Project report. Retrieved from http://e-mobility-nsr.eu/fileadmin/user_upload/downloads/info-pool/Deliverable_3.3_-_E-Mobility_Projects.pdf.

Walker, S. L. (2014). The UK electricity system and its resilience. In W. F. Leal & V. Vasoulous (Eds.), *Global Energy Policy and Security*. Berlin: Springer VS.

Zsilinszky, R. et al. (2011). *E-mobility in Central and Eastern Europe*, Praha. Retrieved from http://www.rolandberger.cz/media/pdf/Roland_Berger_CEE_emobility_study_20111020.pdf.

EV Policy Compared: An International Comparison of Governments' Policy Strategy Towards E-Mobility

Martijn van der Steen, R.M. Van Schelven, R. Kotter, M.J.W. van Twist and Peter van Deventer MPA

Abstract This paper addresses and explores the different strategies governments pursue to support the introduction of plug-in hybrid electric vehicles (PHEVs) and battery electric vehicles (BEVs). This paper presents findings from a European research project that mapped current policies in eight countries, with California as a comparative case to contrast the European findings. The authors analysed the policy strategies that countries have put to practice and analyse how they have performed so far. Arguably, many countries appear to be on track to achieving their short-term goals; in that sense, EV policy is successful. However, once the longer term policy goals for e-mobility are taken into account, it is unlikely that the current policies will be sufficient. Therefore, the authors point out some lessons from current policies that may show a route into the next phase of the introduction of e-mobility. *The paper is part of the Interreg e-mobility North Sea Region (E-Mobility NSR) partnership project, which is co-funded by the EU and participating countries/regions/organisations.*

M. van der Steen (✉)
Netherlands School of Public Administration (NSOB), The Hague, The Netherlands
e-mail: steen@nsob.nl

R.M. Van Schelven
KWINK Groep, The Hague, The Netherlands
e-mail: rvanschelven@kwinkgroep.nl

R. Kotter
Northumbria University, Newcastle upon Tyne, UK
e-mail: richard.kotter@northumbria.ac.uk

M.J.W. van Twist
Erasmus University Rotterdam (EUR), Rotterdam, The Netherlands
e-mail: vantwist@bsk.eur.nl

P. van Deventer MPA
Governor's Office of Planning and Research, State of California & Province of North-Holland, Sacramento, The Netherlands
e-mail: petervandeventer@gmail.com

W. Leal Filho and R. Kotter (eds.), *E-Mobility in Europe*,
Green Energy and Technology, DOI 10.1007/978-3-319-13194-8_2

Keywords Electric vehicle (EV) policy • Plug-In hybrid electric vehicles (PHEV) and battery electric vehicles (BEV) policy strategy • Strategic patterns • Policy innovation

1 Introduction

All over the world, governments attempt to support the transition to e-mobility. The introduction of electric driving is a complex and unpredictable process that is not likely to occur all by itself. Opposition power (such as from the fossil fuel-based value chain connected with motoring and also competing ultra-low-carbon vehicle technology corners, such as hydrogen) is strongly invested. The current (incumbent) market structure benefits continuation of regular cars, and consumers are not yet familiar with e-mobility; many have never driven an EV, let alone have considered buying one. Furthermore, EVs require a substantial investment by consumers. Due to expensive battery packs, the sales price of EVs is higher than those of comparable regular cars. Also, the residual value and life cycle of the batteries is uncertain (although this is tempered currently by manufacturers' warranties on electric batteries, or can be hedged for consumers through leasing models), and any benefits to be gained for consumer from vehicle-to-grid likewise. That makes EVs an *expensive* and *risky* purchase, even though the total cost of ownership is probably competitive to that of a regular car. Also, EVs produce *uncertainty* for drivers. The limited battery range and the uncertain availability of chargers make "carefree" driving difficult. And if there is a charger available, there are issues with *interoperability*, maintenance and the required time to charge. These are all problems that will eventually be solved, but nonetheless are current barriers to consumer take-up (for an overview of EV barriers see Beeton and Butte 2013). There is some momentum for EVs, but it remains a fragile and uncertain venture; the emerging market of EVs can still break down, especially in the early stage that it is in now (once more).

Governmental action is one of the possibilities to overcome the problems of an emerging market. There is a wide array of policy options available to government to support the introduction of EVs and charging infrastructure. Therefore, governmental intervention requires choice; governments wonder which policy to choose, which group or sector to target, what the most effective size and scope of interventions should be and what timing best accommodates the emerging process of the market. Research into the influence of financial incentives and other socio-economic factors on electric vehicle adoption is currently ongoing (see for instance Sierzchula et al. 2014), and there is research into and commentary upon and recommendations towards the effectiveness of EV policy in particular countries (e.g. Green et al. 2014 on the US and Domingues and Pecorelli-Peres 2013 for Brazil). Critical studies attack fiscal subsidies for EVs in the short- and medium term with taxpayers money (Prud'homme and Konig 2012), whilst other authors calculate differently with social/societal lifetime (e.g. public health and atmospheric pollution) costs and come to more favourable results, depending also on the internalisation of the costs by government regulation (Funk and Rabi 1999). Notions of social lifetime costs of battery, fuel cell and plug-in hybrid EVs in relations to conventional vehicles as a more holistic concept may gather more traction in society (Delucchi and Lipman 2010). Not only there are many options to choose from, but there are also many different theories about what to choose for (see Van der Steen et al. 2012; Van Deventer et al. 2011).

Furthermore, it is worth reflecting on the "best" scale of governance for EV policy (see, e.g. Bakker 2014; Bakker et al. 2014). For some, and especially in an EU level, the notion of subsidiarity comes in, to be understood if constructive as a concept "to mean *sharing*, not *shedding*, responsibility in the context of a multi-level policy where the policy process (at least in the European Union) straddles supra-national, national, regional and local levels" (Flynn and Morgan 2004, p. 22). Hierarchically, there is the level of global agreements, e.g. through the International Energy Agency (IEA), which can drive innovation, collaboration and dissemination by a focus on standards and voluntary agreements, realise a policy focus on areas with some impetus funding for research, workshops, training, promotion (IEA IA-HEV 2011), such as through its "Electric Vehicles Initiative," "The Electric Drive Plugs in Implementing Agreement" for cooperation and the "Hybrid and Electric Vehicle Technologies and Programmes" (IEA IA-HEV 2012), as well as the "EV City Casebook" which is a collaboration of the IEA and several partners. There is then the level of trade blocs or integrated markets (such as the European Union, with mandatory standards around emissions for vehicles, urban air pollution, metrics for a New Driving Cycle, labelling and information; and also some US Federal programmes and policy framework initiatives setting the context within which US states operate and can built on); further, there is the national (e.g. EU member state) level, and for the purposes of this paper California as equivalent at that level, which will also have legislation, policy, financial instruments, R&D and demonstration programmes. Then there is the regional (e.g. Electric mobility pilot regions), and not least there is the local level which again has extra policies (e.g. Amsterdam or Utrecht, as cities which offer EV financial incentives on top of what is paid by central government or what the provinces may do, and which will review those extra levels themselves). EV policy is indeed a multi-level policy game, where policy makers continuously have to take into account and operate within frameworks and actions set elsewhere. Governance is *nested*, which is to say that the UK or German or Dutch national level cannot be seen separate from the EU level (see negotiations in the Council of Ministers and the European parliament over emission standards of vehicles, etc.), nor can the regional level be seen as disconnected from the national/Federal or international one in terms of investment, competition, standards (including for charging infrastructure), nor can the local one (e.g. air pollution from the EU one).

The point is that policy initiatives at, e.g. EU or US level do critically rely on dynamism and learning and experimentation and (varied, see the difference between Directive and Framework regulation in the EU) implementation at the national (and arguably regional and local) level, and need the kind of interaction with them (e.g. Plugged-in Places programmes in the UK, which all differed, and were expected and encouraged by the UK government to be varied and different). Nested means there can and would be expected to be variance of policy measures for a variety of reasons and motives, and one should learn from each other, whilst being in the same overall framework which influences what one has to address and to some extent the rules of doing so. A "best practice" example developed and shared (e.g. http://e-mobility-nsr.eu/fileadmin/user_upload/downloads/infopool/Regulation_Subsidy_Fast_Chargers.pdf) should hence be informed and have considered all those connected levels, with subsidiarity being designed, applied and implemented at the most appropriate level/scale.

To fully appreciate this, and for policy makers to best develop, share and apply insights, a relational perspective on space/territory and actors in terms of formal and informal economies and the geographies of knowing and learning is needed (Bathelt and Glückler 2011). In the context of movement in space with electric vehicles, it may be that in the short- to medium term the drive in NSR and EU members states to create interoperability for charging infrastructure in their national territory may well be prioritised over wider cross-border (with exceptions between directly cooperating neighbouring countries and regions) interoperability, thus potentially creating barriers and local lock-in, for both reasons of priority policy targets and also commercial/economic interests of some of the economic agents/actors (Bakker 2013). Also, one needs to clearly consider the motivations and strategies of stakeholders with their commercial interest but also beyond these, with many stakeholders recognise other opportunities presented by EVs: "The most powerful argument in that respect is the potential synergy between EVs and ever increasing renewable electricity shares and many stakeholder activities aim at learning about this opportunity. These activities are however quite limited in scale and mostly focused on off-street charging. Therefore they do not [currently] add, significantly, to the realization of a public recharging infrastructure." (Bakker 2013). However, in the medium term this may change, though, with a likely focus mostly on home charging (Kotter 2013), and with some researchers predicting, e.g. for Germany, that grid-to-vehicle concepts have more of a viable future under the current incentive and policy landscape than vehicle-to-grid concepts (Loisel et al. 2014).

There is a growing literature on EV policy at national, and to some extent regional and local level, and now also supranational level. But only some is of a comparative nature, and usually only between two countries/national levels, other than relatively brief project reports (e.g. Trip et al. 2012) or commissioned consultancy studies undertaking benchmarking at regional level (e.g. E4Tech 2013). Lane et al. (2013), for instance, present a study developing operational definitions of two identified motivations of industrial policy and risk management and uses them to characterise the public policies of six political jurisdictions: California, China, the European Union, France, Germany and the United States. They find that while the European Union is focused primarily on risk management, China, Germany and the United States are primarily engaged in industrial policy. California and France are seen as intermediate cases, with a substantial blend of industrial policy and risk management. Contrast and comment on how California and France both promoted electric and hybrid vehicles to reduce urban air pollution, but differently so and with differeing results at that time for—the authors arguue—differences in the cultural context. Some authors even make policy insight conclusions and recommendations beyond vehicle type (e.g. Yang 2010). Karplus et al. (2010), for instance, undertake an equilibrium-based economic modelling of PHEV penetration in the US and Japan.

Browne et al. (2012) evaluated a range of policies and measures from a range of countries, concluding that developing refuelling infrastructure, supported by tax incentives and awareness campaigns, should be prioritised in the short- to medium term. For them, identified longer term policies and measures could be highly effective include the forced retirement of vehicles that do not adhere to specific fuel economy and emission standards and mandatory import targets (albeit potentially resulting in additional costs for consumers and the domestic vehicle industry, as well as limit consumer choice. Their argument is that "policy-makers have a range of options and should consider the following: (i) develop a transition strategy and engage in scenario planning on a cooperative basis with industry stakeholders; (ii) identify potential "lead adopters" and develop a strategy for strategic niche management; (iii) develop stakeholder partnerships with industry and consumer groups; (iv) promote the adoption of a new socio-technological regime through awareness campaigns and education programmes; (v) change the taxation structure by taxing negative externalities such as [Greenhouse Gas] GHG emissions and creating positive incentives through excise relief and subsidies; and (vi) ensure a consistent mix of policy and regulatory signals, which offer long-term certainty" (Browne et al. 2012, p. 140). They propose that their "evaluation framework" could serve as a useful template for the identification and evaluation of barrier and policy priorities and could be modified depending on the system and/or geographical boundary. In addition, [this framework] can be adapted and used by policy makers in order to guide policy priorities and develop national [Alternatively Fuelled Vehicles] AFV policy strategies or local action plans for strategic niche management. It is sufficiently flexible to be modified for particular jurisdictions, depending on particular consumer choices, policy preferences and the stage of technological innovation. Furthermore, it is suitable for national or cross-country evaluation as particular barriers, policy measures and technologies might be more or less suitable, depending on the jurisdiction. However, as a qualitative tool, it is vulnerable to subjective evaluation and should be supported by empirical analysis, where possible. In addition, this framework should be applied at the particular level of interest and the evaluation should not be construed as universal as it may depend on particular system factors (Browne et al. 2012, p. 140).

A study by Steinhilber et al. (2013) focussing on the socio-technical inertia vis-a-vis the widespread introduction and take-up of electrical vehicles aims to contribute to understanding the key tools and strategies that might enable the successful introduction of new technologies and innovations by exploring the key barriers to electric vehicles encountered in two countries (UK and Germany), where the automobile industry has been historically significant, argues that: Immature developing technology is the major reason behind non-commercialisation of EVs, that EVs currently do not present a significant benefit to the electricity sector, that EVs rely on a mix of regulatory and government measures for their development, that EVs face lock-in problem of unsustainable technologies and related barriers, and that positive "ecosystems" for innovation in vehicle technology and business models are required.

This present chapter adds to this literature and explores the policy options for governments that want to support the further introduction of EVs. The authors aim to provide an empirical answer to that question, based on a study in which they have gathered all of the formally documented policies with regards to e-mobility that a selected group of governments put in place in the period between 2012–2014, to be developed further over time. The project is part of the Interreg *North Sea Region Electric Mobility Network* (E-Mobility NSR) that was launched in April 2011.

1.1 Scope, Methods and Limitations

This research focuses purely on *passenger vehicles*[1] and *multipurpose passenger vehicles*.[2] Furthermore, the present study focuses solely on a specific type of electrified drive trains; of the most commonly used categories—hybrid electric vehicles (HEVs),[3] plug-in hybrid electric vehicles (PHEVs)[4] and battery electric vehicles (BEVs)[5]—the authors take into consideration *only* policies concerning *PHEVs* and *BEVs*. Policy for HEVs is *not* part of the research. Also, the authors did not look at other possible options for clean mobility, such as biofuels, hydrogen or the substitution of cars for public transport (Van der Steen et al. 2014a, b).

In order to collect the data for this study's, the authors have gathered all the documents they could find for the seven case countries in this specific study; the Netherlands, Belgium, Germany, Denmark, Sweden, Norway and the UK. California is added as a comparative case to contrast the European findings. California is widely regarded as a frontrunner in the transition to e-mobility

---

[1]Vehicle with a designated seating capacity of 10 or less (IEA, IA-HEV and AVARE 2013).

[2]Vehicle with a designated seating capacity of 10 or less that is constructed either on a truck chassis or with special features for occasional off-road operation (IEA, IA-HEV and AVARE 2013).

[3]HEV has the ability to operate all-electrically, generally at low average speeds. At high steady speeds such a HEV uses only the engine and mechanical drivetrain, with no electric assist. At intermediate average speeds with intermittent loads, both electric and mechanical drives frequently operate together. (IEA, IA-HEV 2011).

[4]A HEV with a battery pack with a relatively large amount of kWh of storage capability, with an ability to charge the battery by plugging a vehicle cable into the electricity grid. (IEA, IA-HEV 2011).

[5]An BEV is defined as "any autonomous road vehicle exclusively with an electric drive, and without any on-board electric generation capability." (IEA, IA-HEV, 2011).

(Plugincars 2013). To collect the documents, a "snowballing"-method was employed to gather more information about policy. Many documents contained references to other studies and sources that the authors then looked up and included into their model.

The analytical lens—or model—the authors employ is based on, firstly, a value chain approach to e-mobility (Beeton 2014), which the authors here have arranged into three chains—with interactions and interdependencies of the electric vehicle, the charging infrastructure and the (wider) enabling network (the grid, Information and Communications Technologies (ICT) and Intelligent Transport Solutions (ITS) and services etc. Second, the authors adopt Hood and Margetts' (2007) four different categories of tools for government to "steer," and use these four categories as a first lens to organise the policies. In the table below they explain the categories and apply them to policy for EVs. Thirdly, the authors looked at policy as originating from one of four levels of government; policy is *trans-national*, *national*, *regional*, or *local*.

With this first selection of documents the authors populated their database and ran a first scan of results. For each country, they drafted an analysis of its EV policies and asked a local resource colleague to take a critical look at the document; they then asked the local colleagues to correct their document and send them links to relevant documents not yet included in the study. The authors analysed this second set of documents and improved their country analysis on the basis of the feedback from the local colleagues. After that, the authors finalised their findings in a draft report. During 2013 they kept collecting new documents, in order to be able to keep the database up to date with new policies and new data about performances.

As a third round, the authors presented and discussed the draft report in four feedback sessions where expert representatives of the participating countries reflected on their interim findings. Representatives were selected from both the local academic community and the community of EV policy makers from that country, region or municipality. In each session, the authors presented a selection of the findings that were relevant to the particular audience (country). After that, the authors first asked participants if they recognised or could validate the findings and if they had additions or other (critical) remarks about them. Then, there was time for discussion about the more general implications of the findings and possible implications for policy. Each of the feedback sessions resulted in a general recognition for and support of the authors' findings, but also lead to interesting discussions about methodology and about the dilemmas of policy for EVs. In this discussion section of this chapter, the authors present some of those dilemmas and reflect on their implications for the next generation of EV-policy.

1.2 Outline of This Chapter

The authors start with a presentation of the framework used to analyse the policies. After that, they present the assorted variety of policies they found. In the discussion section, some broader observations about general patterns and dilemmas of public policy in the field of E-mobility are presented. Also, the authors reflect on what they think one can learn from these policies for the next phase in the introduction of e-mobility.

2 A Framework for Analysing EV-Policy

2.1 Lens 1: The Value Chain of E-Mobility

"EV-policy" suggests a coherent and single object and objective for policy. However, if one looks closer, e-mobility involves a variety of related but separate elements. Therefore, the authors looked at e-mobility as a value chain (Fournier and Stinson 2011/Squarewise 2010) where the different segments of the chain can each be targeted by policy. Also, e-mobility can be separated into three different value chains (In 't Veld 2005); the value chain of *vehicles*, the value chain of *charging* and the value chain of *surrounding network*. The latter is not so much a chain, but more a third category for policy. For the value chains of vehicles and charging, we see four segments in each chain. Policy can target at least one and possible elements of the chain. For instance, a purchase subsidy targets the vehicles value chain, and within that the consumer segment. Therefore, we categorise that particular policy as a vehicle-consumer-focused policy in our framework. Figure 1 presents the three value chains; Tables 1, 2 and 3 explain the different segments of the value chains.

2.2 Lens 2: Policy as Tools

In their classic *tools of government*-study, Hood and Margetts (2007) distinguish four different categories of tools for government to "steer." We use these four categories as a first lens to organise the policies. In Table 4, the authors explain the categories and apply them to policy for EVs.

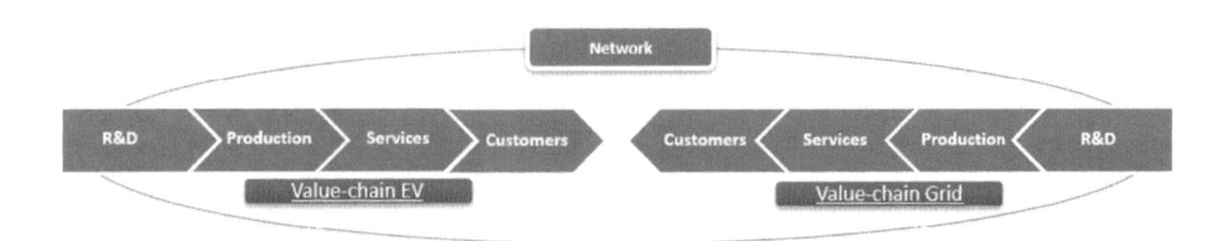

Fig. 1 Three value chains of e-mobility

Table 1 Vehicle value chain

Value chain—electric vehicle	
R&D	• Instruments focused on influencing the research and design of electric vehicles and EV components
Production	• Instruments focused on influencing the production of electric vehicles and vehicle components such as batteries and other hardware (original equipment manufacturers). This segment of the value chain also recognises the software used in electric vehicles
Services	• Instruments focused on influencing service providers for electric vehicles. Different service providers are recognised, such as car dealerships, mechanics, insurance companies, etc.
Customers	• Instruments focused on influencing customers of EVs. The study's methodology recognises individual consumers (end-users), but also fleet-owners (e.g. authorities and leasing companies) and public/governmental agencies (promoting consumerism)

Table 2 Infrastructure value chain

Value chain—charging infrastructure	
R&D	• Instruments focused on influencing the research and design of the complete charging infrastructure
Production	• Instruments focused on influencing the production of charging stations and EV system components such as the electricity network, energy production, etc.
Services	• Instruments focused on influencing service providers for charging stations. Different service providers are recognised, such as energy suppliers, power plants, grid managers, software developers, etc.
Customers	• Instruments focused on influencing customers of charging stations. By "customers" the study refers both to users (consumers) and owners (consumers, companies, public authorities and government). The different types of charging stations (private, public, fast, quick, normal) require different types of steering by governmental units

Table 3 Network value chain

Value chain—Network	
Network	• These are all of the instruments that focus on connecting stakeholders in the EV/ infrastructure value chain. For instance, efforts intended to intensify contacts between different stakeholders, in order to improve value chain alignment and a more efficient functioning of the entire value chain. In addition to the value chain, this includes other policy measures aimed at the e-mobility ecosystem, which are taken into consideration. For instance, policy measures aimed at realising Smart Grids, Smart economies and Smart mobility Beeton (2012)

2.3 Lens 3: Policy at a Certain Level of Government

As a third lens for our analysis, the authors looked at policy as originating from one of three levels of government; policy is *trans-national*, *national*, *regional* or *local*—with a hierarchy but also interactions between levels and a multi-level governance

Table 4 Tools of government

Tools of government	
Legal	• All of the rules and directives designed to mandate, enable, incentivize, limit or otherwise direct subjects to act according to policy goals • E.g. legal requirements, local parking legislation, European legislation for standards for charging station accessibility, limited access to urban areas or roads
Financial	• The policy instruments involve either the handing out or taking away of material resources (cash or kind), in order to incentivize or disincentivize behaviour by subjects. The difference between financial and legal measures is that those affected are not obliged to take the measures involved, but are incentivized to do so by their own choice • E.g. purchase grants, tax benefits for consumers of EVs, government funding for battery research, subsidies on home chargers or free electricity for public charging
Communication	• Instruments that influence the value chain of e-mobility through to the communication of arguments and persuasion, including information and education • E.g. education in schools, government information campaigns
Organisation	• Actions by government that provides the physical ability to act directly, using its own forces to achieve policy goals rather than others. This includes the allocation of means, capital, resources and the physical infrastructure needed to act • E.g. government or public authorities acting as a launching customer, buying an own fleet of EVss, government installing public chargers

nature to it, and competition also between countries, regions and cities (c.f. Bakker 2014). Different countries work with different systems, where other levels of government are responsible for e-mobility. The model takes this into account, in order to be able to analyse the differences in various countries. Some organise policy from the local level, while others have a stronger national policy that is only marginally supplemented by local or regional policies.

3 Findings: An Analysis of EV Policies in Seven EU Countries

In this chapter, the authors compare the variety of policies at different governmental levels in different countries. They present the most important general findings and illustrate them with a range of examples of policies from different countries. The complete results and the total body of policies can be found in the project background report (Van der Steen et al. 2014a, b).

3.1 Finding 1: Most NSR Countries Focus on Financial and Organisational Instruments

The countries in this collated data set primarily focus on financial and organisational instruments. Most policies fall into either one of those two categories of tools.

As for financial instruments, countries adopt very similar policies. They are often conducted by the national government and are mostly fiscal (registrations bonus based on emissions, income tax measures and opportunities for businesses to relieve the cost of an EV against taxable profits). Also, governments apply a considerable number of organisational instruments. Especially at the regional and local levels, the authors observe a lot of "organization tools." Local and regional governments—but also some public–private partnerships—install many local project organisations that, for instance, carry out grant applications and are launching consumer initiatives. This generates extra dynamics to the incentives and benefits set out by the national government.

The focus on legal and communication instruments is limited compared to financial and organisational instruments (Tables 5 and 6).

3.2 Finding 2: Most NSR Countries Initiate Policy from the National Government Level

As summarised in Table 7, in most countries most policy is made at the national level. However, with that said, there are often also very active local and regional communities that provide all sorts of activities to stimulate e-mobility. The main body of policy is national—fiscal, regulation—but that is accompanied by local and regional policy that provides a local colouring and fine-tuning.

Table 5 Type of policy actions (Van der Steen et al. 2014a, b)

Type of policy actions				
NSR countries	Legal	Financial	Communicative	Organisational
Belgium	+	++	+	+++
Denmark	+	+++	+	++
Germany	+	++	+	+++
Netherlands	+	+++	+	+++
Norway	++	+++	+	++
Sweden	+	++	+	+++
UK	0	++	+	++
Comparative case: California	++	+++	+	+

0 = Limited information found/available
+ = Limited focus
++ = Strong focus
+++ = Prevalent focus area

Table 6 Examples of organisational tools used in different countries

Organisational incentives in NSR countries and California (USA)	
Denmark	**Platform** • *Information Centre*. In cooperation with the Danish Energy Agency, the Centre for Green Transport has established (Established in 2011) an information centre to exchange experiences on EVs between local communities in Denmark (Bakker et al. 2012/European Commission 2011/IEA IA-HEV 2014) **Project organisation** • *Copenhagen Electric*. Copenhagen Electric focuses on strengthening the capital region's international competitiveness and ensuring greater cooperation in the Øresund Region and other regions in Europe by providing objective information about electric vehicles to municipalities, companies and private individuals. Also projects, campaigns and partnerships on EVs are initiated (Copenhagen Electric 2014)
Germany	**Project organisation** • **Model regions** – *E.g. Elektromobilitat Model Region Hamburg*. The testing of diesel hybrid buses on lines. Innovative energy storage for rail vehicles. The use and development of electric cars and charging infrastructure. The use of electric vehicles in commercial traffic. These are the priorities of the projects in the model region Hamburg (BMVI—Elektromobilitat Model Region Hamburg 2014) – *E.g. Model region Bremen/Oldenburg*. In the model region Bremen/ Oldenburg, the cooperation between project partners such as the University of Bremen, Bremer Energie Institut and Centre for Regional and Innovation Economics are another building brick in the development in electric vehicle technology. The local Daimler-Benz/Mercedes production plant will use the scientific knowledge to produce these new technologies. The same partnership resulted in plans by the local Daimler-Benz/Mercedes production plant to convert a tractor to an E tractor, to demonstrate the use of commercial Electric Vehicles in daily use (BMVI—Elektromobilitat Model Region Bremen/Oldenburg (2014)
Norway	**Project organisations** • *Gronn Bill*. Set up in 2009 to facilitate the introduction of 200.000 EVs and PHEVs on Norwegian roads by Energy Norway, Novatran, regional authorities and ZERO by 2020 (Bakker et al. 2012) • *Transnova*. Transnova is the public body assigned to reducing CO_2 emissions from the transport sector. Transnova was established in 2007 following a suggestion by ZERO. Today, Transnova has a budget of NOK 75 million per year (Transnova 2014) **Platform** • *Electric Mobility Norway*. The Electric Mobility Norway (EMN) project is being developed in the Kongsberg–Drammen–Oslo region. It is led by Kongsberg Innovation with the support of Transnova (which is managed by the Norwegian Public Roads administration) and Buskerud County Council. The main objective is the "establishment of an innovation and knowledge arena in that region" (Bakker et al. 2012)

(continued)

Table 6 (continued)

Organisational incentives in NSR countries and California (USA)	
Comparative case: California	• *Vehicle Technologies Program (VTP).* Advanced Energy Storage technologies research programmes. Research portfolio is focused on battery module development and demonstration of advanced batteries to enable a large market penetration of Electric Driven Vehicles (EDV) within 5–10 years (EERE 2014a) • *Advanced Power Electronics and Electric Machines.* Subprogramme within the DOE VTP provides support and guidance for many cutting edge automotive technologies now under development. Research is focused on developing revolutionary new power electronics and electric motor technologies that will leapfrog current on-the-road technologies (EERE 2014c) • *LA Cleantech Business Incubator (LACI).* LACI helps accelerate the commercialization of their clean technologies in addition to accelerating new products developed by independent entrepreneurs (LA Cleantech Incubator 2014) • *Clean city.* A national network of nearly 100 Clean Cities coalitions brings together stakeholders in the public and private sectors to deploy alternative and renewable fuels, idle reduction measures, fuel economy improvements and emerging transportation technologies (EERE 2014b)

Table 7 Government level of EV policy (Van der Steen et al. 2014a, b)

Government level			
Country	National	Regional	Local
Belgium	+	+++	+
Denmark	+	+++	+++
Germany	+++	++	+
Netherlands	++	++	++
Norway	+++	+	+
Sweden	+++	+	+
UK	+++	++	+
Comparative case: California	++	++	++

0 = Limited information found/available
+ = Limited focus
++ = Strong focus
+++ = Prevalent focus area

3.3 *Finding 3: In Most NSR Countries Policy Focuses on Vehicles Rather Than Charging*

Policy instruments mostly focus on the vehicle value chain. Within the EV value chain, governments primarily focus policy on consumers. Some countries focus more prominently in R&D and in upstream segments of the value chain.

Table 8 Policy focus on the vehicle value chain (Van der Steen et al. 2014a, b)

Policy focus in the EV value chain				
Country	R&D	Production	Services	Customer
Belgium	+	+	+	++
Denmark	+++	0	+	++
Germany	+++	++	+	+++
Netherlands	+	++	+	+++
Norway	++	+	+	+++
Sweden	++	+	+	++
UK	++	+	+	++
Comparative case: California	+++	++	+	++

0 = Limited information found/available
+ = Limited focus
++ = Strong focus
+++ = Prevalent focus area

Little attention is given to the segment of services, which could be a missing link between the demand of consumers and the supply provided by manufacturers (Tables 8 and 9).

3.4 Finding 4: Policy Mostly Targets the Downstream of the Vehicle Value Chain

Most countries focus their policies downstream in the value chain; they adopt a large number of financial incentives, at different government levels (tax incentives, rebates, subsidies, local benefits, etc.). In Denmark, one-third of the steering instruments in the EV value chain focus on consumers. Different levels of government implement downstream policies. Subsidies and tax incentives are usually implemented at national level. However, local governments also provide financial incentives, often cash but mostly "in-kind." Examples are free or preferential parking, access to toll lanes, free charging, free access to ferries for EVs. At first glance, these are small incentives. However, their impact should not be overlooked. In a recent Californian survey, 59 % of the respondents indicated that access to the high-occupancy vehicle lane (HOV-lane) was extremely or very important in their decision to purchase an EV, making it the most important motivator for purchase found in the survey (CCSE 2014).

Although most countries target the downstream (consumers) of the value chain, some also work more upstream (R&D and production). Most of these instruments are financial (see Table 10 for examples). Germany is one of the countries with a strong focus on R&D in EV policy. This could be explained by the presence of

Table 9 Examples of financial instruments for EVs focused downstream in the vehicle value chain (consumer focused)

Financial incentives—downstream of the value chain (consumer focussed)	
Belgium	**Tax incentives** (ECN 2012) • 120 % of the purchase costs are deductible for companies under a corporate tax system for EVs. 100 % for PHEV with CO_2< 60 g/km (for companies under corporate tax system) • Individuals receive a subsidy of 30 % of the price of the EV up to 9.190 Euros (*by taxes, not directly from invoice*). In Wallonia, the motor vehicle tax for low emission cars is the lowest of all the taxes. In the Flemish region, EVs are exempt from motor vehicle tax **Rebates/subsidies** • *Bonus Malus*. In the Walloon Region, EVs are being promoted through an extra subsidy of 3.500 Euros through a bonus malus system (The New Drive 2014/ECN 2012) • *Subsidy*. Through the subsidy, the city of Gent receives through the CIVITAS demonstration programme. The city grants funds to individuals, taxi and courier services and also to car sharing companies to purchase EVs (CIVITAS 2014)
Denmark	**Tax incentives** (Bakker et al. 2012) • In Denmark, BEVs are exempt from registration tax until 2015. That amount is 105 % of the price of the car for the first 10.000 Euros and 180 % for the rest of the amount • BEVs and fuel cell vehicles are exempted from annual tax until the end of 2015 **Local benefits ('non-fiscal incentives')** • *Parking*. In Denmark, several cities (Copenhagen) have reduced the parking fee for EVs and in some cities EVs are exempt from paying parking fees (Squarewise 2010) • *Toll Roads*. Free use of toll roads for EVs (Bakker et al. 2012)
Germany	**Tax incentives** • Exemption of annual circulation tax for EVs bought during the period of 18 May 2011 until 31 December 2015. The Federal government has decided that the exemption period will be doubled from 5 to 10 years (Spiegel Online 2012) • In Germany, the motor vehicle tax is determined by the amount of CO_2 emissions, which is a pro for EVs (Squarewise 2010) **Rebates/subsidies** • The German government grants subsidies up to 5.000 Euros for EV buyers (Squarewise 2010) **Local benefits ('non-fiscal incentives')** • *Parking*. In several cities in Germany, EVs have privileges for parking (Bakker et al. 2012)

(continued)

Table 9 (continued)

Financial incentives—downstream of the value chain (consumer focussed)	
The Netherlands	**Tax incentives** (IEA IA-HEV 2011) • EVs are exempt from the registration tax and from the annual road tax. Fuel cell EVs follow the same ruling • For leased cars, an income tax measure makes EVs and HEVs attractive. A normal tariff of 25 % of a leased car's value that is added to the annual income tax is eliminated (7 % from 2014) for zero-emission cars (less than 50 g CO_2/km) or will be 14 % or 20 % according to the fuel type and CO_2 emissions if the cars are fuel-efficient • Tax relief regulation for purchasing commercial electric vehicles • Through the MIA and VAMIL regulation of the central government, entrepreneurs can receive a subsidy for purchasing an EV or installing charging infrastructure (RVO NL 2013) **Rebates/subsidies** • The city of Amsterdam grants subsidies up to 5.000 Euros to purchase EVs which are being used for business and up to 10.000 Euros for purchasing electric taxis and courier cars (Programmabureau Luchtkwaliteit 2010)
Norway	**Tax incentives** (WSDOT 2011/Bakker et al. 2012) • EVs are exempt from non-recurring vehicle fees • EVs are exempt from sales tax • EVs are exempt from annual road tax. Tax free allowance given for this tax (calculated as NOK/km) i.e. for trips to/from working places and for business trips is considerable higher for EVs. Reduction for companies: 75 % for EV and 50 % for HEVs • EVs are exempt from taxation for company car benefit tax from 1 January 2009 • Registration tax is calculated according to weight, motor power and CO_2 emissions. The vehicles are classified by groups per CO_2 'tax. EVs are exempt from this tax • Reduced tax for leasing EVs **Rebates/subsidies** (Bakker et al. 2012) • *Grants for individuals*. The Norwegian government grants subsidies (approximately €4.000) to individuals who buy an EV or HEV class N1 or M1 • *Grants for companies*. To purchase EVs, the funding is 50 % of vehicles price; up to 50 % are given to companies **Local benefits** (**'non-fiscal incentives'**) (WSDOT 2011/Bakker et al. 2012) • *Domestic Ferries*. EVs have free use of domestic ferries • *Free access*. EVs have free access to public areas • *Free parking*. EVs can park for free in public parking places. This measure has been in place since the beginning of the 1990s • *Toll roads*. EVs can use the toll roads for free • *Use of Bus and Taxi lanes*. EVs are permitted in bus and taxi lanes. This measure has been in place since 2003

(continued)

Table 9 (continued)

Financial incentives—downstream of the value chain (consumer focussed)	
Sweden	**Tax incentives** (IEA IA-HEV 2012/Bakker et al. 2012) • Taxation is based on the amount of CO_2 emission. This tax has been raised with 33 % in 2011 to stimulate the use of EVs • Hybrid vehicles with CO_2 emissions of 12 G/KM or less and EVs with an energy consumption of 37 kwh per 100 km or less are exempt from the annual circulation tax for a period of 5 years from the date of their first registration starting on 1 January 2010 • For EVs and Hybrid vehicles, the taxable value of the car for the purposes of company car taxation is reduced by 40 % compared with the corresponding or comparable petrol or diesel car **Rebates/subsidies** • A clean vehicle premium of 40,000 SEK (approximately €4.500) has been introduced (from January 2012) for vehicles emitting less than 50 g CO_2 per km **Local benefits ('non-fiscal incentives')** (IEA IA-HEV 2012/Bakker et al. 2012) • *Parking*. In about 50 % of the 70 cities in Sweden where you have to pay to park EVs get a discount or can park for free (Parking. In about 50 % of the 70 cities in Sweden where you have to pay to park, EVs get a discount or can park for free) • *Toll*. EVs bought before 1 January 2009 are exempt from paying toll tax in Stockholm until 2012. Cars bought after 2009 are not exempt. From 1 August 2012, this incentive has been cancelled • *Congestion Charge scheme*. A congestion charging scheme was implemented on a permanent basis in August 2007 in central Stockholm. A fee is charged during times of traffic congestion. PHEVs and EVs are exempt
UK	**Tax incentives** (Bakker et al. 2012) • *Vehicle excise duty or VED (the UK's circulation tax)*. Electric vehicles exempt. VED for other vehicles is graduated by CO_2 emissions (for tailpipe emissions < 100 g CO_2 per km) • *Company car tax*. Employees and employers exempt from income and national insurance contributions • *Van benefit charge*. Exemption for electric vans from income and national insurance contributions (maximum of £3.000) • *Fuel benefit charge*. Electric Vehicles exempt *Enhanced capital allowances*. 100 % first year allowance (FYA): business can relieve entire cost of purchase of an electric car or a van against taxable profits in the year of acquisition for businesses buying low emission cars, a mechanism that effectively allows companies to claim back the cost of the purchase from HM Revenue and Customs (HMRC), which was extended until March 2018 through the 2013 UK budget, with the qualifying threshold will dropping from cars with emissions of less than 110 grams of CO_2/km, to 95 g/km in April and fall again to 75 g/km from April 2015, effectively making it more attractive for companies to purchase the lowest emission vehicles on the market. However, the 2012 budget had *removed* the 100 % FYA for *leasing* vehicles and this was not revised in the 2013 budget. The policy move was nominally designed to counter the possibility of companies

(continued)

Table 9 (continued)

Financial incentives—downstream of the value chain (consumer focussed)	
	leasing low emission cars in the UK and then driving them abroad, which would have no benefit to the country. The British Vehicle Rental and Leasing Association (BVRLA) argues that this threatens to leave leasing companies at a distinct disadvantage when it comes to marketing low emission and electric vehicles, with the risk of so-called cross-border leasing being overstated and that the industry was now being unfairly penalised http://www.businessgreen.com/bg/analysis/2256630/budget-2013-tax-allowances-could-drive-corporate-fleets-away-from-greener-cars **Local benefits** (**'non-fiscal incentives'**) (Bakker et al. 2012) • *Parking Charges*. Some local authorities provide exemptions or a reduced charge for electric cars • *London congestion charge. London congestion charge*. 100 % discount for many types (but not all, e.g. hybrid) EVs, saving up to £2,000 per annum **Rebates/subsidies** (Berkeley 2012/Kotter and Shaw 2013) • *Plug-in car grant*. The purpose of this grant programme is to enable the purchase of ultra-low carbon vehicles. This subsidy programme has a £43 m consumer incentive scheme for EVs and PHEVs, up to 2015. This grant, first available from January 2011, reduces the cost of eligible cars by 25 % up to a maximum of £5,000 for both private and business buyers • *Plug-in van grant*. Aimed at light truck (N1) vehicles that fulfil qualifying criteria; these grants will enable purchasers to receive 20 % off the cost of a van up to a maximum of £8.000 • *Local grants*. Funding through the Local Sustainable Transport Fund (LSTF) will replace the Local Transport Plan funding stream, with £560 m available for 2012–2015
Comparative case: California	**Tax incentive** • Tax credits for purchasing electric vehicle (between $2,500 and $7,500 per vehicle, depending on battery capacity) **Rebates/subsidies** • A credit equal to 10 % of cost up to a maximum of $4,000 is available for kits that will convert a standard vehicle to plug-in EV • Clean Vehicle Rebate Project offers rebates for the purchase or lease of qualified vehicles. Rebates up to $2,500 per vehicle

In Belgium, unlike most of the studied countries, measures such as tax rates are a regional responsibility. Since 2002, the Belgian regions (Flanders, the Brussels Capital and the Walloon Region) are responsible for the vehicle tax base, tax rates and exemptions

major vehicle manufacturers in Germany (which collectively comprise the largest automotive industry in Europe). Sweden also has a strong focus on R&D. Over one-third of the policy instruments found in Sweden focusses on stimulating Research and Development. In France, Renault has teamed up with the CEA (French Alternative Energies and Atomic Energy Commission) to work on electric vehicles, new energies and cleaner combustion engines. Compared to the European cases, California is very upstream (mostly R&D) focused. A lot of programmes fund research activities and experiments.

Table 10 Examples of upstream financial incentives

Financial incentives—upstream of the value chain (R&D and production focussed)	
Germany	**Research funding** (BMWI 2014/Squarewise 2010) • The storage battery programme is founded to build capacities in Germany for implementation throughout the whole supply chain in the production of storage batteries. The programme runs from 2009 until 2012, and the Federal government has granted 35 million Euros to this programme • The third mobility and transport research programme (BMWI) sets out the goals, for instance to research into drive technology. Special importance is attached to developing new vehicle concepts and technologies for reducing energy consumption and pollution by road transport • Through the BMBF ICT 2020 research for innovation, EENOVA receives 100 million Euros for research on energy management in EVs • The Lithium-ion battery alliance is a project to substantially increase the energy and performance density of lithium-ion batteries and to accelerate the possible use in production. The Federal government has granted 60 million Euros to this project
Sweden	**Research funding** • The government invested SEK 240 Million to partially finance research into environmentally friendly vehicles. The Swedish Energy Agency invested SEK 20 Million. One of the projects in which is invested in by the Swedish government is a project that is set up to develop and demonstrate EVs (Government offices of Sweden 2008) • The vehicle strategic research and innovation programme was started in 2009 as a cooperative effort between the government and the Swedish automotive industry. The programme finances common research effort, innovation and development activities. Public funds amount to SEK Million per year (approximately 105 million Euro) (IEA IA-HEV 2011) • The Swedish Hybrid Vehicle Centre Programme focusses on developing a competitive R&D centre for hybrid and electric vehicle technology through continuous cooperation between industry and academia (U.S. Commercial Service Global Automotive Team 2011) • The Environmental Vehicle Development Programme aims to contribute to global leadership within vehicle electronics and software and increase expertise in the efficient design of vehicles (VINNOVA 2013)

(continued)

Table 10 (continued)

Financial incentives—upstream of the value chain (R&D and production focussed)	
Comparative case: California	**Research funding** • Envia Systems Inc. will create a low cost, high energy density, high performance battery system for electric and plug-in hybrid electric vehicles. Grant amount $9 million from CEC and $4 million from American Recovery and Reinvestment Act • Advanced cells and design technology for electric drive batteries. This project will develop next-generation high-energy lithium-ion cells leveraging silicon anodes, doubling the capacity of state-of-the-art vehicle batteries. $4,986,984 • Advanced cells and design technology for electric drive batteries. This project will develop high-energy cells using a lithium metal anode and a proprietary solid polymer electrolyte that will significantly reduce battery cost and size, and improves life and safety. $4,874,391 • Advanced cells and design technology for electric drive batteries. This project will develop next-generation high-energy lithium-ion cells leveraging, high voltage composite cathode materials and silicon-based anodes doubling the capacity of state-of-the-art vehicle batteries. $4,840,781 • Advanced Energy Storage technologies research programmes. Research portfolio is focused on battery module development and demonstration of advanced batteries to enable a large market penetration of Electric Driven Vehicles (EDV) within 5–10 years • Fundamental basic energy research on enabling materials for batteries through the Energy Frontiers Research Centres • Transformational research on revolutionary, "game-changing" energy storage technologies. EDV-related projects include metal–air, lithium–sulphur, magnesium-ion, advanced lithium-ion and solid state batteries, as well as ultra-capacitors • Grid Energy Storage and Battery Secondary Use. The Luskin Centre is developing innovative strategies to enhance PEV value through secondary use of PEV batteries. This includes both vehicle-to-grid power (V2G) and post-vehicle repurposing of used PEV batteries ("second life") into stationary energy-storage appliances (B2G) **Production funding** • *Sales Tax Exclusion*. Advanced Manufacturing (CAEATFA programme). Provides a Sales and Use Tax Exclusion Programme for advanced manufacturing projects. Effective since 1 January 2013 **'Real world testing and experimenting'** • *EV Readiness research*. With funding from the U.S. Department of Energy (DOE) and the Commission for Environmental Cooperation (CEC), California's major regions are assembling PEV Readiness plans. The Luskin Centre is the prime research contractor. This research is aimed at informing the strategic development of public and other charging infrastructure necessary to effectively support a transition to PEVs in Southern California. Additional related projects include examining PEV parking policies • Clean fuel programme provides funding for research, development, demonstration and deployment projects that are expected to help accelerate the commercialization of advanced low emission transportation technologies. South Coast. Approximately $10 million annually

3.5 *Finding 5: Few Countries Focus on Charging Infrastructure. also, Policy in the Infrastructure Value Chain Focuses Less on Downstream and Targets the Upstream Segments (Production and Services)*

In the infrastructure value chain, the focus upstream can be explained by the relatively large number of policies that focus on the installation of (semi)-public charging points (mostly by regional and local governments). Many of those instruments focus on the installation of (semi-) public charging points. Studies show that most EV charging currently takes place at home (Snyder et al. 2012). For instance, the UK national government initiated from 2009 onwards the PIP (Plugged-In-Places) programme. It intended to support the development and consumer uptake of ultra-low carbon vehicles by introducing electric car hubs in six key British cities. Compared to the European cases, California has a lot of rebate/subsidy instruments which focus on the installation of a charging infrastructure. A lot of which are focused on home chargers.

Table 11 shows the focus in policy for the charging infrastructure value chain. Table 12 presents a series of examples of financial incentives that target the downstream of the infrastructure value chain.

Table 11 Policy focus in the infrastructure value chain (Van der Steen et al. 2014a, b)

Policy focus in the charging infrastructure value chain				
Country	R&D	Production	Services	Customer
Belgium	0	+	++	++
Denmark	++	+	+	++
Germany	++	++	+	+
Netherlands	+	+++	+	+
Norway	+	++	+	++
Sweden	++	+	+	++
UK	+	++	++	++
Comparative case: California	+	++	+	+++

0 = Limited information found/available
\+ = Limited focus
++ = Strong focus
+++ = Prevalent focus area

Table 12 Financial incentives downstream in the infrastructure value chain

Financial incentives for charging—downstream of the value chain (consumer focussed)	
Belgium	**Tax incentives** • When a private actor installs a charging point on the outside of his house they are entitled to 40 % tax deduction with a maximum of 260 Euros for the year 2013 (Federale overheidsdienst financiën n.d.) • Additional deductibility of 13.5 % on the investment in charging infrastructure for companies under corporate tax system (ECN 2012)
Netherlands	**Tax incentive** (RVO NL 2013) • Through the MIA and VAMIL regulation of the central government, entrepreneurs can receive a subsidy for installing charging infrastructure **Rebates/subsidies** • Drive4Electric (Province of Friesland) introduced a subsidy on the creation of charging points. Customers and companies that create charging points on private space can get a discount of 500 Euros per charging point (ZERAUTO 2014) • The Rotterdam Electric Programme supports the first 1.000 EV owners with an electric charging point. On private property, a charging point is partly subsided (IEA IA-HEV 2012)
Norway	**Local benefits ('non-fiscal incentives'**) • *Free use of charging infrastructure*. EV users can use the public charging infrastructure for free (ECN 2012) • *Grants*. The Norwegian government has granted 11,9 Million Euro for new recharging stations (Bakker et al. 2012)
UK	*PIP (Plugged-in-places)*. Intended to support the development and consumer uptake of ultra-low carbon vehicles by creating electric car hubs in six key British city or city regions or hubs with the installation of charging point in various locations (Bakker et al. 2012/Kotter and Shaw 2013)
Comparative case: California	**Rebates/subsidies** • *PEV Home Charger Deployment Program*. Provides incentives for up to 2,750 residents who purchase a new plug-in electric vehicle and install Level 2 EVSE from qualifying vendors in Bay Area • *Free charging equipment*. ECOtality offers EV Supply Equipment at no cost to individuals in the Los Angeles and San Diego metropolitan areas. 1,786 EVSE in California installed. 2,785 in total project. The value of the project is $230 million. • *PEV Charging Rate Reduction*. Southern California Edison (SCE) offers a discounted rate to customers for electricity used to charge EVs. Two rate schedules are available for PEV charging during on- and off-peak hours • *Charger Installation Rebate*. The Los Angeles Department of Water and Power (LADWP) provides rebates of up to $2,000 for the first 1,000 residential customers who purchase or lease a qualifying EV and install a rapid, Level 2 charger and a separate time-of-use metre at their home. The programme expires 30 June 2013 • *ChargeUp LA*. LADWP provides rebates to residential customers for the cost of EV chargers and installation. The rebate will cover up to $2,000 of out-of-pocket costs • *PEV Charging Rate Reduction*. In Sacramento Municipal Utility District, this rate option is for residential customers who own or lease EVs • *PEV Charging Rate Reduction*. The LADWP offers a $0.025/kw discount for electricity used to charge EVs during off-peak times. The discount is only applicable for first 500 kWh in month

4 Conclusion

The study finds that EV policy captured here mainly targets the vehicle value chain. Also, most countries adopt policies that target the downstream segments of the value chain, especially consumers. Policy hardly takes into account the segment of services. Within this category of downstream oriented policy, most tools are financial. Especially Denmark, Norway and the Netherlands have strong financial downstream incentives. Three types of financial downstream incentives focusing on EVs are most common: tax incentives, rebates and specific local extra benefits for EV owners (e.g. free parking). The Netherlands and Norway both have a high number of tax incentives that make it very attractive for both businesses and consumers to buy or lease EVs. Interestingly, Denmark has similar financial downstream incentives but has so far seen much lower sales and EV penetration in the market. Only a few countries seem to focus explicitly on charging infrastructure. Also, in most cases infrastructure policies focuses more upstream in the value chain (stronger focus on government purchasing and tenders). In the documents the authors studied there was little clear relation between policy directed at vehicles and those focusing on charging. Although the two are evidently sides of the same coin, policy is often made in two separate silos. A more integral policy strategy could improve the performance of policy.

Given the current phase in the introduction of EVs, the emphasis on financial instruments is understandable. The purchase price of an EV and a private charger are high and this will withhold even the early innovators eager to drive an EV from buying one. Downstream financial instruments can overcome these important barriers and have probably been an important factor for the quite successful penetration of EVs in the market; downstream financial policies have been the backbone of the early market phase of EVs. However, if we take into account the exponential growth in the numbers of sales required for the next phase in the introduction, this policy strategy quickly becomes unsustainable. The exponential growth of the next phase of the introduction of EVs requires a self-enforcing loop in the sales of EV, not government policy that is "pushing" sales by a range of very strong and direct incentives; policy should become more oriented at managing such loops (see: Van der Steen et al. 2013). Already, countries' resources and public support are overstretched and there is societal pressure to downsize financial stimuli. As the quantity of vehicles grows, governments have to look for other tools to stimulate the market for EVs. It is safe to conclude that government policy greatly contributed to the first small but significant steps on the path towards full-scale introduction of e-mobility; however, policy makers will need a different strategy and different policy tools to further the next step in the introduction. This study displays and reviews the policies made to support the small first steps, now policies have to be developed that support the giant leap.

References

Bakker, S. (2013). *Standardization of EV Recharging Infrastructures*. Report written within the framework of activity 4.4 of the Interreg IVB project e-mobility NSR. December 2013. Retrieved from http://e-mobility-nsr.eu/fileadmin/user_upload/downloads/info-pool/4.4_E-MobilityNSR_Recharging_infrastructure_standardization.pdf.

Bakker, S. (2014). *Momentum for electric mobility—dynamics of multi-level governance*. Presentation at conference "Electric Vehicles and Eco Cars: Solutions for Green Growth", 11 April 2014, 9–5 pm, London Metropolitan University, UK. Retrieved from http://e-mobility-nsr.eu/fileadmin/user_upload/events/2014_Final_Conference/Presentations/14_Bakker_EMOB_London_2014.pdf.

Bakker, S., Lima J., & Trip. J. J. (2012). *Electric mobility policies in North Sea Region countries*. NSR E-mobility project report. Technical University of Delft, The Netherlands. Retrieved from http://e-mobility-nsr.eu/fileadmin/user_upload/downloads/info-pool/3.3_-_E-mobility_policies_in_the_NSR_countries.pdf.pdf.

Bakker, S., Maat, L. & Trip, J. J. (2014). *Transition to electric mobility: spatial aspects and multi-level policy-making. Project report NSR E-mobility network*. Retrieved from http://e-mobility-nsr.eu/fileadmin/user_upload/NEWS/Final_report_on_transnational_learning/Transition_to_electric_mobility__3.7_Final_report_.pdf.

Bathelt, H., & Glückler (2011). *The relational economy. geographies of knowing and learning*. Oxford: Oxford University Press.

Beeton, D. and Butte, B. (2013). *Future of markets for electric vehicles. expectations, constraints & long-term strategies*. Report of a roadmapping workshop facilitated by Urban Foresight for the International Energy Agency Hybrid & Electric Vehicle Implementing Agreement and the Austrian Institute of Technology (April 2013). Retrieved from http://www.ieahev.org/assets/1/7/EV_Ecosystems_Future_Markets_Report.pdf.

Beeton, D. (2012). *Electric vehicle cities of the future: A policy framework for electric vehicle ecosystems*. Newcastle upon Tyne: Urban Foresight.

Berkeley, N. (2012). The application of Green Technologies in the automotive industry: An assessment of policy attempts in the UK to stimulate the uptake of alternatively fuelled. *Regions Magazine (Regional Studies Association), 288*(1), 27–28.

BMVI—Elektromobilitat Model Region Hamburg. (2014). Retrieved from http://www.hamburg.de/pressearchiv-fhh/.

BMVI—Elektromobilitat Model Region Bremen/Oldenburg. (2014). Retrieved from http://www.bmvi.de/SharedDocs/DE/Artikel/UI/modellregion-bremen-oldenburg.html?nn=36210.

Browne, D., O'Mahony, M., & Caulfield, B. (2012). How should barriers to alternative fuels and vehicles be classified and potential policies to promote innovative technologies be evaluated? *Journal of Cleaner Production, 35*, 140–151.

CCSE (2014). Center for Sustainable Energy California. *February 2014 PEV Owner Survey Report*. See http://energycenter.org/clean-vehicle-rebate-project/vehicle-owner-survey/feb-2014-survey

CIVITAS. (2014). *City, vitality and sustainability*. Retrieved from http://www.civitas.eu/index.php?id=69.

Copenhagen Electric. (2014). *The regional EV secretariat*. Retrieved from http://www.cph-electric.dk/.

Delucchi, M. A., & Lipman, T. E. (2010). Lifetime cost of battery, fuel-cell, and plug-in hybrid electric vehicles. In G. Pistoia (Ed.), *Electric and hybrid vehicles. power sources, models, sustainability, infrastructure and the market* (pp. 19–60). Amsterdam: Elsevier.

Domingues, S. M., & Pecorelli-Peres, L. A. (2013). Electric vehicles, energy efficiency, taxes, and public policy in Brazil. *Law and Business Review of the Americas, 19*(1), 55–78.

E4Tech (2013). *Low carbon vehicles in the north east—economic impact study. Final Report for North East Local Enterprise Partnership Board*, 11th September 2013. Newcastle. upon Tyne/London.

ECN (2012). *Elektrisch vervoer in Nederland in internationaal perspectief. Benchmark elektrisch rijden 2012*. Retrieved from http://www.rijksoverheid.nl/documenten-en-publicaties/rapporten/2012/07/23/elektrisch-vervoer-in-nederland-in-internationaal-perspectief.html.

EERE (2014a). US department of energy. *Energy efficiency & renewable energy. vehicle technologies program*. Retrieved from http://www1.eere.energy.gov/library/.

EERE (2014b). US Department of Energy. Energy efficiency & renewable energy. *Clean Cities program*: http://www1.eere.energy.gov/cleancities/about.html.

EERE (2014c). US Department of Energy. Energy efficiency & renewable energy. *Advanced Power Electronics and Electric Machines Program*. Retrieved from https://www1.eere.energy.gov/vehiclesandfuels/pdfs/program/2010_apeem_report.pdf.

European Commission. (2011). Eco-innovation action plan. *Danish green transport plan to get the environment back on track*. Retrieved from http://ec.europa.eu/environment/ecoap/about-eco-innovation/policies-matters/denmark/388_en.htm.

Federale overheidsdienst financien (n.d.). *Belastingvermindering bij aankoop van een elektrisch voertuig of installatie van een laadpaal*. Retrieved from http://www.minfin.fgov.be/portail2/nl/themes/transport/vehicles-electric.htm.

Flynn, A. and Morgan, K. (2004). Governance and sustainability. Chap. 2, pp. 21–39, In M. Thomas & M. Rhisiart (Eds.), Sustainable regions. Making sustainable development work in regional economies. Vale of Glamorgan: Aureus Publishing Ltd.

For North East Local Enterprise Partnership Board. (11th September 2013). Newcastle upon Tyne/London.

Fournier, G., & Stinson, M. (2011). The future thinks electric. Developing an electric mobility value chain as a foundation for a new energy paradigm. *Interdisciplinary Management Research, 7*, 867.

Funk, K., & Rabi, A. (1999). Electric versus conventional vehicles: social costs and benefits in France. *Transportation Research Part D: Transport and Environment, 4*(6), 397–411. November 1999.

Government Offices of Sweden (2008). *Joint initiative to present Swedish electric cars*. Retrieved from http://www.government.se/sb/d/10123/a/100866.

Green, E. H., Skerlos, S. S., & Winebrake, J. J. (2014). Increasing electric vehicle policy efficiency and effectiveness by reducing mainstream market bias. *Energy Policy, 65*, 562–566. Feb. 2014.

Hood, C., & Margetts, H. (2007). *The tools of government in the digital age* (2nd ed.). Basingstoke: Palgrave Macmillan.

http://energycenter.org/clean-vehicle-rebate-project/vehicle-owner-survey/feb-2014-survey.

http://luskin.ucla.edu/sites/default/files/Non-Residential%20Charging%20Stations.pdf.

http://www.plugincars.com/six-bills-would-ensure-californias-ev-future-128410.html.

IEA IA-HEV. (2011). International Energy Agency, Electric Vehicles Initiative. Hybrid and Electric Vehicles. *The electric drive plugs in. implementing agreement for co-operation an hybrid and electric vehicle technologies and programmes*. www.ieahev.org.

IEA IA-HEV. (2012). International Energy Agency, Electric Vehicles Initiative. *EV City Casebook*. Retrieved from http://www.cleanenergyministerial.org/Portals/2/pdfs/EV_City_Casebook_LR.pdf.

IEA IA-HEV. (2014). International Energy Agency, Electric Vehicles Initiative. The International Energy Agency (IEA) Hybrid & Electric Vehicle Implementing Agreement. Retrieved from http://www.ieahev.org/by-country/demark-on-the-road-and-deployments/.

IEA IA-HEV & AVERE. (2013). International Energy Agency, Electric Vehicles Initiative and AVERE (2013) *Global EV Outlook 2013*. International Energy Agency, Electric Vehicles Initiative. Retrieved from http://www.iea.org/publications/globalevoutlook_2013.pdf.

Karplus, V. J., Paltsev, S., & Reilly, J. M. (2010). Prospects for plug-in hybrid electric vehicles in the United States and Japan: A general equilibrium analysis. *Transportation Research Part A: Policy and Practice, 44*(8), 620–641.

Kotter, R. (2013). The developing landscape of electric vehicles and smart grids: A smart future? *International Journal of Environmental Studies, 70*(5), 719–732.

Kotter, R. and Shaw, S. (2013). *Work Package 3. Activity 6: Micro to macro Investigation. Project report, April 2013*. North Sea Region Electric Mobility Network. http://www.northsearegion.eu/files/repository/20130716114213_E-mobility_3.6._Main_Report_April_2013.pdf.

Lane, B. W., Messer-Betts, N., Hartmann, D., Carley, S., Krause, R. M., & Graham, J. D. (2013). Government promotion of electric car: risk management or industrial policy? *European Journal of Risk Regulation, 4*(2), 227–246.

Loisel, R., Passaoglu, G., & Thiel, C. (2014). Large-scale deployment of electric vehicles in Germany by 2030: An analysis of grid-to-vehicle and vehicle-to-grid concepts. *Energy Policy, 65*, 432–443. Feb. 2014.

Los Angeles Cleantech Business Incubator. (2014). Retrieved from http://laincubator.org/about/.

Plugincars. (2013). *Six Bills That Would Ensure California's Electric Car Future*. http://www.plugincars.com/six-bills-would-ensure-californias-ev-future-128410.html

Programma bureau Luchtkwaliteit. (2010). Amsterdam Elektrisch Actieplan.

Prud'homme, R., & Konig, M. (2012). Electric vehicles: A tentative economic and environmental evaluation. *Transport Policy, 23,* 60–69 (September 2012).

R.J. in 't Veld (2005). Zonneklaar. *Onderzoek naar de rol van de overheid bij de introductie van zonnestroom in Nederland* (English—A research on the government role in introducing solar energy in the Netherlands).

RVO NL. (2013). Tax Measures in the Netherlands. Retrieved from http://www.agentschapnl.nl/programmas-regelingen/mia-milieu-investeringsaftrek-en-vamil-willekeurige-afschrijving-milieu-invest.

Sierzchula, W., Bakker, S., Maat, K., & Wee, B. van (2014). The influence of financial incentives and other socio-economic factors on electric vehicle adoption. *Energy Policy*, *68*(C), 183–194.

Snyder, J., Chang, D., Erstad, D., Lin, E., Falkan Rice, A., & Tzun Goh et al. (2012). *Financial Viability of Non-Residential Electric Vehicle Charging Stations*. UCLA Luskin Center for Innovation.

Spiegel Online. (2012). Gesetzentwurf der Regierung: Elektroautos fahren zehn Jahre steuerfrei. Retrieved from http://www.spiegel.de/auto/aktuell/keine-kfz-steuer-fuer-elektroautos-zehn-jahre-lang-a-834800.html.

Squarewise. (2010). *Elektrisch Rijden: internationale stand van zaken*. (English—E-mobility: international overview). For the Ministry of Economic Affairs in The Netherlands.

Steinhilber, S., Wells, P., & Thankappan, S. (2013). Socio-technical inertia: Understanding the barriers to electric vehicles. *Energy Policy*, *60,* 531–539 (September 2013).

The New Drive. (2014). Retrieved from http://thenewdrive.be/nl/elektrisch-rijden/wat-kost-een-elektrische-auto/subsidies.

Transnova. (2014). Retrieved from http://www.transnova.no/english/.

Trip, J. J., Lima, J., & Bakker, S. (2012). *Electric mobility policies in the North Sea Region countries*. Project report NSR e-mobility. Retrieved from http://e-mobility-nsr.eu/fileadmin/user_upload/downloads/info-pool/3.3_-_E-mobility_policies_in_the_NSR_countries.pdf.pdf.

U.S. Commercial Service Global Automotive Team. (2011). *Electric Vehicles. Europe in Brief.* Retrieved from http://export.gov/build/groups/public/%40eg_main/%40byind/%40autotrans/documents/webcontent/eg_main_035287.pdf.

Van der Steen, M., van Deventer, P., de Bruijn, J. A., van Twist, M. J. W., ten, E. F., & Heuvelhof, E. et al. (2012). *Governing Innovation: The Transition to E-Mobility-A Dutch Perspective*. Paper presented at the AAG Annual Meeting, Paper Session 'Electric Vehicles', Saturday 25 February 2012, in NY, NY.

Van der Steen, M., Van Schelven, R., Mulder, J., & Van Twist, M. (2014a). *Introducing e-mobility: Emergent strategies for an emergent technology. Ambition, Structure, Conduct and Performance. Background report*. The Hague: Dutch School for Public Administration (NSOB). Retrieved from http://e-mobility-nsr.eu/info-pool/.

Van der Steen, M., Van Schelven, R., Mulder, J., & Van Twist, M. (2014b). *Introducing e-mobility: Emergent strategies for an emergent technology. Ambition, Structure, Conduct and Performance. Main report*. The Hague: Dutch School for Public Administration (NSOB). Retrieved from http://e-mobility-nsr.eu/info-pool/.

Van der Steen, M., van Twist, M., Fenger, M., & LeCointre, S. (2013). Complex causality in improving underperforming schools: a complex adaptive systems approach. *Policy & Politics, 41*(4), 551–567.

Van Deventer, A. P., van der Steen, M., De Bruijn, J. A., ten Heuvelhof, E. P., & Heynes, K. E. (2011). *Governing the transition to e-mobility: small steps towards a giant leap*. The Hague: Netherlands School of Public Administration.

VINNOVA. (2013). Vehicle Development Program. Retrieved from http://www.vinnova.se/en/FFI—Strategic-Vehicle-Research-and-Innovation/Vehicle-Development/.

WSDOT. (2011). *Electric Vehicle Policies, Fleet, and Infrastructure: Synthesis*. Retrieved from http://www.wsdot.wa.gov/NR/rdonlyres/5559AE0E-8AB5-4E6B-8F8B-DEAA7ECE715D/0/SynthesisEVPoliciesFleetandInfrastructureFINALRev112911.pdf.

Yang, C.-Y. (2012). Launching strategy for electric vehicles: Lessons from China and Taiwan. *Technological Forecasting and Social Change, 77*(5), 831–834. June 2010.

ZERAUTO (2014). *Elektrisch Vervoer in Nederland. EV in Friesland*. Retrieved from http://91.205.33.8/Agora/index.php?page=EV_in_Friesland&pid=319.

An Analysis of the Standardization Process of Electric Vehicle Recharging Systems

Sjoerd Bakker and Jan Jacob Trip

Abstract Electric vehicles of various manufacturers are being deployed throughout Europe. To recharge these vehicles, an infrastructure of rechargers is needed to enable charging at both private and public parking facilities. Throughout Europe, different charging protocols, plug designs and billing systems have been developed and introduced. In this chapter, the authors describe these standards and analyse the current situation in north-western Europe regarding the installed equipment and initiatives to realize national and international interoperability between currently isolated networks of chargers. The authors conclude that there is a problematic tension between early attempts to define national standards and the eventual need for international interoperability to enable cross-border travel with electric vehicles.

Keywords E-mobility · Vehicle · Recharging systems · The Netherlands · Infrastructure

1 Introduction

Electric vehicles (EVs) are quickly entering the European car market and they have reached market shares of up to 1.5 % in countries such as the Netherlands and Norway (Sierzchula et al. 2014). Even though most users are able to use their cars at home or at their offices, a complementary infrastructure of vehicle rechargers is needed nevertheless to extend the practical range of these vehicles and thereby to increase their value to their users (Dimitropoulos et al. 2013). Such an infrastructure can be realized on private grounds, such as offices and shops, but also in public space (Bakker and Trip 2013). To maximize the utility of the recharging infrastructure, all EVs would have to be able to recharge at all available charging

S. Bakker (✉) · J.J. Trip
Faculty of Architecture and the Built Environment, Delft University of Technology, Jaffalaan 9, 2628 BX Delft, The Netherlands
e-mail: s.bakker-1@tudelft.nl

W. Leal Filho and R. Kotter (eds.), *E-Mobility in Europe*,
Green Energy and Technology, DOI 10.1007/978-3-319-13194-8_3

stations. To achieve this goal, standardization of recharging plugs, charging protocols and payment systems is thus necessary. First and foremost, EV drivers would be served by regional and national standards, but at some point international standardization is necessary as well to facilitate cross-border trips.

Besides the practical value of interoperability to EV drivers throughout Europe, one could also argue that standardization would be beneficial to equipment manufacturers and charging network operators as it would provide much needed clarity and bring about positive scale effects (Brown et al. 2010). Furthermore, standardization could also take away some of the uncertainties about charging among potential EV drivers and, in a broader rhetorical sense, position the EV as a viable option today instead of presenting it as an underdeveloped future option.

Attempts to standardize EV plugs are as old as the EV itself and go back to the early 1900s (Mom 2004), but a global standard has not emerged yet. In practice, as the authors will show in this chapter, there are still many competing standards in use worldwide. In fact, even within many (European) countries, multiple networks of chargers have emerged that do not (yet) allow EV drivers to roam between these networks. The variety and incompatibility among these networks mean that EV drivers cannot use their EVs to their full potential and that cross-border trips are virtually impossible. The EU has called for standardization of EV charging systems on several occasions, but only in January 2013 has it published a clear draft Directive, agreed in amended form in March 2014, that provides clarity on the charging systems and plug designs that are to become the new European standard (European Commission 2013).

One could however argue that this Directive is issued too late and that the new EU standard will have to compete with incompatible standards already installed in many regions and nations. To make matters worse, the now agreed Directive only deals with harmonization of the hardware, e.g. the plugs and sockets used, but it does not address the interoperability between recharging networks in terms of customer identification and payment systems (Knox 2013). Much more thus needs to be done before true interoperability of European charging networks is realized and to ensure that EV drivers can actually charge their cars anywhere in the EU.

In this chapter the authors provide an overview of available standards worldwide and the status quo regarding the different standards in use in the seven countries in the North Sea Region (NSR). The authors thereby limit themselves to a general description of the plugs standards in use and the charging protocols that guide the actual charging of the vehicles.[1] For each country, the authors then describe the most common standards in use and the differences among the major recharging networks. Such networks may be regional, but they may also be defined by a specific network operator. In the first case, charging and travelling may be limited to a specific region, in the latter case, charging is typically limited to a (national) network of a specific operator. Whereas as the overview of available standards is

[1]Technical details of the various standards can be found, amongst others, in publications from the Green eMotion project: http://www.greenemotion-project.eu.

drawn from the literature, the analysis of charging standards in use in the NSR is based on a series of (telephone) interviews with stakeholders in the individual countries.

2 Electric Vehicle Recharging Standards: The Options

There are three basic options for recharging an electric vehicle: wired, wireless by means of induction,[2] and by swapping its batteries.[3] The remainder of this chapter deals exclusively with wired charging since this is currently the only option that is used in practice and for which both cars and charging equipment are commercially available.

For wired charging, two options can be distinguished: AC and DC charging. Charging with alternating current (AC) is used for conventional and semi-fast charging at homes and offices and the majority of public recharging stations. Direct current (DC) is used for fast charging. Since all batteries require DC power to be charged, the AC power that is delivered by the electricity grid needs to be converted to DC at some point. An AC/DC convertor is thus needed between the grid and the battery. In the case of AC charging with regular mains power, power levels are low enough to install a small converter on-board the vehicle. For fast charging with higher power levels, a bigger and more expensive converter is needed that would not easily fit in a typical car. These high-power converters are therefore incorporated in the charging station and DC power is delivered to the car (see Fig. 1 for a schematic drawing of these differences). Also, because of the higher power levels and related safety concerns, DC charging cables are always fixed to the charging

[2]A battery can be charged inductively without using cables. Instead, an electromagnetic field is used to transfer energy to the vehicle. This way of charging is tested in numerous experiments and may very well become the charging mode of the future. Today however, no cars are ready for conductive charging nor are there any standard chargers available for this. Still, it is an attractive option because it does not involve any cumbersome equipment or cables and thereby does not spoil the streetscape in inner cities for instance. Also, and perhaps more importantly, it could be used for charging while driving by means of inductive road surfaces or for instance to quick-charge buses at bus stops.

[3]An empty battery can be swapped with a fully charged one. Potentially, this is the fastest way of 'recharging' an EV, but most likely also the most expensive way since it requires the construction of (automated) swapping stations. Additional batteries are needed as well. Furthermore, standardization of EV batteries would be needed to some extent (to prevent stocking a wide variety of battery types and sizes), and EVs need to be specifically designed to be suitable for battery swapping (most of them currently are not). The company Better Place was well known for developing and deploying swapping stations in, amongst others, Israel, Denmark, Japan, and the Netherlands, but went bankrupt in May 2013. Renault was the only car manufacturer that cooperated with Better Place and has produced about a thousand Fluences with switchable batteries. The dedicated EV manufacturer Tesla Motors has also announced that it is testing battery swapping options, but this has not been done outside its factory gates yet (http://www.teslamotors.com/batteryswap).

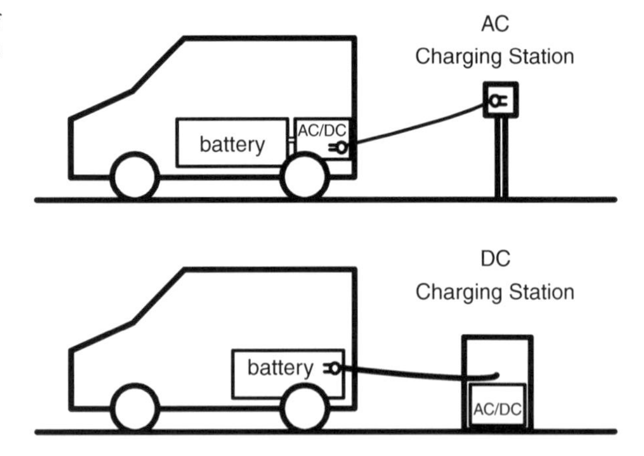

Fig. 1 Schematic drawing of AC and DC charging stations

station and there is thus only one plug that needs to be standardized. AC charging cables are often, but not always, loose cables with plugs on both ends and standardization may thus be necessary for both plugs.

In the following sections, the authors will present and discuss the emergence of the various AC and DC standards separately and also address the need for standardization on the car- and wall-side.

2.1 Specific Standards: Modes, Types and Identification Systems

For both AC and DC charging, multiple plug designs and charging modes have been developed and have been deployed throughout the world. Next to that, an even wider variety of identification and billing systems have been developed. In order to enable EV drivers to roam between networks and ultimately between countries, interoperability, and thus standardization, is necessary between the various modes, plugs, and identification and billing systems. Below the authors introduce these concepts and explain the major differences between the various options.

2.1.1 The Charging Modes

The charging mode refers to power levels that charger and its connectors are rated for and the control and safety features that guarantee safe and efficient charging. The International Electrotechnical Commission (IEC) has recognized four different charging modes that vary in terms of complexity of the system and the speed with which a vehicle can be recharged (Van den Bossche et al. 2012; IEC 2014a, b).

Mode 1 charging encompasses charging from regular mains sockets (up to 16 Amperes) and is done without any specific safety or communication features. This mode by definition requires the usage of a loose cable with plugs that match the car-side as well as the wall-side (e.g. a home socket or charging equipment).

Mode 2 charging encompasses charging from regular mains sockets as well, but features a special cable with a so-called in-cable-control-box that controls the power level and thereby protects the user and the vehicle. Both Modes 1 and 2 are used in situations where a dedicated infrastructure is lacking (e.g. at home) or where the network operator has decided to offer a rather uncomplicated system. For instance, in Norway, most of the 'regular' recharging networks consist of basic sockets that can be used by any EV driver who has a key to unlock the charger. Because Mode 1 and 2 charging make use of regular sockets, plug designs vary per country and cross-border trips would require the use of several cables with varying plugs on the wall-side.

Mode 3 charging, which is to become the European standard, makes use of dedicated charging equipment which guarantees safe usage and also enables communication between the charging equipment and the vehicle. Because of these additional features, a special cable and plug and socket combination are necessary.

Finally, *Mode 4* charging entails the use of an AC/DC converter and charger in the charging equipment (instead of on-board the vehicle) and DC power is delivered to the vehicle. Mode 4 is typically used for fast charging with power levels starting at 50 kW.

2.1.2 The Plug Types

The plug type refers to the physical design of the plug(s) with which the vehicle is connected to the charging equipment. There are three officially recognized plug designs for Mode 3 charging, these are designated as Types 1, 2 and 3 (IEC 2011).

Type 1 (Yazaki) is used mainly in the US and Japan and is supposed to be used on a cable that is fixed to the charging equipment. In other words, the Type 1 plug is used specifically to plug into the car and therefore requires a car with a compatible inlet (the vehicle inlet).

The *Type 2* plug (Mennekes), the new European standard, is used on loose cables and connects the cable to the charging equipment. On the car-side, the cable can have any plug that matches the vehicle's inlet, but this is often a Type 1 design because many cars have a Type 1 inlet anyway. Type 2 plugs are rated for higher power levels than Type 1 plugs and can therefore be used for semi-fast charging with chargers that make use of three-phase power connections.

The *Type 3* plug (Scame) is mostly the same as the Type 2 plug, but its use is limited to several countries in southern Europe (i.e. Italy, France). These countries prescribe the use of so-called safety shutters on power outlets that are installed outside and the Type 3 socket features such shutters. Because the Type 2 and 3 plug and socket combinations are not compatible, travelling between, for instance, Germany and France would require an additional cable.

As for mode 4, DC fast charging, there is currently only one design that is used in practice. This is the CHAdeMO standard and this standard specifies the charging protocol as well as the physical design of the plug and vehicle inlet. This implies that a CHAdeMO charger, like all mode 4 chargers, makes use of a fixed cable. As the authors will describe further on in this chapter, a large consortium of car manufacturers has agreed on a competing standard in which either a Type 1 or Type 2 plug is combined with additional pins for DC power. These are the Combo 1 and 2 plugs and are meant to be used on vehicles with a matching vehicle inlet that is also compatible with Type 1 and Type 2 plugs.

No actual standardization has yet taken place in Europe and the various plug types are still in use and most countries have in fact not even agreed on a national standard, despite the aforementioned now agreed Directive (European Commission 2013; Council of the European Union 2014). In this now agreed Directive two plug designs are selected to become the EU standard (Type 2 and Combo 2). To quote the now agreed draft Directive (where the following was not amended):

- *Alternating Current (AC) slow* recharging points for electric vehicles shall be equipped, for interoperability purposes, with connectors of *Type 2* as described in standard EN62196-2:2012.
- *Alternating Current (AC) fast* recharging points for electric vehicles shall be equipped, for interoperability purposes, with connectors of *Type 2* as described in standard EN62196-2:2012.
- *Direct Current (DC) fast* recharging points for electric vehicles shall be equipped, for interoperability purposes, with connectors of *Type 'Combo 2'* as described in the relevant EN standard, to be adopted by 2014.

In addition, this now agreed draft Directive which at the time of writing is still to be adopted formally approved by the European Parliament and the Council of Ministers (agreement at first reading, entering into force 20 days after its publication in the EU Official Journal), with Member States having 2 years to adopt national provisions to comply with the Directive, in July 2014 the European Commission issued a draft request to standardization bodies (which also contained a request relevant to electric buses, and also wireless charging) to develop a supplementary standard for the Type 2 plug over the next 2 years. This standard should specify how Type 2 plugs can be used on outdoor sockets with the mechanical shutters such as prescribed in a number of Member States (European Commission 2014) (Table 1).

So far the authors have described the various options for the charging hardware and the standardization of plugs and sockets. Just as important, to realize true interoperability between various charging networks, is the harmonization of driver (or vehicle) identification and billing systems. In the next subsection the authors briefly introduce this challenge.

2.1.3 The Identification and Billing System

The identification and billing system is the system that identifies the driver (or the car) and allows him or her to charge a car at a given charging station. This system thus comprises the identification of the driver by the charging station, but also the communication with the network's back office to secure payment or to verify that

Table 1 Plug types and maximum current and voltage levels, based on Van den Bossche (2010) and SAE, GreenEmotion

	Current (A)	Voltage (V)	Power (kW)	Charging point	Vehicle
Type 1	32	250	7.2	Cable with plug	Standard inlet today
Type 2	63[a]	480	44	Cable with plug or socket	Announced by consortium of OEMs
Type 3	63	480	44	Socket	N/A
CHAdeMO	125	500	62.5	Cable with plug	Standard inlet today
Combo 1 and 2	200	500	100	Cable with plug	Announced by consortium of OEMs

SAE, http://www.sae.org/mags/aei/11005/
GreenEmotion, http://www.greenemotion-project.eu/upload/pdf/deliverables/D7_2-Standardization-issuses-and-needs.pdf
[a]In case of a 3-phase power line

the driver is otherwise allowed to make use of the charger. In other words, this system provides the link between the driver and the network operator and/or service provider. Standardization of these systems, or compatibility between them, is crucial to realizing interoperability or roaming between the different recharging networks, both for national as well as international interoperability. Interoperability between networks also requires a higher level system (a clearing house) that connects the back offices of individual networks and that takes care of financial transactions between the network operators (or service providers). This chapter does go into the technical details of these systems, but there are several options for identification of the driver and to communicate with the back office(s). The most popular identification method today is the use of smart Radio-Frequency Identification (RFID) cards. Identification by means of a mobile phone or direct communication between the car and the charger are also in use or under development. All of these options can be used in subscription-based systems in which the system 'knows' the driver and can allow the driver to charge the EV. The driver is then also billed for the charging time or energy use (when the membership does not include energy use on a flat-rate basis). In case of roaming between networks, the system should be able to acquire information about the user from its own network and send a bill to that network. Mobile phone identification can also be used for ad hoc charging by means of an SMS payment and this option therefore provides, theoretically at least, most flexibility.

3 An Overview of the Plugs in Use in the NSR Region

Despite the fact that some countries in the north of Europe such as Germany and the Netherlands have adopted the Type 2 plug as their national standard, different plugs and sockets are still in use as well and roaming between different networks is by and far still impossible. Below the authors describe the status quo situation in the individual NSR countries and highlight initiatives that specifically aim to realize true international interoperability between the charging networks.

The findings are based on 30 interviews with representatives of stakeholders in these countries.[4] These stakeholders include local, national, and regional governments, electric utilities, electricity suppliers and dedicated recharging network operators. Except for the interviews in the Netherlands, all were conducted by telephone. Where appropriate, references to (online) written sources are provided.

3.1 United Kingdom

In the UK, about half of the recharging points have been realized as part of the Plugged-in-Places initiative (Kotter and Shaw 2013). These PiP's have been set up in eight different regions and in each of the regions, separate systems have been developed and roaming was by and large impossible. The problem was however not so much with the sockets on the chargers. Many, but not all, chargers in the UK offer both a regular British three-pin socket (BS 1363) and a Type 2 socket. In an attempt to further reduce the variety in sockets, national government decided that from April 2012 onwards, all chargers that are (partly) funded with public money have to offer at least a Type 2 socket. Insofar as DC fast chargers are installed, these are all CHAdeMO (among others at Nissan dealerships).

The real issue with interoperability in the UK, as in many other countries, is that the different PiP networks use different identification and billing systems and that roaming is virtually impossible between these networks.

A first exception was the harmonization of the networks in London and East of England (north of London) where EV drivers can use their identification cards on both networks. Another initiative to develop a nationwide network is the Charge Your Car (CYC) initiative. CYC started as the Plugged-in-Places project in the north-east of England, but has since strived to become the national recharging network for the UK. It was joined by Scotland's publicly funded charging stations[5] and later by other regional networks in the west[6] and north-west of England.[7] Still, various identification and billing protocols are in use in the individual PiP's and so far, the two initiatives mentioned above do not allow roaming between them.

It is also noteworthy that in the UK there is a trend to move away from flat-rate subscription-based systems to pay-per-charge systems in which drivers either pay

[4]The number of interviews per country: the UK 3, the Netherlands 7, Belgium 3, Germany 7, Denmark 4, Sweden 2, Norway 5.

[5]http://chargeyourcar.org.uk/news/transport-scotland-joins-national-electric-vehicle-charge-point-network/.

[6]http://chargeyourcar.org.uk/news/west-of-england-joins-charge-your-car/.

[7]http://chargeyourcar.org.uk/news/north-east-england-joins-charge-your-car-2/.

with their debit or credit card that is connected to their RFID charging card or by paying with their mobile phone. This move can be interpreted as a step towards commercialization of EV recharging and the development of viable business models.

Four major networks have emerged in recent years. In London there is the abovementioned Source London network that offers a variety of 3-pin sockets, Type 1 plugs and Type 2 sockets.[8] There is currently one CHAdeMO fast charger in the city centre. These chargers can be used in combination with a Source London membership card (RFID) that costs £10 per year.

As noted, the Charge Your Car network was based primarily in the North-East and started off as the regional PiP. This network consists of many local hosts that own the actual stations. CYC provides interoperability and roaming by providing a RFID card that works on all connected stations. Local hosts can either offer free electricity or can charge the EV driver via the RFID card and the driver's debit or credit card. The network includes both AC and DC CHAdeMO chargers. The AC chargers differ in terms of the sockets (British 3-pin, Type 2). Some fast-charge stations offer both CHAdeMO (50 kW) and AC Type 2 (22 kW).[9]

The POLAR network offers a subscription-based RFID card that can be used for POLAR's own network and in the future also for the Source London network. POLAR's chargers are mainly found at strategic locations between the already existing PiP networks. In addition, POLAR will also install several DC fast chargers throughout the country.[10] Semi-fast AC Type 2 chargers are offered nationwide by ECOTRICITY[11] in collaboration with and located at Welcome Break service stations. These will be complemented with CHAdeMO fast chargers in the future.

3.2 Belgium

In Belgium, the fast majority of EV chargers have been installed in pilot projects of the Flemish Living Lab initiative. These projects were explicitly meant to develop new technologies and systems and to learn about their usage. Technological variety was therefore fostered and standardization was never a priority. Agreements have been made however to use Type 2 sockets (possibly in combination with other sockets) and RFID cards for identification. Talks are ongoing to realize interoperability among the pilot projects and the networks of several commercial operators. These include the networks of Blue Corner, The Plug-in Company and The New Motion from the Netherlands that also operates several chargers in Flanders.[12]

---

[8]https://www.sourcelondon.net/map.php.

[9]http://chargeyourcar.org.uk/#map.

[10]http://www.polarnetwork.com/.

[11]http://www.ecotricity.co.uk/for-the-road.

[12]www.bluecorner.be.

BlueCorner is Belgium's biggest operator with a network of 50 charging stations which are produced by the company itself (under the Enovates brand). This is a subscription-based network and its members use an RFID card to access the chargers, but they do pay for the electricity on a time basis. BlueCorner's chargers offer a Type 2 and CEE 7/5 (French standard) socket. The Plug-in Company[13] offers 16 stations[14] with Type 1 and Type 2 connections and these are found among others at IKEA stores. The Plug-in Company uses an SMS payment system for its chargers in public space.

In terms of international interoperability, Blue Corner and The New Motion have both embraced the e-clearing.net initiative to enable interoperability with the Dutch and parts of the German network[15] (stations that are part of the Ladenetz initiative). Interestingly, Blue Corner has also signed an agreement with Hubject, the German commercial initiative to realize interoperability and payment services.[16]

3.3 The Netherlands

Of the roughly 2500 charging stations in the Netherlands, about 2000 have been installed by the E-Laad foundation. E-Laad was founded by (almost all of) the grid operators and offered free chargers to municipalities.[17] Other chargers were installed by a number of large cities (e.g. Amsterdam, Rotterdam and Utrecht) and some as part of individual pilot projects. All of the E-Laad stations and the large majority of municipal stations are fitted with Type 2 sockets. In fact, it was decided early on by a broad range of stakeholders that this was to be the Dutch standard for EV plugs and socket. Only some older charging stations still make use of the standard three-phase power socket (IEC 60309 industrial plug). These are for instance part of an early network that was installed in the city of Amsterdam.

Today, several companies (among others The New Motion) offer RFID cards that allow usage of all E-Laad chargers and the networks that were commissioned by several municipalities (e.g. Amsterdam, Rotterdam and Utrecht). These cards are offered for free, but EV drivers pay for the electricity on a charging-time basis.

There are about 50 DC fast chargers in the Netherlands, all of which use the CHAdeMO protocol and plug. These are all paid for on a pay-per-charge (time) basis using a specific RFID card of the fast-charger operator. The E-Laad network developed the OCCP (open charging point protocol) for communication between

---

[13] http://www.theplugincompany.com/en/.

[14] These are for electric vehicles (according to www.openchargemap.org), the network further consists of hundreds of chargers for e-bikes.

[15] http://www.e-clearing.net/news.php.

[16] http://www.hubject.com/pdf/PM_hubject_20130522_EN.pdf.

[17] http://www.e-laad.nl/.

individual charging stations and the network's central system[18] that is now in use in several countries. And it is also involved in the development of the OCHP (open clearing house protocol) for communication between multiple networks to allow roaming of customers and billing across networks. The OCHP forms the basis for the international roaming initiative e-clearing.net.

3.4 Germany

The German situation is comparable to the UK and Belgium. As a result of regionally oriented test and demonstration projects, separate charging networks with different designs have emerged. In Germany, in contrast to the UK and Belgium, the plug itself was standardized early on and the Type 2 plug (a German design) was selected. Roaming between the different networks is however not possible due to differences in the identification and billing systems. Besides local charging networks that have installed as part of the model regions (Modellregionen) initiatives, charging stations have been installed by the large energy companies such as Vattenfall (mainly in Hamburg and Berlin) and RWE (throughout the country, but mostly in North Rhine-Westphalia). The RFID passes both of these private networks that are not interchangeable and in a city like Berlin, roaming between the separate networks is not possible. There are currently no plans to realize interoperability between them. Further to this, RWE is co-founder of the joint venture Hubject[19] (together with the BMW Group, Bosch, Daimler, EnBW and Siemens), while Vattenfall has joined the Ladenetz.de initiative[20] (a cooperation of 21 municipal utilities and several international stakeholders).

Fast chargers have been installed on only a few locations in Germany and these are all CHAdeMO chargers. New fast chargers are likely to feature the new COMBO 2 plug to fit with the vehicles that have been announced by German car manufacturers for the coming years.

3.5 Denmark

The Danish recharging infrastructure consists of multiple networks that were set up by private companies, utilities and local governments. Today there is agreement

[18] http://www.ocppforum.net/sites/default/files/ocpp%201%205%20-%20a%20functional%20description%20v2%200_0.pdf.

[19] https://www.press.bmwgroup.com/pressclub/p/pcgl/pressDetail.html?title=hubject-aims-at-connecting-public-charging-infrastructure-for-electric-vehicles-across-european&outputChannelId=6&id=T0134530EN&left_menu_item=node__2379.

[20] http://www.pressebox.de/pressemitteilung/vattenfall-gmbh/Kooperation-fuer-barrierefreie-Elektromobilitaet/boxid/571560.

that all chargers should be equipped with Type 2 sockets, but in practice many chargers only offer Schuko or CEE industrial sockets.[21] The most prominent network operator in Denmark is CLEVER which operates both regular as well as DC quick chargers in public space. Its members make use of an RFID card with which they can charge on a pay-per-kWh basis. Until its bankruptcy, in May 2013, Better Place was the other major operator in Denmark with both its swapping stations as well as 770 regular chargers (Type 2), which suddenly closed. Better Place also worked with a membership scheme, but non-member EV drivers were able to use the network on an ad hoc basis after a phone call to Better Place's service centre. The charger was then opened remotely and electricity was paid for by credit card. In September 2013, EON made an investment in purchasing this charging station network as part of their strategic focus on green transportations in Denmark (having already being involved in heavy transportation).[22]

As the core of the business model of the individual operators is mostly with home charging services, realizing interoperability is not too high on their agendas. However, a Clean Charge Solutions, one of the smaller operators with only a couple of chargers in public space, is linking up with the German Hubject initiative to realize international interoperability and also to connect with Hubject mobile phone-based payment system.

3.6 Sweden

Sweden is one of the countries where the recharging infrastructure emerges relatively slowly, possibly due to the fact that the Swedish national government is still defining its position in relation to electric mobility. The resulting lack of direction and coordination between the various initiatives has probably also caused the wide variety of plug and socket types that are in use. Sweden is one of the few countries in Europe where Type 1 plugs are in use[23] (often next to Type 2 sockets) and continued installation of Type 1 plugs (on fixed cables) is being considered despite the likelihood of Type 2 becoming a European standard. This is especially the case in the Jämtland region (in the centre of Sweden) where the local utility Jämtkraft has installed a small network of chargers. In the city of Gothenburg, most chargers still offer Schuko sockets.

Identification and billing is not an issue in Sweden and chargers in public space are by and large not equipped for this. Since most electricity in Sweden is generated by hydropower and nuclear power plants, electricity is relatively cheap and elaborate billing systems are therefore not worthwhile. In Gothenburg for instance, the

---

[21]http://www.uppladdning.nu/.

[22]http://www.investindk.com/News-and-events/News/2013/EON-enters-Danish-EV-market-with-Purchase-of-Better-Place.

[23]http://www.uppladdning.nu/.

price of electricity is simply part of the regular parking tariffs (and for long-term parking the parking tariffs are a bit higher for EVs).

In addition to the 'real' EV chargers, there are hundreds of thousands of engine preheater sockets in Sweden that can be used to charge an EV. In some instances these have been upgraded with additional safety features to make EV charging safer from these sockets.[24]

3.7 Norway

Norway is one of the countries with the highest number of charging stations in Europe. Many of these are basic Schuko sockets (CEE 7/4) that are installed in public space. Only the newest chargers are equipped with Type 2 sockets, for instance in the city of Oslo, but always next to a Schuko socket. A key is needed to access the chargers and the chargers are free to use for members of the Norwegian EV Association. Similar to Sweden, power in Norway is predominantly generated at hydropower plants and billing of the electricity is thus not a priority and identification and billing systems are not an issue.

Norway also has quite a number of DC fast chargers that make use of the CHAdeMO protocol and plugs. These can be used in combination with an RFID card. The several network operators, all connected to regional energy companies, do allow roaming between the networks, but this is done on an ad hoc basis. This means that an EV driver needs to call the network operator and is subsequently granted (one-time or permanent) access ('added to the white list') to the network. So far, no billing takes place as the number of EV drivers is still fairly limited and any billing system would be more expensive than the actual charging costs. Also, it is thought that roaming is spread evenly among the operators, so there is no direct need to settle costs among the operators.

4 Initiatives to Realize International Interoperability

In this section, the authors briefly describe four initiatives which are explicitly aiming for harmonization of both plugs and identification and billing systems.

4.1 Ladenetz

Ladenetz originates from the cooperation between the local utilities (Stadtwerke) of Aachen, Duisburg and Osnabruck in 2010. A further 18 utilities have joined them

[24]Elforsk, Laddningsinfrastruktur—Marknadsinventering och rekommendationer, Lennart Spante och "Arbetsgrupp P5" Juni 2010.

later on.[25] The focus of the initiative is to enable roaming between the charging networks of the individual utilities and in practice this means that drivers are able to charge off any charger with a single RFID card. The protocol do so is the so-called Open Clearing House Protocol (OCHP) with which the individual networks communicate with each other to exchange user data and to take care of the financial transactions. Having started as a national platform of semi-public organizations, Ladenetz is now expanding the use of the protocol and was joined by Vattenfall for instance. Internationally, Ladenetz has initiated the e-clearing.net platform on which the OCHP is used to enable cross-border roaming.[26] In March 2012 the so-called 'Treaty of Vaals' was signed[27] to confirm this international cooperation and the treaty was signed by Ladenetz.de, the E-Laad foundation from the Netherlands, BlueCorner and Becharged from Belgium as well as Estonteco from Luxemburg, Vlotte from Austria, ESBeCars from Ireland, and Inteli from Portugal. All these network operators will use the OCHP for both national as well as international roaming. In an earlier stage, The New Motion, the largest service provider of the Netherlands, was one of the first to join the e-clearing.net initiative. The entire Dutch network is therefore open to drivers from abroad using an e-clearing.net compatible RFID card.

4.2 Hubject

Whereas Ladenetz and e-clearing.net are not-for-profit attempts to realize interoperability, Ladenetz's spin-off Hubject tries to do the same on a commercial basis. Hubject is a joint venture of BMW, Bosch, Daimler, EnBW (the regional utility company of Baden-Württemberg), RWE and Siemens.[28] It develops the so-called eRoaming platform that acts as a clearing house for network operators and service providers. Any network operator or service provider can join the platform and from there on allow customers of other associated operators to charge at its stations. Charging stations that take part in the Hubject system feature a QR code that can be scanned by an app on the phone. The app then takes care of the identification of the driver and the subsequent financial transaction between the user's own provider and the local host.

Hubject is quite similar to Ladenetz protocol, but the two systems are not directly compatible. It is however thinkable that both systems are used on top of each other and Ladenetz (or any other initiative of multiple network operators)

[25] http://ladenetz.de/index.php?id=partner.

[26] http://www.e-clearing.net/news.php.

[27] http://ladenetz.de/index.php?id=35&tx_ttnews%5Btt_news%5D=1070&cHash=277c3081cbe3b83b3b7f5017ea9d5ab7.

[28] https://www.press.bmwgroup.com/pressclub/p/pcgl/pressDetail.html?title=hubject-aims-at-connecting-public-charging-infrastructure-for-electric-vehicles-across-european&outputChannelId=6&id=T0134530EN&left_menu_item=node__2379.

could for instance be coupled to the Hubject platform (‘hubbing the hubs’). True ad hoc roaming is however not possible since Hubject only acts as a platform for other operators/service providers and not directly for customers.

4.3 Crome

One especially interesting project is CROME (Cross-border Mobility for EVs). This is a German–French cooperation to enable cross-border travel in the Alsace and Moselle regions in France and Baden-Württemberg region in Germany. Within this project, charging stations are installed that can be used with a single RFID card on both sides of the border. Strikingly, these stations will offer both Type 2 and Type 3 sockets so that cross-border travellers do not have to carry along additional cables.[29]

4.4 Green eMotion

Finally, the EU funded Green eMotion project also aims to develop standard for interoperability between charging networks. Green eMotion’s standard is currently in the research and design phase and no implementation has taken place yet. Interestingly, many participants in the project are also active in the other initiatives and especially in the CROME project and Hubject (e.g. Siemens and Bosch).

5 Conclusions: Looking Forward

The development and acceptance of a European standard for recharging finds itself at an interesting intersection. Most countries are still struggling to define their national standard to enable their EV drivers to charge throughout the country. At the same time, there are several initiatives to realize a European network of chargers with compatible plugs and especially with interoperable identification and payment systems. In theory, this could be the ideal point in time to push for international standardization as the need for international standards is recognized by many and national standards are still fluid. From the analysis however, it follows that the individual countries have prioritized their national standardization process over international efforts to define a true European standard. In other words, those countries that have not realized domestic interoperability seem eager to do this on the short term. As a possible consequence, some of these national networks will

[29] http://crome-projekt.de/index.php?id=312.

create a barrier, resulting from a local lock-in, to international interoperability. As noted in relation to the individual countries, the plugs and sockets are not likely to be the real problem. Most countries are indeed moving towards the Type 2 plug as encouraged by the proposed 2013 EU Directive, sometimes side-by-side with prevailing local standards such as the Schuko or the British 3-pin plug.

In the meantime, in case an older infrastructure is still in use or when a local stakeholder does not subscribe to the emerging international standard, EV drivers still need to carry multiple plugs to be sure they can charge their car during a trip. This practical solution does go a long way and may be acceptable to today's early adopters of EVs. However, it may not be acceptable to 'regular' drivers in the near future. Furthermore, this solution works only for (most) regular chargers and will not be an option for DC fast charging.

The real challenge however, as noted several times in the chapter, is with the identification and billing system. The Netherlands is the only country in which roaming is possible between the regular AC charging networks. In other countries like Norway and Sweden 'roaming' is possible because there is no billing system at all. In the rest of the countries, there are at least two networks with their own identification and billing systems. In the case of DC fast charging, roaming is only possible on an ad hoc basis. The two major initiatives to realize international roaming, e-clearing.net and Hubject, are incompatible and it is unlikely that they can exist side-by-side at individual charging stations or locations.

Given the unlikeliness of the emergence of a single European system for roaming between the networks, it would be best if charging station operators would be able to include ad hoc and payment systems in their chargers (using SMS or credit card payments for instance) and that the location of such 'open' chargers are available through comprehensive and up-to-date charge point maps (Lilley et al. 2013). This would be especially relevant for DC fast chargers to facilitate cross-border trips. Alternatively, in those countries where a national roaming system emerges, it would be convenient for foreign EV drivers to be able to purchase a (prepaid) card with which they can charge from the local network(s) during their cross-border trip. It is also recommended that key stakeholder websites and online and offline regional Electric Mobility Information Centres provide detailed enough information on charging an EV (Kottász 2014).

References

Bakker, S., & Trip, J. J. (2013). Policy options to support the adoption of electric vehicles in the urban environment. *Transportation Research Part D: Transport and Environment, 25*, 18–23.

Brown, S., Pyke, D., & Steenhof, P. (2010). Electric vehicles: The role and importance of standards in an emerging market. *Energy Policy, 38*(7), 3797–3806.

Council of the European Union (2014) *Clean fuel infrastructure rules agreed by the Council and the European Parliament*, Press release 8203/14. Brussels, 26 March 2014.

Dimitropoulos, A., Rietveld, P., & van Ommeren, J. N. (2013). Consumer valuation of changes in driving range: A meta-analysis. *Transportation Research Part A, 55*, 27–45.

European Commission. (2013). *Proposal for a Directive of the European Parliament and of the Council on the deployment of alternative fuels infrastructure*. Brussels: CEC.

European Commission, Enterprise and Industry Directorate-General. (2014b). A Notification under Article 12 of Regulation (EU) No 1025/2012. Subject matter: Possible future standardisation requests to the European standardisation organisations (Art. 12, point b). A draft standardisation request addressed to the European standardisation organisations in support of the implementation of the proposal for a directive of the European Parliament and of the Council on the deployment of alternative fuels infrastructure COM(2013) 18final, Brussels. Retrieved July 2, 2014.

IEC—International Electrotechnical Commission. (2011). *Plugs, socket-outlets, vehicle connectors and vehicle inlets—Conductive charging of electric vehicles—Part 2: Dimensional compatibility and interchangeability requirements for a.c. pin and contact-tube accessories*. IEC 62196-2.

IEC—International Electrotechnical Commission. (2014a). *Plugs, socket-outlets, vehicle connectors and vehicle inlets—Conductive charging of electric vehicles—Part 1: General requirements*. IEC 62196-1.

IEC—International Electrotechnical Commission. (2014b). *Plugs, socket-outlets, vehicle connectors and vehicle inlets—Conductive charging of electric vehicles—Part 3: Dimensional compatibility and interchangeability requirements for d.c. and a.c./d.c. pin and contact-tube vehicle couplers*. IEC 62196-3.

Knox, J. (2013). Cooperation on eMobility standardization. *Automotive Industries*.

Kottász, R. (2014). *Comparative analysis of electric car (EV) information available on key stakeholder websites: Application of the Danish framework*. NSE E-mobility Project Report, London Metropolitan University. Retrieved from http://e-mobility-nsr.eu/fileadmin/user_upload/NEWS/New_report__Recommendations_for_setting_up/Appendix_Comparative_analysis_EV_websites_Danish_Framework-13.06.14.pdf.

Kotter, R., & Shaw, S. (2013). *A micro to macro investigation on electric vehicle policy in the UK: Work package 3 activity 6 report*. NSR E-mobility Network. Newcastle upon Tyne/London. Retrieved from http://e-mobility-nsr.eu/fileadmin/user_upload/downloads/info-pool/E-mobility_3.6._Main_Report_April_2013.pdf.

Lilley, S., Kotter, R., & Evett, N. (2013). *A review of electric vehicle charge point map websites in the NSR—Interim report*. NSR E-mobility Network, Newcastle upon Tyne/London. Retrieved from http://e-mobility-nsr.eu/fileadmin/user_upload/downloads/info-pool/Review_of_NSR_charge_point_maps_interim_report_June_2013_final.pdf.

Mom, G. (2004). *The electric vehicle: Technology and expectations in the automobile age*. Baltimore: Johns Hopkins University Press.

Sierzchula, W., Bakker, S., Maat, K., & van Wee, B. (2014). The influence of financial incentives and other socio-economic factors on electric vehicle adoption. *Energy Policy, 68*, 183–194.

Van den Bossche, P., Omar, N., & Van Mierlo, J. (2012). *A tale of three plugs: Infrastructure standardization in Europe*. Paper presented at Electric Vehicle Symposium 26, Los Angeles.

Van den Bossche, P. (2010). Electric vehicle charging infrastructure, Chap. 20. In G. Pistoia (Ed.), *Electric and hybrid vehicles—Power sources, models, sustainability, infrastructure and the market* (pp. 517–543). Amsterdam: Elsevier.

Addressing the Different Needs for Charging Infrastructure: An Analysis of Some Criteria for Charging Infrastructure Set-up

Simon Árpád Funke, Till Gnann and Patrick Plötz

Abstract Electric mobility is an important means to decarbonise the transport sector. Especially in cities, the use of zero-emission vehicles like electric vehicles is favourable, as emissions of conventional cars cause severe air pollution. Besides CO_2, the most important emissions are nitric oxides, particular matter and noise. Given the trend of urbanisation, the problem of air pollution in large cities will rather grow than diminish. Although electric vehicles are an infrastructure-dependent technology, one important advantage of plug-in electric vehicles (EV) compared to hydrogen-powered vehicles is the possibility to use the existing electricity infrastructure in households for charging. While additional public charging infrastructure is also needed for interim charging or overnight charging for the so-called 'on-street parkers' without own garage, the majority of vehicles could be operated as EVs without additional public charging infrastructure. However, public charging infrastructure is an important component for the large-scale diffusion of electric vehicles and political action seems necessary since no business models are presently available. In the present paper the authors combine different data sets concerning German charging points and mobility patterns to describe the different needs for charging infrastructure, and provide an overview of the underlying different technical options. Based on the current charging infrastructure stock, the set-up methodology and the impact of user needs on charging infrastructure, the authors compare a coverage-oriented and a demand-oriented approach. The authors also estimate the number of public charging points for those two approaches. Finally, criteria for charging infrastructure are categorised and related to the different approaches. It results that the number of charging stations needed for the two

S.Á. Funke (✉) · T. Gnann · P. Plötz
Fraunhofer Institute for Systems and Innovation Research ISI, Breslauer Strasse 48, 76139 Karlsruhe, Germany
e-mail: simon.funke@isi.fraunhofer.de

T. Gnann
e-mail: till.gnann@isi.fraunhofer.de

P. Plötz
e-mail: patrick.ploetz@isi.fraunhofer.de

W. Leal Filho and R. Kotter (eds.), *E-Mobility in Europe*,
Green Energy and Technology, DOI 10.1007/978-3-319-13194-8_4

different scenarios and the actual distribution of this predefined number of charging stations are answers to fundamentally different questions. As one consequence, an explicit statement on the number of charging stations needed on large scale (such as Germany) is difficult to make on the basis of (local) user demand.

Keywords Electrification of transport • Electric vehicles • Charging infrastructure set-up • Coverage-oriented and demand-oriented approaches

1 Introduction

Electric mobility is widely recognised as an instrument to fulfil the greenhouse gas emission targets in the transport sector (c.f. German Federal Government 2011). Electric vehicles (EVs) help to reduce both global, European, national and local emissions. Locally, in contrast to conventional vehicles, EVs have no pollutant emissions. Therefore, EVs are particularly suited to improve air quality in cities (especially with high smog, like, e.g. Beijing, c.f. AQI 2014; OECD 2014 and UN 2012). Additionally, their noise emissions are vanishingly low at lower speeds. Finally, EVs could help to reduce the dependence on fossil fuels. Although some EVs are on the roads already, there are still obstacles to overcome for wide market diffusion. Consumer surveys often find purchase price reductions and the installation of charging infrastructure to be means of supporting a large-scale diffusion of electric vehicles (see, e.g. Dütschke et al. 2012). It is often postulated (NPE 2012), and a common (but perhaps not psychological) sense approach, to install charging infrastructure in line with demand for it driven by electric vehicles. But the determination of actual demand for the different types of charging infrastructure is difficult and knowledge about it is rare. This might be one reason for the European parliament to engage national governments to build up an 'appropriate number of electric recharging points accessible to the public' until 2020 (European Voice 2014) instead of setting specific targets as suggested by the European Commission (European Commission 2013). For Germany, the European Commission suggested a target of 1.5 million charging points in total until 2020, 150,000 of them in public (cf. ibid.). The proposal is based on a supranational top-down approach to address geographical coverage as well as user demand. User demand is derived from the forecasted number of electric vehicles (NPE 2012). However, the general aim of public charging infrastructure is to provide a social infrastructure, i.e. to guarantee a minimum standard of service at low cost to the widest possible public (cf. Wirges and Fulda 2010). A demand-oriented installation of charging infrastructure might not be in line with this task of building up the aforementioned social infrastructure. A comparable conflict of interests can be found in public transport. Scarcely used railway lines are operated to guarantee mobility for the widest possible public, although these lines are operated at a loss. In conclusion, for a holistic view, the construction of public charging infrastructure has to be regarded from different levels of perspectives. Consequently, literature on charging infrastructure set-up is very heterogeneous. Approaches range from the discussion of location criteria to the estimation of charging infrastructure demand based on complex mathematical models (see, e.g. BMVI 2014a, b; Lam et al. 2013; NPE 2013; Sandin 2010; Siefen 2012; Stroband et al. 2013; TU Berlin 2011). Therefore, an overview of the different options could serve stakeholders to better decide which approach to apply for the installation of public charging infrastructure in the special case. However, to the authors' best knowledge, a holistic overview including the main different approaches does not exist. To address this gap, this paper categorises public charging infrastructure into different options and show the different stakeholders involved in the set-up of the different options. The paper combines the qualitative discussion of different types of public charging infrastructure with quantitative models used to estimate demand as well as with the authors' calculations. The work is intended to answer the question whether generally applicable criteria exist for the set-up of charging infrastructure or whether their application depends on different types of charging infrastructure.

Thus, the authors provide a brief overview of the different types of charging infrastructures and relate them to user needs as well as to the underlying technical options. The paper is structured as follows. In Sect. 2, it provides background information about public charging infrastructure as well as the methodology used in the following approach. To be more precise: An overview of the most relevant characteristics and design possibilities of charging infrastructure is supported by a discussion about the current state of infrastructure construction (Sect. 2.1). The authors describe their methodology and identify the main user needs for public charging infrastructure (Sect. 2.2). In Sect. 4.1 the authors use available data on user needs to roughly estimate the number of public charging points based on a geographical coverage and on a demand-oriented coverage for EV drivers without a garage and fast-charging options for rare long-distance trips. In Sect. 4.2 the authors derive general criteria from current approaches for building up charging infrastructure to be able to compare them to the results given in Sect. 4.1 (Sect. 5). Finally, the authors conclude with an outlook for future set-up of public charging infrastructure (Sect. 6).

2 Background and Methodology

2.1 Background and Current Status of Public Charging Infrastructure

Charging infrastructure can be distinguished in many ways, including according to its accessibility, its power, connection type and many more (see, e.g. Kley et al. 2011 or Michaelis et al. 2013 for a detailed description). In the following, the most relevant characteristics of charging infrastructure are presented. Table 1 provides an overview of the criteria for differentiation. Accessibility to EV charging infrastructure is

Table 1 Characteristics and design possibilities of charging infrastructure (adapted from Kley et al. 2011, p. 3396)

<table>
<tr><th>Characteristic</th><th colspan="6">Design possibility</th></tr>
<tr><td>Accessibility</td><td colspan="2">Private</td><td colspan="2">Semi-public</td><td colspan="2">Public</td></tr>
<tr><td>Power connection</td><td>1-phase
(level1; 3.7 kW)</td><td colspan="2">3-phase
(level 2; 11-22 kW)</td><td colspan="2">High voltage AC
(level 3; >22 kW)</td><td>High voltage DC
(level 3; >50 kW)</td></tr>
<tr><td>Connection type</td><td colspan="3">Unidirectional</td><td colspan="3">Bidirectional</td></tr>
<tr><td>Information flow</td><td colspan="2">None</td><td colspan="2">Unidirectional</td><td colspan="2">Bidirectional</td></tr>
<tr><td>Type of billing</td><td colspan="2">No fee</td><td colspan="2">Fixed rate</td><td colspan="2">Pay-per-use</td></tr>
<tr><td>Metering</td><td colspan="2">No metering</td><td colspan="2">At charging station</td><td colspan="2">In vehicle</td></tr>
</table>

distinguished as private, semi-public or public charging infrastructure. While private charging infrastructure is only accessible to one person, vehicle or household and thus it is the most restricted option (e.g. a garage), public charging stations are open to everybody (e.g. at a public parking spot). Semi-public charging points are restricted to a certain group of people, e.g. the member's of a sports club or the paying users of a car park. From here on, the paper focuses on public charging stations.

The power connection of charging points ranges from 1-phase AC charging, up to 50 kW high-voltage DC charging, while most public charging options are expected to offer at least 11 kW (level 2). The connection as well as the information flow may be uni- or bidirectional, while for billing feeless systems, fixed rates and pay-per-use options can be distinguished. Finally, the metering differs: there are systems without any metering and others that use metres either in the charging station or in the vehicle (see Dallinger et al. 2013).

The current status of public charging infrastructure in Germany is quite diverse. Currently, there are about 2,900 locations for public charging with approximately 4,900 charging stations and 8,400 sockets in Germany (Lemnet 2014). Almost half (45 %) of the stations are equipped with simple type 1 sockets, another 38 % with type 2 (three phase). The majority of the charging stations (51 %) have a power of up to 3.7 kW. Another 30 % of the charging stations have a power higher than 16.7 kW, predominantly operated by 3-phase alternating current. In total, only 68 of the charging stations reported are operated with direct current (DC). Until today, it is not clear whether DC charging or 3-phase AC charging will become standard for fast charging (NPE 2013). These numbers support the assumption that public charging infrastructure is needed primarily for charging at low power. Fast charging points are also needed, but only in a limited number. Concerning the geographical location, we find about 60 % of all charging stations in Bavaria, Baden-Württemberg and Northrhein-Westfalia, the federal states with the highest populations and also the highest number of EVs in Germany. Among them, Baden-Württemberg has the highest charging infrastructure density per capita (0.1 charging stations per 1,000 inhabitants). It is not surprising that the city states Berlin and Hamburg have a charging infrastructure density per square kilometre that is about 15 times higher than the German average (0.22 charging stations per square kilometre for both city states compared to 0.015 on average) (see also Schneider et al. i.p.). They have a high population density and, furthermore, these states were part of the publicly funded project family German Pilot Regions (BMVBS 2011). Compared to this, the third city state in Germany, Bremen, has a relatively low number of charging stations resulting in an infrastructure density of 0.09 charging stations per kilometre. The market of charging station operators is very concentrated. About 25 % of the charging stations are operated by one of the four large energy providers, with RWE (ca. 11 % of all charging stations) being the most prominent. The highest proportion of charging stations, 12 %, is operated by local energy providers. Altogether, public utilities and local energy providers operate around 35 % of all charging stations in Germany (Lemnet 2014). Another very important part of charging infrastructure is formed by privately operated infrastructure. The access to private charging points is restricted to private persons and members, respectively (see, e.g. Park&Charge 2014 and Table 1). This limitation allows for a simple and therefore cheap infrastructure. Private infrastructure is also installed for marketing reasons, e.g. by restaurants. Ca. 50 % of non-public charging infrastructure can be used free of charge (Lemnet 2014).

The installation of further charging points in different projects has already been announced. In context of the project family E-mobility Show Case Regions (see, e.g. E-mobil BW 2014) charging infrastructure is one focus. In Baden-Württemberg, the construction of 1,000 additional charging points until 2015 is planned (E-mobil BW et al. 2013). Within the publicly funded project “SLAM—Schnellladenetz für Achsen und Metropolen” another 400 DC fast charging points are about to be installed by 2017 (see, e.g. Fraunhofer IAO 2014).

2.2 *Impacts of User Needs on Charging Infrastructure*

The usage of passenger cars is highly heterogeneous and the daily driving of passenger cars is not very regular (Wietschel et al. 2013). However, vehicles return home overnight for the vast majority of days per year, i.e. a long-distance travel with overnight stay is exceptional (Pasaoglu et al. 2013; Axhausen et al. 2002). Thus, for EV users with own garage, ‘at home’ is the location that could be used to charge most of the time. In small villages (less than 20,000 inhabitants) a high percentage of 82 % does have a garage or a carport at its disposal, in big cities (more than 1 million inhabitants) this proportion is at 44 % (Nagl and Bozem 2014). However, there is a certain number of potential users without a permanent parking space, the so-called ‘on-street parkers’. For these potential users public infrastructure is needed to compensate the lack of a possibility to charge at home. After charging at home, the highest increase in electrification of driven kilometres is achieved by charging at work (Santini 2013). For company cars, the same conclusions as for private cars can be applied. If company cars cannot be charged at the company ground, public infrastructure is needed. This could be the case for small companies without own parking areas. Due to the long parking periods overnight for both cases, low power charging is sufficient.

Nonetheless, on some days per year users drive long distances and spend the night away from home. In this case, when the daily driving distance exceeds the technical range of a BEV, fast charging infrastructure is needed for interim charging during rare long-distance trips, though arguable hotels should also be encouraged to supply them. Plug-in hybrids (PHEV) do not necessarily need infrastructure for interim charging, since they can go long distances with an internal combustion engine that serves as range extender. Additionally, interim charging can generally be applied at low power during longer parking periods. This might be the case for longer trips to, e.g. shopping outlets where the total daily trip is interrupted by the parking time at the shopping centre.

To conclude, different user needs and their impact on public charging infrastructure are summarised in Fig. 1. Generally, public charging infrastructure can be distinguished into (1) infrastructure for 'on-street parkers' and (2) for 'interim charging'. The need for 'on-street parking' infrastructure arises from (a) private drivers without own garage and (b) workplace charging for employees at companies without own parking spots. The latter one is not absolutely necessary but offers large benefits for many private users. 'Interim-charging' is needed for long-distance trips with BEVs for both private and company cars. Depending on the trip purpose, interim charging can further be distinguished into (c) charging with low power in the proximity of shops, restaurants, etc. and (d) fast charging. This categorisation follows the Swiss Forum for Electromobility (Schweizer Forum Elektromobilität 2012) (see also Schatzinger and Rose 2013) as, in the authors' view, it represents an appropriate categorisation of the different types of charging infrastructure in an easily comprehensible way by using the activities probably carried out during charging. The categories of activities used are the most likely during parking time (Follmer et al. 2010b). For other classifications see, e. g. Botsford (2012) and Sandin (2010). However, the above-mentioned classification is not without overlap. On-

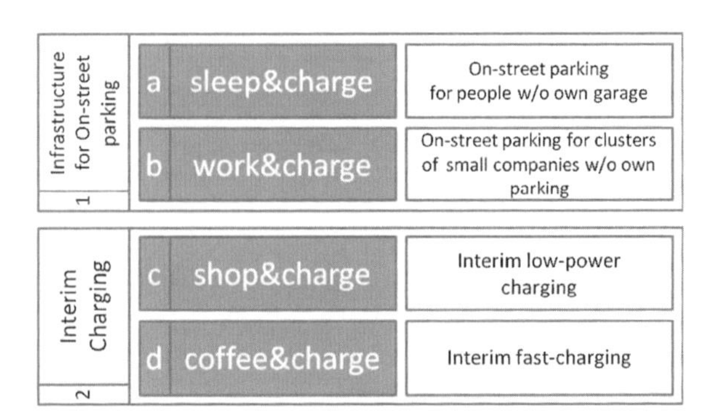

Fig. 1 Types of public charging infrastructure by user needs (own illustration, categorisation based on the Swiss Forum for Electromobility—Schweizer Forum Elektromobilität (2012))

street parkers (case a, Fig. 1) could use the same infrastructure that is used for interim low-power charging (case c). While the infrastructure is used by the first group overnight, the latter application would be during the day. This overlap for the different types of infrastructure has to be considered when analysing the involved stakeholders (see Sect. 4.2) for the different types of public charging infrastructure.

The described use cases are determined by user needs and are analysed below to estimate the demand for public charging infrastructure from the user perspective (Sect. 4.1).

3 Methodology

Diverse factors make the decision about the construction of charging infrastructure complex. On the one hand, charging infrastructure should guarantee a minimum standard of service implying the need for a dense charging infrastructure. On the other hand, a demand-oriented construction of charging infrastructure is desirable. In this work, the authors analyse and compare both different approaches. Together with the different types of public charging infrastructure and the different approaches for their set-up, a holistic view is given. For the analysis of the different approaches, the authors use different methods of technology assessment. An overview of these methods is given in Tran and Daim (2008). First, in a kind of a top-down approach, the authors estimate the number of charging stations needed on the basis of a predefined geographical coverage. Different data sets on vehicle registrations and demography are combined. Secondly, the authors estimate the number of charging stations on the basis of user behaviour and user need. As the authors consider the single user as point of reference for this approach, the analysis is characterised by a bottom-up view. A mathematical model comprises the use of different scenarios on mobility patterns, the prediction of potential buying decisions based on a total cost of ownership (TCO) analysis (Wietschel et al. 2013). The aforementioned approaches are supplemented by user-specific requirements on public charging infrastructure in the form of general criteria. Literature and general information about ongoing projects is reviewed thoroughly to provide a holistic view. The data obtained is clustered and categorised systematically to make a differentiated understanding of the needed information possible.

4 Results

4.1 Estimation of Real Needs for Charging Infrastructure

As we mentioned before, charging infrastructure can be set up according to a broad coverage of all regions (geographical coverage) or according to users' needs (user-oriented coverage) (see Ball and Wietschel 2009, pp. 415–417). In this section, the authors estimate the number of charging infrastructure stations that would result from both approaches and discuss their usefulness.

4.1.1 Geographical Coverage with Charging Infrastructure

For a geographical coverage, the population density can be analysed. The present number of refuelling stations in Germany (currently about 14,700) leads to one charging station every 3.4 km.[1] As it is known that there is a higher refuelling stations density in cities than in rural areas, one can be a little bit more precise for this estimation. There is a common differentiation between area types into:

- core cities: cities with more than 100,000 inhabitants
- condensed areas: areas with a population density >150 inhabitants/km^2
- rural areas: areas with a population density <150 inhabitants/km^2

The number of charging stations for a geographical coverage depends on the area and the maximum distance between two charging stations as shown in Fig. 2.

It seems clear that a small distance between two charging stations leads to a higher number of charging points, but also the total numbers are important as they are connected to the areas. At a first glimpse, one would argue that the charging network should be denser in core cities than in rural areas. Thus, assuming that a charging station every 500 m is sufficient in core cities, one obtains a result of about 25,000 charging stations as the surface area is only about 13,000 km^2. The authors used a separating distance of 2 km between charging stations in condensed areas and 5 km for rural areas (Table 2) which return a total of about 51,500 charging stations.

Thus, although the largest share of surface area is in rural areas, the number of charging stations is small if one assumes that the charging stations density can be low. The assumed distances also point out that even if the surface area is low in core cities, a dense network of charging infrastructure results in more than 50 % of charging stations in core cities with this geographical approach.

An estimate based on vehicle registrations in the area types yields a different distribution of charging points as more vehicles are registered in the rural areas. If one studies only those vehicle owners that do not own a garage or parking spot close by and 2.5 % thereof (equivalent to 1 million EVs of 40 million vehicles in total), one finds about 41,300 charging stations that would be necessary for overnight charging in Table 3.

Although the total number of charging stations is fairly equal to the first estimation, it neglects the fact that some users are more likely to buy EVs than others, which is subject to the following subsection.

[1]One has to divide the area of Germany (357,097 km^2) by the number of refuelling stations multiplied by $\sqrt{3}$ (intersection point of three equal circles).

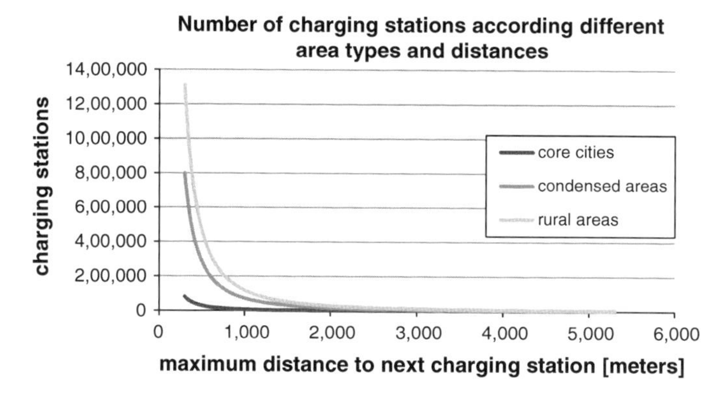

Fig. 2 Number of charging stations according to different area types and distances between charging stations

Table 2 Charging stations in different area types (1) (Destatis et al. 2013)

Area types	Core cities	Condensed areas	Rural areas
Surface area (km^2)	13,086 (3.9 %)	130,181 (36.5 %)	213,831 (59.9 %)
Maximum distance between two charging stations (m)	500	2,000	5,000
Charging stations hence to area	28,900	17,900	4,700

Table 3 Charging stations in different area types (2) (Follmer et al. 2010a, b; Destatis et al. 2013)

Area types	Core cities	Condensed areas	Rural areas
Vehicles	11,364,366	21,457,184	12,949,325
Vehicles not parking on own ground or close by	863,692	536,430	297,834
Charging stations weighted with vehicle stock (share of 2.5 % EVs 2020)	19,800	13,600	8,000

4.1.2 User Need for Charging Infrastructure

The paper now turns to the estimate of required public charging infrastructure based on user behaviour as discussed above. The authors estimate the order of magnitude for the years 2020 and 2030 based on the market evolution results contained in Wietschel et al. (2013).

Since on-street parking infrastructure serves as an alternative to home charging, the need for it results both for BEVs and REEVs or PHEVs respectively. Of these EVs in Germany about 1.5 % is expected to be on-street parkers (Wietschel et al. 2013). For the potential one million BEVs in 2020 and six million BEVs in 2030 the authors thus arrive at 15,000 charging points in 2020 and 90,000 in 2030 for on-street parkers. The demand for fast public charging infrastructure can be estimated by the number and distances of rare long-distance driving that BEV would take in 2020 or 2030. The average number of days per year with more than 100 km of driving (requiring interim recharge for BEVs) has been estimated by Plötz (2014) based on the assumption of log-normal distributed daily vehicle kilometres travelled and is given by about 30 days. The typical distance on these days is given by the mean excess function of the log-normal distribution and is (for typical parameter values as displayed in Plötz (2014)) in the range 160–220 km, so that one recharging per long-distance driving day seems typical. The number of required fast charging points is determined by the number of EV users multiplied by the number of long-distance driving days per year divided by the number of days per year. The result for assumed one million EV users in 2020 would be 20,500 recharging events per day in 2020 and 125,000 in 2030 if charging events are distributed equally over the year. If each charging point can on average serve 10 users per day once finally arrive at a user demand of 2,000 fast charging points in 2020 and 12,500 fast charging points in 2030. These numbers are linear in the assumptions for the EVs in the respective years and the number of users that can be served by a charging station. However, this does not take into account that drivers might use other alternatives for these long-distance trips (e. g. rental cars, car sharing or public transport) which would decrease the number of required public fast charging stations. Please note that these rough estimates provide an order of magnitude for the public charging infrastructure demand that are intended to help analyse the different criteria for charging infrastructure set-up discussed in the next section.

4.2 *Current Heuristics for Charging Infrastructure Set-up*

After the estimation for the numbers of charging stations needed for the two different approaches, geographical coverage and demand-oriented coverage, the paper now takes a look at strategies or heuristics that are used in projects today and compare them to the above-described approaches. The need for the development of specific strategies for the construction of public charging infrastructure results from the fact that existing refuelling infrastructure cannot be used to charge electric vehicles. Besides, the integration of charging stations into the existing gas station infrastructure is not viable due to the long charging periods (Lam et al. 2013). Car sharing stations are not suitable either for charging stations since they are designed for round trips (Wirges and Fulda 2010). In Germany, apart from private initiatives, infrastructure is installed in politically funded projects (e.g. the German Pilot Regions and the Show Case Regions, see e.g. BMVI (2014a) and E-mobil BW (2014)).

4.2.1 Characterisation of Stakeholders Involved in the Construction of Charging Infrastructure

Building up infrastructure means to take the fundamentally different primary objectives of the different stakeholder groups into account. We identify four different stakeholder groups: (1) EV drivers, (2) the operator of the charging infrastructure and the electric grid respectively, (3) the municipality in which charging infrastructure is installed (local authority) as well as (4) the national or supranational (e.g. EU) authority. While the driver of an electric vehicle will accept limitations neither in mobility patterns nor in high costs, the operator of the charging station pursues the development of a business model. However, the aim of the local and the national authority is to ensure a minimum standard of service. Concerning charging infrastructure in particular, the public supply mandate may imply the integration of its set-up into a general urban plan (Rothfuss et al. 2012), especially as the target group of charging infrastructure, the electric car drivers, is still very small (KBA 2014) and public space is a scarce resource. As a summary of this section, Table 4 displays criteria for the involved stakeholder groups. The list comprises criteria concerning the visibility, the handling and cost of the charging stations. They are divided into categories described in the following:

Table 4 Categorisation of criteria for charging infrastructure (on basis of BMVBS 2011; Wirges and Fulda 2010; Hoffmann 2013)

		Electric Vehicle Driver	**Local/ national Authority**	**Infrastructure Operator**
Primary Objective		No limitation in mobility at limited additional cost	Charging Infrastructure as consequence of the public supply mandate	Charging Infrastructure as Business Model
Basic/Excluding Criteria		Charging infrastructure must be: • Fully accessible • Unrestricted • Safe	Obeying of different regulations: • Fire prevention • Protection of historical monuments	Profitability and Grid Stability
Target Criteria on micro level Detailed Characterization of Location	Visibility	• Visibility as a pioneer • Station easy to find & recognize	• Inconspicuous Integration into Cityscape vs. Image as Green City	• Visibility to reach high utilization rate
	Site/Handling	• Simple access & declaration • Safety & weather protection	• Non-discriminatory access of public charging infrastructure • Safety & ease of traffic (safety obligation)	• Extendibility • Low cost for installation & maintenance
Target Criteria on macro level General Overview		Comfort & Practicability: • Demand-oriented • High availability	Integration into Urban Development: • No shortage of parking space • BUT: Limitation of land use for parking • Support of intermodality	Utilization: • High utilization vs. Grid stability
	Cost	Limited additional cost for charging	Parking fees as income	Low cost e.g. for parking space and grid connection

4.2.2 Categorisation of Criteria by Decision Level

As shown, diverse criteria at different decision levels make the set-up of charging infrastructure complex. Decisions have to be taken by different stakeholders and the criteria for evaluation are heterogeneous as they have, e.g. technical or legal character (Wirges and Fulda 2010). Therefore, a categorisation of the criteria used for the location of charging infrastructure by its decision level into (1) micro level and (2) macro level is practical. The specific distinction into those two categories depends on the actual level of decision. Thus, criteria on the macro level can affect either a nation[2] or a city as a whole. However, criteria on macro level affect the charging infrastructure in general, whereas criteria on micro level affect the specific location and the detailed realisation of the charging points. Thus, recommendations concerning the realisation of charging infrastructure given in the actual publicly funded projects often affect the micro level. They emphasise the need for further standardisation of technical equipment of billing and communication infrastructure. In Europe, the combined charging system is announced to become the standard plug for electric vehicles (ENS 2014). However, different infrastructure operators use different unharmonised billing systems that still make the use of public charging infrastructure uncomfortable (WIWO 2014). Additionally, possibilities to integrate charging infrastructure into existing neighbourhoods as in parking metres and bollards are presented (BMVBS 2011). On this level, a harmonised approach is difficult to implement. On the micro level, a further distinction of criteria into excluding and non-excluding criteria or into (1) basic and (2) target criteria (BMVBS 2011) is practicable. Basic conditions comprise, e. g. the protection of historical monuments or fire prevention regulations. In general, charging stations should not be built in historical view centres, nor on public places or in the proximity of public listed buildings (Wirges and Fulda 2010).

4.2.3 Fundamental Distinction of Approaches into Maximum Coverage and Demand-Oriented Coverage

On the macro level, different approaches for the set-up of charging infrastructure are possible. The distinction used in Sect. 4.1 into (1) an approach to reach the maximum possible coverage and (2) a demand-oriented approach has fundamental character. While the latter takes an economic or a user-specific view, respectively, the aim of this approach is to reach a high utilisation rate of the charging infrastructure. The first approach, however, takes a more social view: a broad reachability of charging infrastructure (Sect. 1). The remarkable difference in the numbers of charging stations for the two different approaches underlines their contradictory character (Sect. 4.1). Social infrastructure could also help reducing range anxiety of EV drivers. In Tokyo,

[2]Or supranationally if agreed at that level, or if devolved to federal state level at that level.

e.g. the technical range of electric vehicles was fully used only after the installation of public charging infrastructure (E-mobil BW 2013).

An optimal geographical coverage of charging infrastructure can be determined with methods of operations research. Depending on focus, evaluation criteria of infrastructure set-up are (besides others): (1) the number of demand sides covered, (2) the average number of reachable charging points or (3) the mean minimal distance between the supply sides as an indicator for infrastructure density (Siefen 2012). In general, centre problems and covering problems can be distinguished (Hoffmann 2013). While centre problems intend to minimise the distance between the supply and the different demand points, covering problems suppose a given maximal reachability to ensure a given minimum standard of service. Mainly, location problems are formulated as covering problems. If the number of charging stations p to be installed is predefined, the problem is called a maximum-p-covering problem (Siefen 2012). Overall, these methods provide optimal results, but are solvable with reasonable effort only for a limited number of demand points. A viable way is to use heuristics[3] (Lam et al. 2013). The described methods of defining a maximum possible coverage can be combined with methods to determine user demand for infrastructure (Stroband et al. 2013). One possible way to determine demand is to divide the city into cells and analyse detailed data on travel behaviour (see, e.g. Wirges and Fulda 2010; Vélib 2014). A huge amount of data makes this approach time-intensive and probably costly. Alternatively, demand requirements can be estimated. For charging infrastructure, e.g. population size and the penetration rate of electric vehicles could be used (Lam et al. 2013) (see Sect. 4.1). In studies for practical application, a combination of both perspectives, the maximum coverage and demand-oriented approach, can be found (Hoffmann 2013, Wirges and Fulda 2010; TU Berlin 2011). Taking a macro view first, a rough estimate on needed charging stations (see Sect. 4.1) is a useful starting point to predefine a number of charging stations p to be installed. In a second step, this number can be used for solving a maximum-p-covering problem. Depending on the defined criteria, suitable locations for the predefined number of charging stations can be identified. Finally, the selected areas can be analysed in more detail.

5 Summary and Discussion

Comparing the different types of approaches for infrastructure set-up, the authors distil two different main questions: (1) The number of charging stations needed in total and (2) the actual distribution of this predefined number within a certain area. The decision concerning these questions has to be taken by different authorities on

[3]The word heuristics in this context is used as the description of a mathematical method to approximate the optimal solution. In contrast, the word heuristics in the title is used to describe generally an experience-based approach for infrastructure set-up strategies.

different decision levels. Therefore, the different approaches and methodologies for the analysis of charging infrastructure reach from rough estimates to complex models suitable for the different decision levels. On the macro-level rough estimates using different data sets are a viable way to estimate the number of charging stations needed on national scale. In contrast, a demand-oriented approach on a national view is difficult to implement. Due to a detailed analysis of local traffic volumes, this approach is data-intensive and therefore particularly suitable for a local analysis with a predefined limited number of charging stations. Concerning the distribution of the predefined number of charging stations, different stakeholder interests have to be taken into account. As an example of different interests, the authors compare an approach to reach a maximum coverage and a demand-oriented approach. In the early phase of the electric vehicle market, a demand-oriented approach will lead to a lower number of charging stations than a maximum coverage approach (see Sect. 4.1). For 2020, the authors estimate the number of charging stations for a maximum coverage-oriented approach in the range of 50,000 and for the demand-oriented approach in the magnitude of 17,000 charging stations. Although it is complex to estimate a demand-oriented need for charging stations taking a macro view, these numbers underline the contradictory character of the two approaches. The postulated number of 75,000 public charging stations[4] in European Commission (2013) is even higher. The high number of charging stations could be an indicator for the installation of a dense social infrastructure with the aim to push market penetration, although this might not trigger the diffusion by itself (Gnann and Plötz 2015).

6 Conclusions and Outlook

In the set-up of public charging infrastructure different stakeholder groups with different interests are involved. Furthermore, decisions about the installation of public charging infrastructure are taken on different decision levels. The authors find that, depending on the level of the decision, e.g. nationwide or locally, the analysis of needs for public charging infrastructure has different implications. The analysis presented here shows that the estimation of the total number of charging stations for a specific area and the decision about the distribution of a predefined number of charging stations are two different questions. While the use of detailed mobility data to estimate the number of charging stations needed on national scale is not viable due to high data intensity, rough estimates based on data sets are imprecise on local scale. Therefore, the involved stakeholders first have to become clear about the form and main target of the respective infrastructure to be installed.

[4]In the document 150,000 charging points are postulated. For the estimation of the resulting number of charging stations, the authors assume two charging points per charging station (Lemnet Europe e.V. 2014).

On basis of this information a suitable approach can be applied by comparing the different approaches. To do so, the authors conducted a holistic analysis by integrating all the different approaches at the different levels into one approach. The analysis in this paper focuses on public charging infrastructure. However, charging infrastructure is expected and recommended to emerge first in private and semi-public areas (Rothfuss et al. 2012; BMVI 2014b; Kley 2011) and is sufficient for a large number of car owners (Wietschel et al. 2013). For local authorities a possible way is to partnership with private utilities to build up publicly planned infrastructure on private ground, e.g. in car parks (i.e. semi-public charging infrastructure, see Sect. 2.1).

The overview provided in this paper allows for a better understanding of the underlying assumptions and targets of the different approaches for public charging infrastructure set-up as well as of their implications. Nevertheless, further research is needed to determine the impact of public charging infrastructure on EV purchase decision. This is key for the understanding and prediction of real user demand for charging infrastructure. Besides the psychological effect, the impact of technical development on the need for public charging infrastructure is important. An extended driving range, e.g. due to a higher energy density of a new battery technology, probably will affect substantially the need for public charging infrastructure. Finally, for infrastructure operators the rentability of the stations is essential. Part of further research thus should contain the development of a detailed model to determine the utilisation rate and potential business models for charging points at different locations.

Acknowledgments The research was made possible as part of the REM 2030 project, which is funded by the Fraunhofer Society and the federal state Baden-Württemberg.

References

Air Quality Index AQI. (2014). *Beijing air pollution: Real-time air quality index (AQI)*. Required May 14, 2014, from http://aqicn.org/city/beijing/.

Axhausen, K. W., Madre, J. L., Polak, J. W., & Toint, P. (Eds.) (2002). *Capturing Long Distance Travel*. Baldock: Research Studies Press.

Ball, M., Seydel, P., Wietschel, M., & Stiller, C. (2009). *Hydrogen-infrastructure build-up in Europe*. In M. Ball & M. Wietschel (Eds.), *The hydrogen economy: Opportunities and challenges* (pp. 385–453). Cambridge: Cambridge University Press.

Botsford, C. W. (2012). *The Economics of Non-Residential Level 2 EVSE Charging Infrastructure*, Proceedings of EVS26 International Battery, Hybrid and Fuel Cell Electric Vehicle Symposium, Los Angeles, California.

BMVBS (German Federal Ministry for Transport, Building and Urban Development) (2011). *Elektromobilität in Deutschland—Praxisleitfaden; Aufbau einer öffentlich zugänglichen Ladeinfrastruktur für Genehmigungsbehörden und Antragsteller*. Retrieved April 11, 2014, from http://www.electrive.net/wp-content/uploads/2011/12/Praxisleitfaden-Ladeinfrastruktur.pdf.

BMVI, German Federal Ministry for Economic Affairs and Energy. (2014a). *Modellregionen Elektromobilität*. Retrieved April 11, 2014, from http://www.bmvi.de/SharedDocs/DE/Artikel/UI/modellregionen-elektromobilitaet.html?nn=36210.

BMVI, German Federal Ministry for Economic Affairs and Energy (BMVI) (2014b). *Genehmigungsprozess der E-Ladeinfrastruktur in Kommunen: Strategische und rechtliche Fragen—eine Handreichung*. Berlin.

Dallinger, D., Funke, S., & Wietschel, M. (2013). *On-board metering*. Working paper sustainability and innovation. Karlsruhe: Fraunhofer Institute for Systems and Innovation Research ISI. Retrieved April 10, 2014, from http://publica.fraunhofer.de/documents/N-246997.html.

Destatis (Federal Statistical Office), Wissenschaftszentrum Berlin für Sozialforschung (Eds.) (2013). *Datenreport 2013. Ein Sozialbericht für die Bundesrepublik Deutschland*. Retrieved April 11, 2014, from https://www.destatis.de/DE/Publikationen/Datenreport/Downloads/Datenreport2013.pdf?__blob=publicationFile.

Dütschke E., Schneider, U., Sauer, A., Wietschel, M., Hoffmann, J., & Domke, S. (2012). *Roadmap zur Kundenakzeptanz: Zentrale Ergebnisse der sozialwissenschaftlichen Begleitforschung in den Modellregionen*. Karlsruhe: Fraunhofer Institute for Systems and Innovation Research ISI. Retrieved April 10, 2014, from http://publica.fraunhofer.de/documents/N-192915.html.

E-mobil BW GmbH (State Agency for Electric Mobility and Fuel-Cell Technology Baden-Wuerttemberg GmbH) (Eds.) (2013). *Systemanalyse BWe Mobil 2013; IKT- und Energieinfrastruktur für innovative Mobilitätslösungen in Baden-Württemberg*. Retrieved April 29, 2014, from http://www.e-mobilbw.de/files/e-mobil/content/DE/Publikationen/PDF/Systemanalyse_BWemobil_IKT_Energie_2013.pdf.

E-mobil BW GmbH (State Agency for Electric Mobility and Fuel-Cell Technology Baden-Wuerttemberg GmbH) (2014). *Schaufenster LivingLab Bw^e mobil. Das baden-württembergische Schaufenster*. Retrieved April 11, 2014, from http://www.e-mobilbw.de/de/aufgaben/schaufenster-livinglab-bw-emobil.html.

ENS (Environment News Service) (2014). *Europe sets common standard for electric vehicle charging*. Retrieved May 23, 2014, from http://ens-newswire.com/2013/01/28/europe-sets-common-standard-for-electric-vehicle-charging/.

European Commission (2013). *Proposal for a directive of the European parliament and of the council on the deployment of alternative fuels infrastructure*. Retrieved April 11, 2014, from http://eur-lex.europa.eu/LexUriServ/LexUriServ.do?uri=COM:2013:0018:FIN:EN:PDF.

European Voice (2014). *Deal on electric car recharging, but no EU targets*. Retrieved March 21, 2014, from http://www.europeanvoice.com/article/2014/march/deal-on-electric-cars-but-no-eu-targets/80166.aspx.

Follmer, R.; Gruschwitz, D., Jesske, B., Quandt, S., Nobis, C., & Köhler, K. (2010a). *Mobilität in Deutschland (MiD) 2008—Tabellenband*. Infas Institute for Applied Social Sciences GmbH, German Aerospace Center (DLR), Bonn, Berlin.

Follmer, R., Gruschwitz, D., Jesske, B., Quandt, S., Lenz, B., Nobis, C., Köhler, K., & Mehlin, M. (2010b). *Mobilität in Deutschland (MiD) 2008—Ergebnisbericht*. Infas Institute for Applied Social Sciences GmbH, German Aerospace Center (DLR), Bonn, Berlin.

German Federal Government. (2011). *Regierungsprogramm Elektromobilität*. Retrieved May 20, 2014, from http://www.bmwi.de/BMWi/Redaktion/PDF/Publikationen/regierungsprogramm-elektromobilitaet.

Gnann T., & Plötz, P. (2015). *A review of combined models for market diffusion of alternative fuel vehicles and their refuelling infrastructure*. Renew. Sustain. Energy Rev. *47*, 783–793.

Hoffmann, J. (2013). *Raumplanungskonzept für Fahrzeugwechselstationen und Ladeinfrastruktur zur Nutzung der Elektromobilität im intermodalen Personennahverkehr*. Bonn: Geographisches Institut der Rheinischen Friedrich-Wilhelms-Universität.

IAO, Fraunhofer Institute for Industrial Engineering (2014). *Projekt-Kick-off SLAM—Schnellladenetz für Achsen und Metropolen auf der Hannover Messe*. Retrieved April 11, 2014, from http://www.iao.fraunhofer.de/lang-de/presse-und-medien/1332-start-fuer-flaechendeckendes-schnellladenetz.html.

KBA (German Federal Motor Transport Authority) (2014). *Jahresbilanz des Fahrzeugbestandes am 1. Januar 2014*. Retrieved April 11, 2014, from http://www.kba.de/nn_125264/DE/Statistik/Fahrzeuge/Bestand/bestand__node.html?__nnn=true.

Kley, F. (2011). *Ladeinfrastrukturen für Elektrofahrzeuge—Entwicklung und Bewertung einer Ausbaustrategie auf Basis des Fahrverhaltens*. Karlsruhe: Karlsruhe Institute for Technology (KIT).

Kley F.; Lerch, C., & Dallinger, D. (2011). *New business models for electric cars—a holistic approach*. Energy Policy, *39*(6), 3392–3403 (doi: http://dx.doi.org/10.1016/j.enpol.2011.03.036).

Lam, A.; Leung, Y. -W., & Chu, X. (2013). *Electric Vehicle Charging Station Placement*, Proceedings of IEEE SmartGridComm 2013 Symposium—Smart Grid Services and Management Models. Retrieved April 11, 2014, from http://ieeexplore.ieee.org/stamp/stamp.jsp?tp=&arnumber=6688009.

Lemnet Europe e.V. (2014). *Data on German Charging Stations*. Lemnet.org.

Michaelis J., Plötz P., Gnann T., & Wietschel, M. (2013). *Vergleich alternativer Antriebstechnologien Batterie-, Plug-in Hybrid- und Brennstoffzellenfahrzeug*. In P. Jochem (Ed.), *Alternative Antriebskonzepte bei sich wandelnden Mobilitätsstilen: Tagungsbandbeiträge vom 08. und 09. März 2012 am KIT* (pp. 51–80). Karlsruhe: KIT Scientific Publishing.

Nagl, A., & Bozem, K. (2014). *Elektromobilität: Erwartungen des Marktes an das Laden*. Repräsentative Ergebnisse der Studie Future Mobility, Zeitschrift für die gesamte Wertschöpfungskette Automobilwirtschaft (ZfAW) 01/2014, Bamberg.

NPE. German National Platform for Electric Mobility (NPE) (2012). *Fortschrittsbericht der Nationalen Plattform Elektromobilität (3. Bericht)*. Retrieved April 11, 2014, from http://www.bmub.bund.de/fileadmin/bmu-import/files/pdfs/allgemein/application/pdf/bericht_emob_3_bf.pdf.

NPE. German National Platform for Electric Mobility (NPE) (2013). *Technischer Leitfaden Ladeinfrastruktur. Arbeitsgruppe 4 Normung, Standardisierung und Zertifizierung*. Retrieved April 11, 2014, from http://www.vda.de/de/publikationen/publikationen_downloads/detail.php?id=1186.

OECD (2014). *Rising air pollution-related deaths taking heavy toll on society, OECD says*. Retrieved May 23, 2014, from http://www.oecd.org/greengrowth/rising-air-pollution-related-deaths-taking-heavy-toll-on-society.htm.

Park&Charge e.V. (2014). *Ladeinfrastruktur für Elektrofahrzeuge*. Retrieved April 11, 2014, from http://www.park-charge.de/.

Pasaoglu, G., Fiorello, D., Martino, A., Zani, L., Zubaryeva, A., & Thiel, C. (2013). *Travel patterns and the potential use of electric cars–Results from a direct survey in six European countries*, Technological Forecasting and Social Change. Retrieved from http://www.sciencedirect.com/science/article/pii/S004016251300276X.

Plötz, P. (2014). *How to estimate the probability of rare long-distance trips*. Working Paper Sustainability and Innovation, S 1/2014, Fraunhofer Institute for Systems and Innovation Research ISI, Karlsruhe. Retrieved April 10, 2014, from http://www.isi.fraunhofer.de/isi-de/x/publikationen/workingpapers_sustainability_innovation.php.

Rothfuss, F., Rose, H., Ernst, T., & von Radecki, A. (2012). *Strategien von Städten zur Elektromobilität—Städte als Katalysatoren auf dem Weg zur Mobilität der Zukunft*. Stuttgart: Fraunhofer Institute for Industrial Engineering IAO.

Sandin C. -O. (2010). *Developing Infrastructure to Promote Electric Mobility*. Royal Institute of Technology, Stockholm.

Santini, D. J. *IEA*-HEV (2013). *Task 15 Report: Plug-in Hybrid Electric Vehicle*. IEA energy technology network.

Schatzinger, S., & Rose, H. (2013). *HafenCity Hamburg Praxisleitfaden Elektromobilität*. Retrieved May 26, 2014, from http://www.hafencity.com/upload/files/files/HafenCity_Praxisleitfaden_Elektromobilitaet.pdf.

Schneider, U., Gnann, T., Plötz, P., & Dütschke, E. (i.p.). *Public charging infrastructure for electric vehicles—wishes and reality*. Fraunhofer Institute for Systems and Innovation Research ISI: Karlsruhe.

Schweizer Forum Elektromobilität (2012). *Schweizer Road Map Elektromobilität*. Retrieved April 11, 2014, from http://www.forum-elektromobilitaet.ch/home/engagieren/e-road-map.html.
Siefen, K. (2012). *Simulation und Optimierung der Standort- und Kapazitätsauswahl in der Planung von Ladeinfrastruktur für batterieelektrische Fahrzeugflotten*. Paderborn.
Stroband A., Koopmann S., Sowa, T., & Schnettler, A. (2013). *Platzierung und Dimensionierung von öffentlicher Ladeinfrastruktur für Elektrofahrzeuge—Sitzing and Sizing of public charging infrastructure for electric vehicles*. International ETG-Congress: Berlin.
TU Berlin (Technische Universität Berlin) (2011). *Teilvorhaben Nutzerverhalten und Raumplanung regionale Infrastruktur. Final report*. Berlin.
Tran, T. A., & Daim, T. (2008). *A taxonomic review of methods and tools applied in technology assessment*. Technol. Forecast. Soc. Chang. *75*, 1396–1405.
UN. United Nations (UN), Department of Economic and Social Affairs (2012). *World Urbanization Prospects, the 2011 Revision*. Retrieved May 14, 2014, from http://esa.un.org/unpd/wup/pdf/FINAL-FINAL_REPORT%20WUP2011_Annextables_01Aug2012_Final.pdf.
Vélib (2014). *Vélib*. Retrieved April 11, 2014, from http://www.velib.paris.fr/.
Wietschel, M., Plötz, P., Kühn, A., & Gnann, T. (2013). *Market evolution scenarios for electric vehicles*. Summary report. Fraunhofer Institute for Systems and Innovation Research ISI: Karlsruhe. Retrieved April 11, 2014, from http://www.isi.fraunhofer.de/isi-de/e/projekte/npetco_316741_plp.php.
Wirges, J., & Fulda, A.-S. (2010). *MeRegioMobil. Standortplanung von Ladestationen für Elektrofahrzeuge*. European Institute for Energy Research (Eifer): Karlsruhe.
WIWO. Wirtschaftswoche (2014). *Die Ladesäule, an der Sie stehen, gibt es gar nicht*. Retrieved April 23, 2014, from http://www.manager-magazin.de/unternehmen/autoindustrie/fahrbericht-bmw-i3-rwe-eon-vattenfall-etc-verschlafen-e-mobilitaet-a-955489.html.

Results of the Accompanying Research of the 'Modellregionen Elektromobilität' in Germany for Charging Infrastructure

Robert Kuhfuss

Abstract It has been the German approach so far to not promote electromobility by direct subsidies but to organise large research projects, which should enable electromobility on Germany's way to a technological leader in this field. A part of this is the 'Modellregionen Elektromobilität' programme, which is funded since 2009 by the Federal Ministry of Transport and Digital Infrastructure (BMVI). The main focus is to make electromobility visible in everyday life, therefore funding fleets and testing a bunch of solutions for different fields in electromobility on the street. Since there are a lot of different projects in the eight regions, an accompanying research programme merges the results and thus can be seen as a multiplier. In the field of infrastructure, there is a lot of ongoing research. A research topic from the programme was the access, billing and technical requirements for the electric infrastructure. In addition, there is a big question the accompanying research addresses. The issue is, how to build up additional infrastructure on a demand-based basis, since it is still hard for manufacturers and service companies to find a business case—and it remains uncertain how much infrastructure will be needed in the future. This article provides some theoretical and practical rudiments based on the report of the Federal Ministry of Transport and Digital Infrastructure (BMVI 2014a) to answer this, which are also tested in the shown demonstration projects. Generally spoken, the results are in consistency with other big funded programmes in Germany like first results of the Schaufenster-programme (BMVI 2014b) or the Fraunhofer Systemforschung (FhG).

Keywords Electromobility · Infrastructure · Best practice · Demand-based construction

R. Kuhfuss (✉)
Department Electrical Systems, Fraunhofer Institute for Manufacturing Technology and Advanced Materials IFAM, Wiener Straße 12, 28359 Bremen, Germany
e-mail: Robert.Kuhfuss@ifam.fhg.de

W. Leal Filho and R. Kotter (eds.), *E-Mobility in Europe*,
Green Energy and Technology, DOI 10.1007/978-3-319-13194-8_5

1 Introduction: Electromobility in the Demonstration Projects 'Modellregionen Elektromobilität' and in North Germany

As described briefly in the abstract, a strong motivation on Germany's way to a technological and market leader in electromobility is to demonstrate the technologies to the user. Therefore, a big governmental funded project, the 'Modellregionen Elektromobilität' started in 2009 with eight regions competitively selected from all over Germany, where electric vehicles made it on the street and charging infrastructure was built up. To transfer the gathered knowledge to the enablers of electromobility [which for the programme mainly are the following stakeholders: (local) energy provider, local authorities], an accompanying research was set up for a range of topics such infrastructure, safety, regulatory law, fleet management, user acceptance and traffic development (Braune 2014).

In the field of infrastructure, most of the federally funded projects from the BMVI programme with around 2500 charging points (mainly with alternating current) were identified and evaluated. The main research topics addressed were the technical and information technological aspects interoperability and demand-based construction.

This article gives a summary for the first results of the accompanying research regarding the field of infrastructure according to (BMVI 2014a). Because of the innovation of the whole topic electromobility, most results can be considered to be relatively new and topical, since for example all individuals, organisations or companies who want to set up new infrastructure still face a lot of difficulties. Especially, the question as to where the infrastructure should be build was seized upon during the projects which are discussed here.

In recent years, when talking about development of charging points, one encounters the term 'demand-based' (NPE 2012a). This is perceived as the availability of a sufficient number of charging points that cover the actual demand on electricity for 'refuelling' existing and future created electric powered vehicles (BMVI 2014a). 'Demand-based' charging infrastructure is to be distinguished from the widespread existence of charging infrastructure, where the charging points are spatially, more or less equally, distributed, and present even in places where there is no need to be. It is most likely, that an area-widespread charging infrastructure exceeds (so far and future) demand which can be considered also from a user perspective (Peters and Dütschke 2010). The desire for an area-widespread charging infrastructure also arose due to the existed range anxiety (European Commission 2013; NPE 2012a), and it seems to be more a 'need' than a necessity (BMVI 2014b; Peters et al. 2013).

A delimitation of the term 'demand-based' can be achieved using the following key-point questions.

2 Key Questions for the Building of Infrastructure

2.1 *Quantity Structure—How Many?*

Considering the subject demand, and in order to determine the quantity structure, not only the question about the number of the charging possibilities has to be answered, but also all the additional questions arising. In addition, here mainly in Germany, a quantity structure can be derived based on the expected development of the market, as shown by the EU and the NPE statistics. Closely related to the precise number of charging points is the development of use-cases and business models, in order for the developing charging infrastructure to operate within the costs. For this reason also, a more modest, based on the actual demand development is reasonable. A forced privately financed infrastructure development is rather not to be expected, due to the current costs. Basically, however, the question is raised whether newly constructed public infrastructure could be an incentive for vehicle purchases (the 'chicken-or-the-egg-problem').

3 Facilities e.g. Positioning—Where?

Closely connected to the number of charging points are certainly the spatial aspects, namely the concrete spatial positioning of the facilities, at first.

Based on whether they are publicly, semi-publicly or privately positioned, they make for different demand requirements. The private charging today makes up the biggest part of the charging activities in Germany with expected 90 % of all charging events (Peters et al. 2013; NPE 2012a). With some exceptions of street parking without charging possibilities in the garage, the carport or also in home parking spaces, the charging in public places is a complementary infrastructure, meaning a supplementary opportunity to home charging. Public charging can normally be found in the driveway of the destination areas. Accordingly, for a demand-based development it is reasonable to consider the traffic to the destination, taking into account at the same time the origin of the traffic. Related to this, it is common to refer to the possibility for charging activities during work. However, this is still legally obstructed and connected to questions which reflect the current (political) discussions; as charging activities during work can constitute financial benefits for the employee, and thus this constitutes a reason for this benefit to be taxed. Thereby, if the employer provides electricity to an employee by itself, under actual legal circumstances he faces a treatment as an energy supplier which will raise energy-related financial questions, as well as legal ones. Intermodal links are also under consideration in this context. Therefore, charging infrastructure should be developed where electromobility is offered as a systems element in the integrated urban transport system, as for example in 'mobility points', in connection to the public transport stops or e-car sharing offers. These points are primarily well visible and highly visited. Aspects of the urban structure should be also considered.

During the discussions, several general factors for suitable and some exclusion criteria for the construction could be identified. First, the following aspects have to be considered, while thinking about a location for a charging station (EMiS 2014; BMVI 2011):

from a general point of view:

- the availability of the spot (also urban land-use planning)
- the structural and technical applicability
- traffic safety
- building law, cityscape, city design guidelines
- the preservation of sites of historic interest, conservation, other specific regulations

from the provider point of view:

- the effort needed for structural integration and electric connection to the grid
- attractiveness/representativeness of the location, noticeability
- expandability

from the user point of view:

- accessibility, cognizability
- attractiveness and centrality
- expressions of preference for specific locations from users
- points of intermodal transport connections

There are several spots which are in general well suited for the building of infrastructure:

- clinics and health centres
- large retail sector clusters in business parks (shopping malls, centres, hardware stores, etc.) and other retail concentrations
- Sustenance stations with good transport/traffic link (e.g. roadhouses at the motorway)
- event halls, congress centres or sports stadium
- centres of tourism and leisure (funfair, special excursion destinations)
- educational centres: (high-)schools, colleges and universities
- hubs of public transportation (especially train stations)

Exclusions should be considered the avoid mistakes at the placing of infrastructure if one operates from a demand-based perspective, leading to exclusion criteria for the construction of charging infrastructure (BMVI 2011, 2014a):

Factual obstacles in relation to the concrete regions of the concerned location-areas, are the lack of availability during the working time of the model project, their constructional or technical suitability, such as lack of size or incompatibility with constructional or (electro-) technical requirements, or city-constructional issues, especially aspects of city picture and structure.

Legal obstacles can be present when the status of the regions, due to planning laws or similar contexts, does not allow the construction of charging columns. Alongside this, limitations can be posed by legal obstacles through conflicting protection standards, concerning monument protection laws, nature protection laws or due to specific area rules.

Furthermore, it has to be said that cities often lack the human and financial resources required for specific on-site (model supported) demands of their charging infrastructure. At the same time, the models can depict the reality only in a limited way. The demand is determined theoretically by scientific or consultancy institutions; therefore, the limits of the models impinge on the reality of the community. At the same time, questions of city planning and actors' interests are not mapped into the models. For public infrastructure, therefore, a well-established communication at the connection between the infrastructure stakeholders and the community where the infrastructure is planned, is necessary and important. Together with questions such as the detailed position, integration in the city area, demand-oriented dimensioning, issues such as aspects of maintenance security, as well as further local political interests are arising from the interface of the infrastructure with the community which follow the development of charging infrastructure. For the city, except for the maintenance aspects, the representation of their specific innovation activities can be in the foreground. However, also of importance are questions regarding political and social acceptance of the charging infrastructure. Therefore, for the political level it is important to secure the acceptance of the city, or of the district representatives, or to establish the approval within the city council. For the social level, the acceptance of the citizens and the other users is important. Citizens have an increasing interest for participation; they want information and to have a voice. This increasing importance of the public participation, together with the introduction of the electromobility, or the development of charging infrastructure, constitute a comprehensive management task which needs appropriate organisation and ways to cooperate, both on the administration, and in cooperation with the legitimised decisions of the political bodies, as well as with the citizens (BMVI 2014c). Below examples of good practices of cities, communities, energy suppliers and other clients that follow different approaches for the development of charging infrastructure are presented. They provide methodological tips for the determination of the deposited quantity structure, and serve—at different spatial levels—the discovery of appropriate location.

Regardless of the methodological approaches, it is advised to consider the processes of demand-based infrastructure development from at least two points: First, the determination of local demand; and second, the subsequent implementation phase, which considers—further to the quantitative measurements—qualitative aspects (e.g. aspects of the monument protection, civic participation, etc.).

3.1 What Technology? For Whom?

Another question related to demand-based development is, which technology should be developed further? The currently available charging possibilities have different charging power and different technologies associated with them regarding access and payment, and fulfil different requirements and needs. Also, the choice of power classes belongs, in particular cases, to the needs dimension. The time advantage of fast charging—in contrast to the slower AC charging—for the equally different power classes that come with different requirements—is to be compared with the cost disadvantages. The installation of fast-charging stations can also refer to different use and business-cases, than the slower AC charging infrastructure (see for example the ideas of fast-charging stations at motorways, to bridge longer distances (ChargeLounge) *and* (Fastned). In principle, a mix-form of fast charging and AC charging infrastructure, depending on particular use-cases in the cities or communities, seems meaningful.

During the research undertaken, some minimum requirements were defined for the construction of the infrastructure:

- the use of a consistent plug (EN 62196-2; Type 2, so called 'Mennekes' plug)
- the infrastructure should be able to communicate with an IT-back end in order to make the connection to a network (e.g. roaming network) between different infrastructures possible
- a billing possibility which enables spontaneous charging (e.g. without an extended provider contract)
- an RFID authentification possibility

3.2 User Groups and Use-Cases—For Whom?

In the system of electromobility, involved actors (including users and providers, communities and approved authorities) can basically pursue different, and partly also conflicting, interests. Therefore, for example, the desire of the users for an area-wide infrastructure can be in contradiction with that of the operation company, which in principle—based on a business model—would establish only a limited and profitable number of charging opportunities.

Also, a demand-based development requires a differentiation by different target or user groups (e.g. private/commercial customers, community/private fleets), which have different charging needs and, therefore, requirements in a demand-based development.

3.3 Time Aspects—When?

Another important point concerns the time aspects; so, on the one hand, the question for which time horizon the demand is being determined, has to be answered. If it is planned for the current status, the requirements and the number in a demand-based infrastructure is different to that of infrastructure which is planned for future-development, which is based on market start-up-scenarios. On the other hand, the duration-of-parking aspect plays some role (night charging, charging during the labour time, fast charging, etc.).

As mentioned before, the subject of demand-based infrastructure has, for the user, psychological components with the so-called 'range anxiety', which should be considered separately from the real demand (effective driving performance and mobility patterns). To what extent the development of infrastructure affects the purchase and mobility behaviour of the different user group cannot be answered at this point. A demand-based installed infrastructure should be—regardless of its quantity structure—accessible, reachable and visible notable.

4 Best Practice for the Building of Demand-Based Infrastructure

The following project represents one of many possible methods to build up infrastructure systematically.

4.1 Best Practice Local SIMONE in the Project Metropol-E

The subject of demand is being addressed with the help of the questions outlined above already and contained in the second NPE-report (NPE 2011) as a projection of the national total demand for charging infrastructure using, as a meaningful regional distribution. A solution approach for a demand-based distribution presented here is the 'settlement-oriented model for sustainable development and support of e-charging infrastructure' (in German: **Si**edlungsorientierte **Mo**dell für **N**achhaltigen Aufbau und Förderung der **E**-Ladeinfrastruktur—SIMONE) (metropol-E 2014).

With the SIMONE concept, three main questions, in terms of the allocation of the publically accessible charging infrastructure, should be basically answered:

- How high is the demand for publically accessible charging infrastructure?
- Which incentive systems are being applied for the public support of the development of the charging infrastructure?
- How are and should demand and support mechanisms of charging infrastructure be distributed in communal areas?

To answer these questions, a settlement-structured approach should be used. Therefore, it is assumed that inside the community different demands for public accessible charging infrastructure exist. The different area types of a community (e.g. city-centre area, city-centre area near the old-buildings area, pure residential areas, commercial and industrial areas, etc.), is associated, with the help of location criteria, with the demand for publically accessible charging infrastructure. At the same time, the area standardisation of the model offers an approach to structure the financial support of the publically available AC-charging infrastructure. Basically, it is assumed that a joint engagement between public sector and economy is important in order to ensure a demand-based development.

The local authorities determine the demand for publically available AC charging infrastructure over the zones. The identified demand defines at the same time the maximum eligible quantity of charging infrastructure per area type—accumulated for local authorities. The concrete arrangement of the funding rates (amount of the funding rates and allocation from the grant scheme awarding source to Federal, state, district, community) accordingly follow the defined area types. The SIMONE concept sets a rule framework which the communities can apply for the identification of the demand for charging infrastructure, as well as the amount of the funding then needed. The local authorities take the responsibility for the organisation of this, in order to integrate the demand and the positioning of the charging infrastructure into the concept as a tool. With the SIMONE approach, the Federal (or state) government should obtain a possible simple method with verification mechanisms in order to assign the funding rates based on the local authorities.

The aim of the work in the Metropol-E project is to design the SIMONE-concept in an exemplary fashion for the city Dortmund (local SIMONE). The knowledge gained from this exercise should provide approaches which, in the future, can be transferable to other communities. One central task of the PTV Group (PTV) before using was to develop a planning approach for the identification of the demand for publically accessible charging infrastructure per area type. The development of incentive systems for the efficient establishment of charging infrastructure in community areas occurs through the TU Berlin—Workgroup for Infrastructure Policy (WIP).

As part of the work of the PTV on the identification and the distribution of the publicaly available infrastructure demand, an assessment process for an overall assessment or overall result, consisting of two components, namely the location criteria as a measurement for the potential and a synthesis method for the preparation of the detailed results obtained with the location criteria, is developed.

After the definition of the analysis basis, such as user groups, vehicle type and user terms, the overall publicaly accessible charging infrastructure demand for the city Dortmund was calculated. The resulting demand of 800–1100 publically available charging points is placed—by means of traffic-related location indications in the level of traffic cells (TC)—into the context of a small-part distribution model in the area of Dortmund. For the efficiency of the charging points, connection points are being developed, which recognise as relevant the level of the destination-traffic volume in the traffic cells as well as the proportion of vehicles with a charging-related duration of stay in the traffic cell for (a) all and (b) specifically the electro vehicles and the proportion of the publically and semi-publically parking places in the total number of the parking places.

In the first step, the results identified in different units of the location indicators (e.g. destination traffic volume in vehicles/24 h; proportion of charging-related parking in %) are being transferred, for every traffic cell, with the help of a function, into dimensionless use-points. For every traffic cell, a specific value is derived through the addition of weighted use-points (potential value) which can be placed in the potential value of the other traffic cells. The demand in charging infrastructure of the whole city can be distributed by this potential value.

Particularity in the context of the developed methods represents the location criteria of 'intermodal connection points'. A transformation of the results of those indicators into use-points, or utility points, is considered not effective, since the location areas can be determined exactly. In this respect, the charging points at the connection points are being 'located' through the community, and their number will be subtracted from the overall demand of charging infrastructure. The aforementioned utility-analytical methods will be then applied for the remained located charging points.

4.2 Best Practice Demand-Based Development of Charging Infrastructure in Berlin

In the project 'Concept for the identification of charging station locations for station-unbounded car sharing e-vehicles' (in German: 'Konzept zur Identifizierung von Ladesäulenstandorten für stationsungebundene Carsharing-E-Fahrzeuge'), in which the German Centre for Aeronautics and Space Research—Institute of Transport Research (in German: DLR, Institut für Verkehrsforschung—IVF) and the VMZ Berlin Betreibergesellschaft mbH is involved, a scientifically consolidated approach was developed and implemented, in order to determine future locations for charging possibilities for electrical vehicles in the city area of Berlin (DLR 2013). Within the scope of this project, urban areas were identified, for which a demand for charging possibilities for a specific use-profile is expected. Preliminary analysis revealed that for these areas, at least, for all the operations of the big car-sharing fleets with electrical vehicles, a demand is to be expected. The determined charging stations are based strongly on this scenario. The identification of the demand per city area serves as basis for the on-site testing and implementation through the relevant authorities (Figs. 1, 2, 3 and 4).

Figure 5 shows the course of action pursued in the model. The essential initial size was as the definition of the user profile describes, with the result to concentrate on e-vehicle-sharing-flex vehicles (in German: E-Carsharing-Flex-Fahrzeuge, ECSF), which are the electrical car-sharing vehicles without a fixed location. The next step in the model is the analysis of empirical traffic data which was conducted. Those are in particular the mobility survey SrV from the year 2008 (Ahrens 2009)

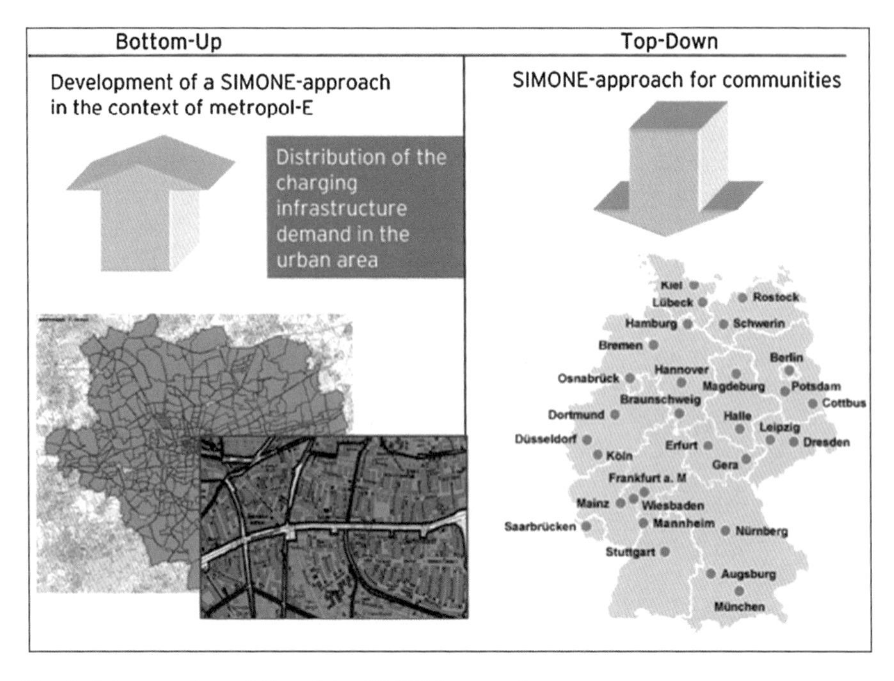

Fig. 1 Schema and portability of the SIMONE-approach

and data from the MIV (motorised individual traffic, in German: Motorisierter Individualverkehr)—Matrix from the traffic model Berlin 2008. Examples are the data on the origin and destination traffic for the ECSF-related routes across the urban area of Berlin. The intensity of the colours is at the same time information about the clustering of the sources or sinks (Fig. 6).

The localisation of this related traffic leads to a fine distribution in the urban area, which is being evaluated with a final simulation. This serves the derivation of the spatial focus-points of the demand, as well as the required number of charging possibilities. An essential aspect of the simulation is the change in perspective of the user (derived from the mobility restrains) towards the vehicle (simulation). This is because a car-sharing vehicle usually experiences a much more frequent use by more than one user. Part of the simulation is also the depiction of the single routes per day which afterwards will be linked as route chains. This allows planners to draw conclusions about the charging demand of the vehicles. Important input parameters for the simulation are also the fleet size, the condition terms and the operation models of the potential ECSF-fleet operators. From the analysis of the simulation derives, for example, the average daily charging demands per traffic area for different scenario and thus, the total demand on charging possibilities.

Figure 7 shows how, based on this model, the existed infrastructure of Berlin, in a first stage of construction, should be expanded with additional charging points in a further step in order to cover the demand on charging infrastructure. A final result of

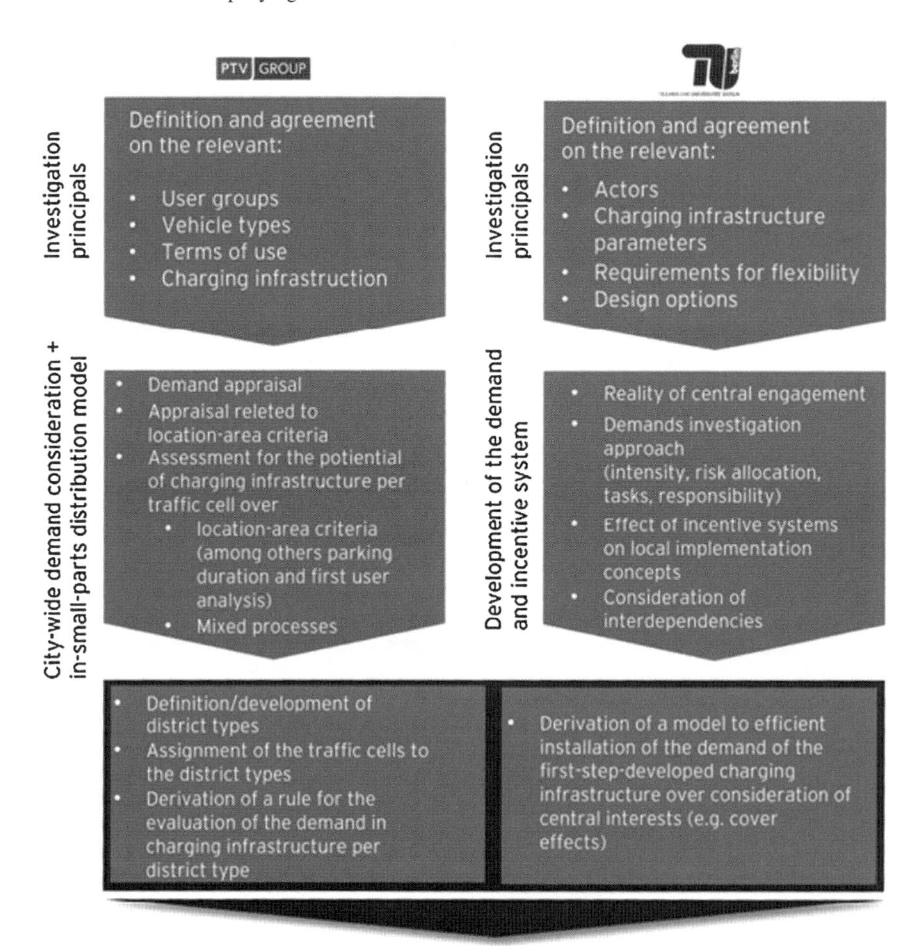

Fig. 2 Overview of the research aspects of the project

the project was the recommendation of the expansion of the charging infrastructure in about 264 charging stations (242 traffic districts) to a total of 534 charging points (366 traffic districts). As a further assistance in reaching the decision about the development of charging infrastructure the public transport division and parking space management, for example, can be consulted with.

As mentioned earlier, the Federally funded 'Modellregionen Elektromobilität' programme was introduced to make electromobility visible in everyday life, and hence attract a large variety of users for electric vehicles. In order to assure a smooth fleet testing, the need of training according to the safety standards in Germany on how to operate electric vehicles and to respond in critical situations became obvious. Hence, some participants of electric vehicles received training courses (offered by Fraunhofer IFAM). In these seminars, not only safety aspects

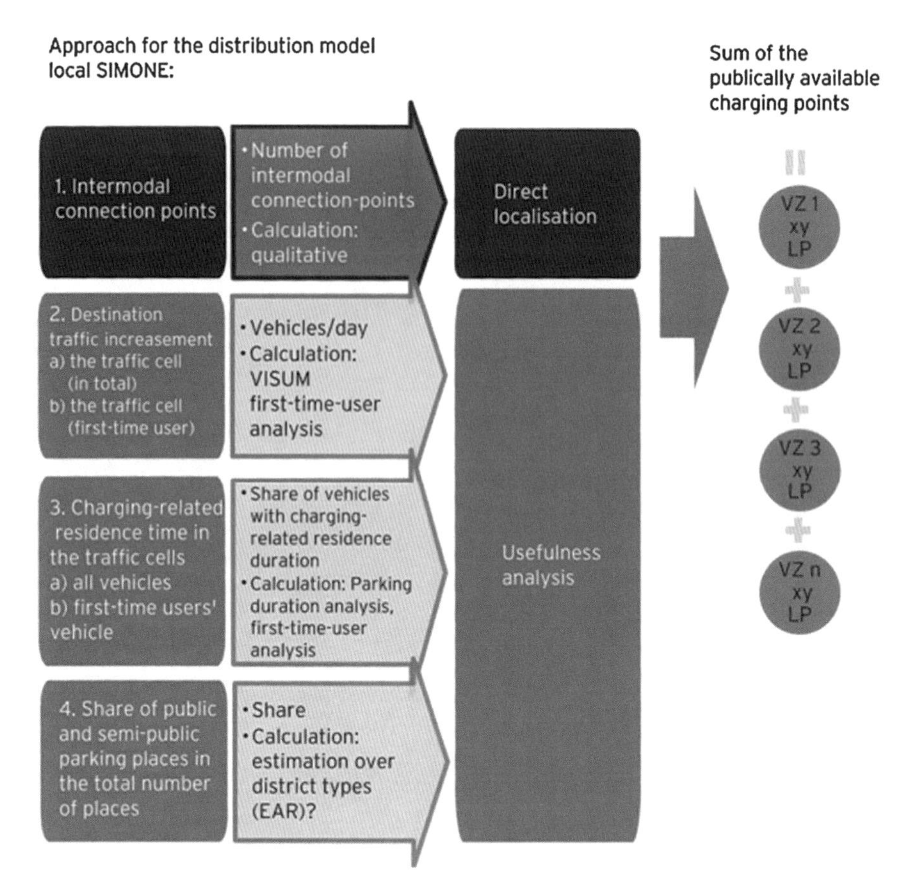

Fig. 3 Approach of the SIMONE-model

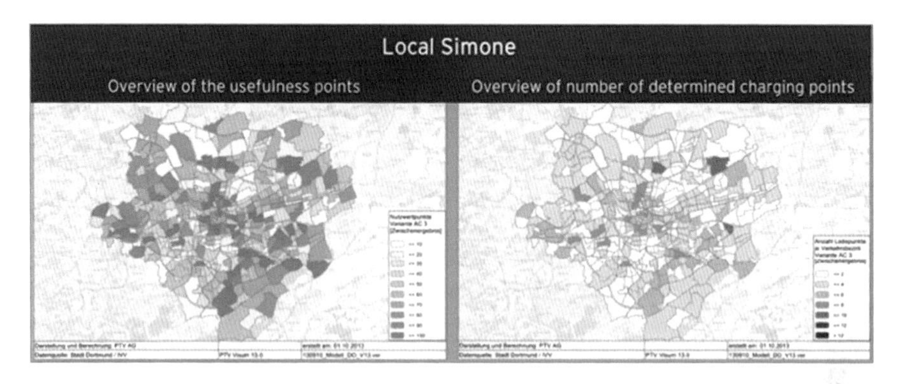

Fig. 4 Principle representation for the identification of the number of charging points pro traffic cells through utility-analytical methods

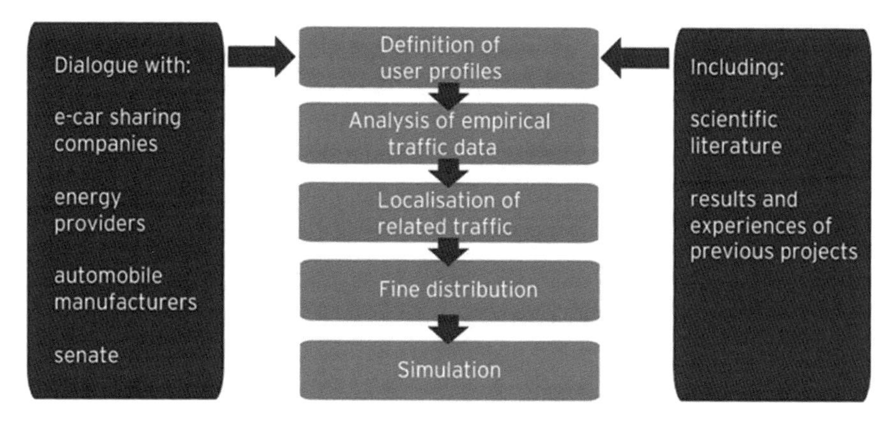

Fig. 5 Flowchart for the determination of the required infrastructure for the Berlin project

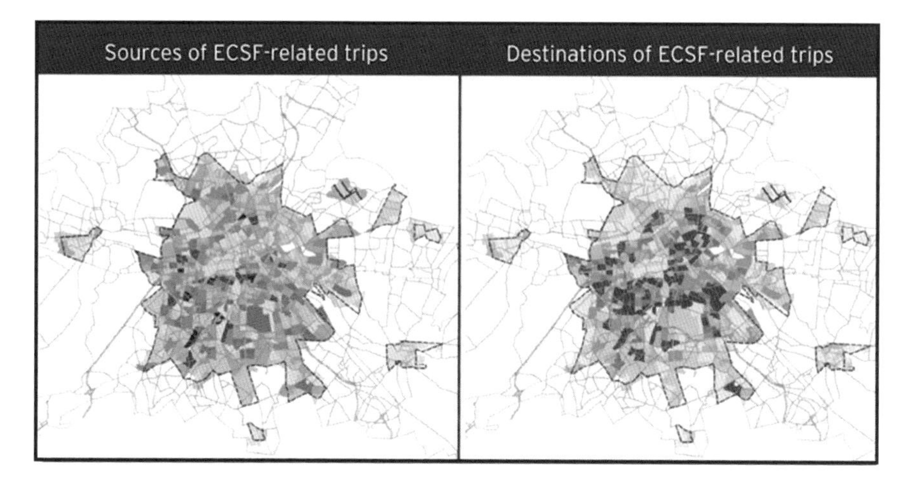

Fig. 6 Overview of the source and destination routes of the related user groups

but also recent developments in the field of electromobility technologies were introduced. Thus, the trainings further supported the aim of the 'Modellregionen Elektromobilität' by informing potential users of electric vehicles and by raising awareness for electromobility in Germany.

5 Conclusion

In this article, results of the accompanying research for a big Federally funded governmental programme regarding electromobility in the field of infrastructure are discussed. If talking about the building up of charging infrastructure, the key

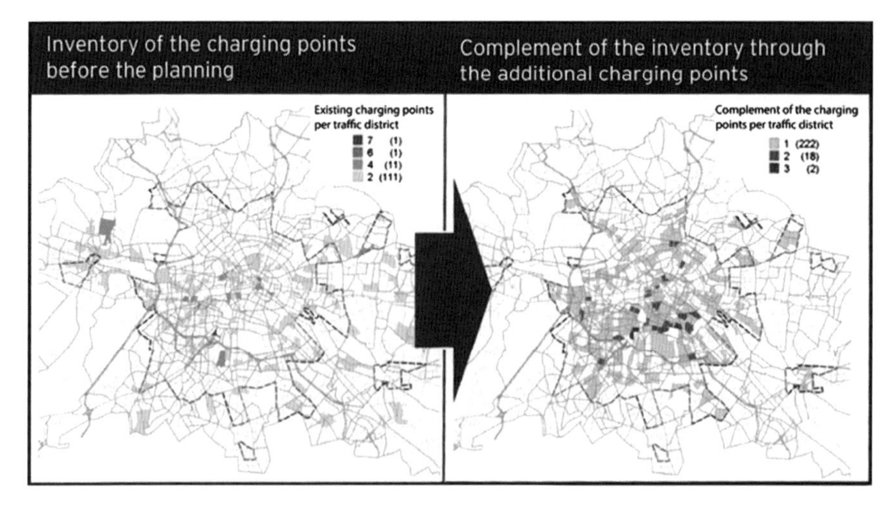

Fig. 7 Overview of 'before' and 'after' (existed charging points at the *left* and accordingly determined charging points for the city-area on the *right*)

questions set out of the article should be faced by the individual, organisation or company responsible for the construction. Some regulations are outlined regarding technology development and several hints are given for the strategic location of the infrastructure based on demand. Furthermore, two scientific approaches are shown which address a demand-based build up of charging infrastructure.

References

Ahrens, G.-A. (2009). *Endbericht zur Verkehrserhebung Mobilität in Städten – SrV 2008 und Auswertungen zum SrV-Städtepegel*. TU Dresden, Lehrstuhl Verkehrs- und Infrastrukturplanung; PDF. Retrieved June 17, 2014 from http://www.mil.brandenburg.de/media_fast/4055/Potsdam_TEXT.pdf.

BMVI - Bundesministerium für Verkehr und digitale Infrastruktur – Federal Ministry of Transport and Digital Infrastructure. (2014b). Förderprogramm. *Schaufenster Elektromobilität*. Retrieved June 17, 2014 from https://www.bmvi.de/SharedDocs/DE/Artikel/IR/elektromobilitaet-steckbriefe.html.

BMVI - Bundesministerium für Verkehr und digitale Infrastruktur – Federal Ministry of Transport and Digital Infrastructure, and NOW GmbH. (2011). *Elektromobilität in Deutschland – Praxisleitfaden - Aufbau einer öffentlich zugänglichen Ladeinfrastruktur für Genehmigungsbehörden und Antragsteller*. Berlin.

BMVI - Bundesministerium für Verkehr und digitale Infrastruktur – Federal Ministry of Transport and Digital Infrastructure *and* NOW GmbH. (2014a). *Öffentliche Ladeinfrastruktur für Städte, Kommunen und Versorger - Kompendium für den interoperablen und bedarfsgerechten Aufbau von Infrastruktur für Elektrofahrzeuge*. Berlin. [covering several projects from the federally funded program „Modellregionen Elektromobilität“ – Model Regions Eectromobility].

BMVI - Bundesministerium für Verkehr und digitale Infrastruktur – Federal Ministry of Transport and Digital Infrastructure *and* NOW GmbH. (2014c). *Elektromobilität in der Stadt- und Verkehrsplanung – Praxiserfahrung aus den Modellregionen und weitere Wissensbedarfe*. Berlin.

Braune, O. (2014). *Elektromobilität in Modellregionen - Begleitforschung als Erfolgsfaktor*; contribution for conference „Elektromobilität vor Ort: Fachkonferenz für kommunale Vertreter", Bremen, organized by the Bundesministerium für Verkehr und digitale Infrastruktur (BMVI) - Federal Ministry of Transport and Digital Infrastructure; PDF. Retrieved June 17, 2014 from http://bit.ly/UHAEZF.

ChargeLounge. *Elektromobilität ist die Zukunft – concept page for fast charging network in Germany*. Retrieved June 17, 2014 from http://www.chargelounge.de/.

EMiS – Elektromobilität im Stauferland. (2014). Retrieved February 07, 2014 from http://www.emis-projekt.de/.

DLR, Institut für Verkehrsforschung – IVF. (2013). *Konzept zur Identifizierung von Ladesäulenstandorten für stationsungebundene Car-Sharing-E-Fahrzeuge*. Retrieved June 17, 2014 from http://www.dlr.de/vf/desktopdefault.aspx/tabid-958/4508_read-36923/4508_page-2/

European Commission. (2013). *Proposal for a Directive of the European Parliament and of the Council on the deployment of alternative fuels infrastructure*. COM (2013) 18 final. Brussels.

Fastned. *Concept page for fast charging network in Netherlands*. Retrieved June 17, 2014 from http://www.fastned.nl/.

FhG - Fraunhofer Gesellschaft. *Systemforschung Elektromobilität*. Retrieved June 17, 2014 from http://www.elektromobilitaet.fraunhofer.de/

Metropol-E – Elektromobilität Rhein Ruhr. (2014). Retrieved April 24, 2014 from http://www.metropol-e.de/.

NPE - Nationale Plattform Elektromobilität – National Platform Electromobility. (2011). *Zweiter Bericht der Nationalen Plattform Elektromobilität*. Berlin.

NPE - Nationale Plattform Elektromobilität – National Platform Electromobility. (2012a). „Fortschrittsbericht der Nationalen Plattform Elektromobilität (Dritter Bericht)".

NPE - Nationale Plattform Elektromobilität – National Platform Electromobility. (2012b). *Ladeinfrastruktur bedarfsgerecht aufbauen*. Berlin.

Peters, A., Dütschke, E. (2010). *Zur Nutzerakzeptanz von Elektromobilität - Analyse aus Expertensicht*. Results from the project Fraunhofer Systemforschung Elektromobilität (FSEM). Retrieved June 17, 2014 from http://www.forum-elektromobilitaet.ch/fileadmin/DATA_Forum/Publikationen/FSEM_2011-Ergebnisbericht_Experteninterviews_t.pdf.

Peters, A., Doll, C., Plötz, P., Sauer, A., Schade, W., Thielmann, A., et al. (2013). *Konzepte der Elektromobilität: Ihre Bedeutung für Wirtschaft, Gesellschaft und Umwelt*. Berlin.: Edition Sigma.

Large-Scale Deployment of Public Charging Infrastructure: Identifying Possible Next Steps Forward

Peter van Deventer, Martijn van der Steen, Rogier van Schelven, Ben Rubin and Richard Kotter

Abstract This paper presents the next steps forward for a large-scale deployment of public charging infrastructure after the first round of infrastructure was mainly financed by government agencies over the last 3–5 years. In order to create a sustainable and market-driven public charging network, governments are increasingly looking for strategies to support the next generation of public charging infrastructure with creative financing mechanisms and limited public funding. The primary goal is to review and analyze the different models that are currently being tested in early adopter markets such as Norway, the Netherlands, California, and United States. Based on this early learning, identify possible business models for large-scale deployment of public charging infrastructure. This paper describes the challenges and opportunities in these early markets, identify six different (international) models for investing into public charging infrastructure and describe their individual advantages and disadvantages. By applying these models to California, a state that is actively involved in public policy development and introduction of electric vehicles, this paper identifies preferred financing models applying three different scenarios. The research provides insights into international comparison of deployment of public charging infrastructure and possible financial models. Based on a case study, the various advantages and disadvantages of these models are exemplified. Finally, suggestions are made for further research and modeling.

P. van Deventer (✉)
Coast to Coast e-Mobility, One Montgomery Street, San Francisco, CA 94104, USA
e-mail: petervandeventer@gmail.com

M. van der Steen
Netherlands School of Public Administration (NSOB), The Hague, Netherlands

R. van Schelven
Kwink Group, The Hague, Netherlands

B. Rubin
M Public Affairs (Past: California Governor's Office/OPR), New York, USA

R. Kotter
Northumbria University at Newcastle upon Tyne, Newcastle upon Tyne, UK
e-mail: richard.kotter@northumbria.ac.uk

W. Leal Filho and R. Kotter (eds.), *E-Mobility in Europe*,
Green Energy and Technology, DOI 10.1007/978-3-319-13194-8_6

Keywords Electric vehicles · Evs · E-mobility · EVSE · EV-policy · Policy implementation · Charging infrastructure · Innovation · Financial strategies · The Netherlands · California · Coast-to-Coast partnership

1 Introduction

This paper looks at the next steps forward for large-scale deployment of public charging infrastructure for plug-in electric vehicles (PEV's). The first round of infrastructure has been financed by government agencies typically for at least 5 years. The main problem with introducing PEV's is not to start but how to grow from small numbers to large-scale market adaptation (Van Deventer et al. 2011).

The essence of this "how to get from small steps to a giant leap" is basically a chicken-egg problem; in order to get more people interested in buying PEV's more chargers are needed (among other arrangements); however, the reverse is true as well, in order to get more chargers more EV's are needed.

The question is how to break through this classical public policy problem with many barriers both for battery electric vehicles (Schot et al. 1994; Diamond 2008; Steinhilber et al. 2013) as well as fuel cell electric vehicles (Debe 2012). If chargers are installed at great cost to taxpayers but no one uses them, public support for clean transportation may be at risk (InsideEVS 2013b). On the other hand, if governments have been supporting the introduction of PEV's[1] and the early adopters have profited but charging becomes nearly impossible due to a lack of charging infrastructure, public support may turn around quickly if politicians, the public, and the media become critical of progress in this area. (Telegraaf 2014; Autoweek 2014).

For nearly a decade, much research has been undertaken how to successfully introduce EV's and provide for the necessary public infrastructure (Ahman 2006; Morrow et al. 2008; Sperling and Gordon 2009; Schiffer 2010; Snyder et al. 2012). Also, attention has focused on consumer perspectives (Luis et al. 2013), relevance of smart grids (Kotter 2013), connection with Intelligent Transport Systems (Hübner et al. 2012), and commercial viability (Quandt 1995; Kley et al. 2011). These research efforts have all mainly focused on situations where numbers of EV's were low or governmental policies had recently been put in place (e.g., Boonen 2012; Williams 2013)

Although excellent research was executed comparing policy effectiveness for 25 + municipalities in the Netherlands (Boonen 2012), hardly any research has focused on the next steps forward for those areas in the world who have been leading both in terms of available supportive policies, EV's on the road and infrastructure put in

[1]In addition to the term "PEV's," also terms like BEV's (battery electric vehicles) and PHEV's (plug-in hybrid) electric vehicles are often times used. However, PEV's includes them both: BEV's representing EV's with only batteries to propel the vehicle and PHEV representing EV's that also have a small combustion engine to support/charge the battery. This uses the term PEV's.

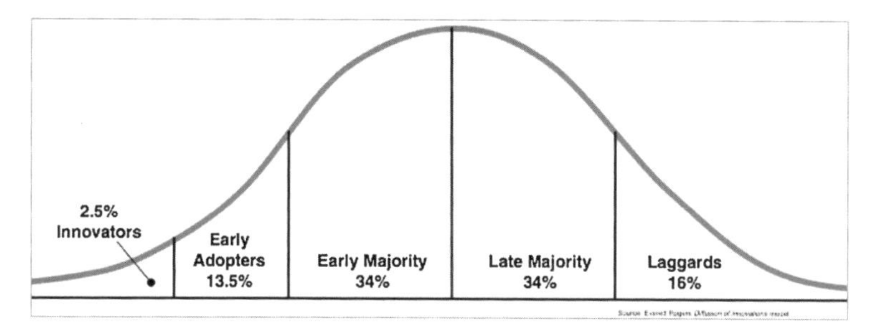

Fig. 1 Innovation adoption curve (Rogers 2003)

place. It is assumed that these areas will first run into the above described problems and risks when the market moves from the innovator stage (sales share[2] of EV's < 2.5 %) to early adopter stage (2.5–16 %) and later to early majority stage (16–50 %); see Fig. 1.

Since these areas have already supported the introduction with large sums of public funding, it is assumed that governments of leading states will increasingly look for strategies to support the next generation of public charging infrastructure with creative financing mechanisms mainly due to decreasing public funding. In other words, how then to create a sustainable and market-driven public charging network?

2 EV Market Leadership

EV adoption rates differ between countries. Norway, the Netherlands, and California are clearly leading the way in terms of PEV sales shares (ICCT 2014). Figure 2 presents a breakdown of worldwide sales of new cars with leading states/ countries on top.

Looking at Fig. 2, Norway, the Netherlands, and California are clearly leading in terms of PEV's[3] sales shares (Cleantechnica 2014; The Foreigner 2014). As of April 2014, Norway has about 20.000 PEV's on the road and more than 6 % of all cars sold in 2013 were PEV's. More importantly, the market share of EV's in Norway is currently about 10 % or higher on a monthly basis (Myklebust 2013; EVObsession 2014); in fact, in October 2013 the Nissan Leaf was the best sold car in Norway beating the VW Golf.[4]

---

[2]PEV sales share is the percentage of PEV's sold relative to all cars sold over a given period of time. The ambition for some of the leading states such as California, Norway, and the Netherlands is to reach 10–25 % over the next decade.

[3]PEV's include here both BEV's and PHEV's.

[4]Best-selling car in Europe since many years.

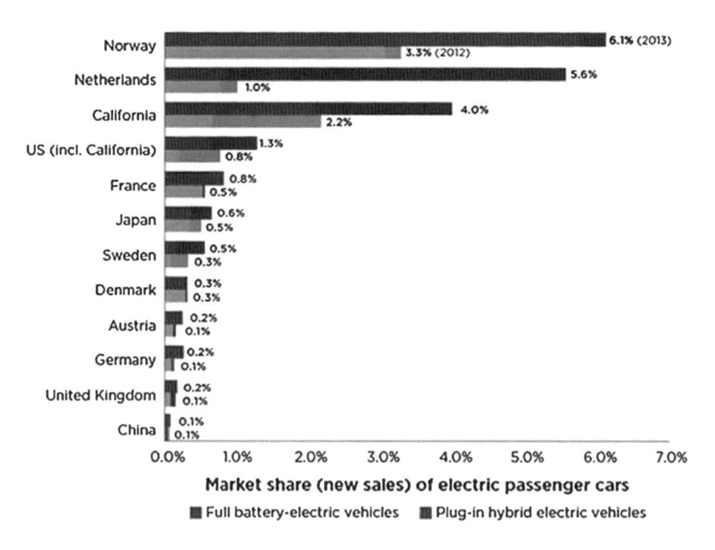

Fig. 2 Market share of EV's relative to all vehicles sold in various countries including the State of California (as of April 2014); *Source* ICCT (2014) http://www.theicct.org

The Netherlands has similar numbers: nearly 20.000 PEV's were sold in 2013, representing over 5 % of the total market; in fact, the Mitsubishi Outlander (PHEV) is making up more than 40 % of all PEV's sold, and ranks in the top 10 of all cars in 2013 (RAI 2014). And, December 2013 was even more impressive: 25 % of all cars sold were electric cars, mainly due to a (slight) change in tax regime.

California is leading the way in the US with currently 65.000 EV's, being about 0.5 % of all cars on the road. Looking at the number of PEV's sold in California on a monthly basis (approximately 3000), the monthly market share would be about 2.5 % (Bizjournals/EV Inside 2013). This results in a third place position behind Norway and the Netherlands.

In terms of market share, all these countries and the state of California have just moved into the early adopter stage and possibly Norway and the Netherlands will be moving toward "early majority stage" in the coming 2–3 years based on their rapid EV sales growth. California is assumed to be following the same pace as Norway and the Netherlands but arriving a few years later at the "early majority stage"; however, individual regions within the state like the Bay Area, San Diego, and Los Angeles, County could possibly be in the "early majority stage" at the same time as Norway and the Netherlands. Studies comparing the introduction of the Toyota Prius a decade ago with the first 3 years of PEV sales in California indicates that PHEV's are selling at a higher rate (Turrentine et al. 2013). The question then becomes: will charger deployment be able to keep up with that pace?

3 EVSE Market Development

The number of chargers seems to lag behind in the countries were EV adoption is high. The ability to charge an electrical vehicle at home, work or in the public domain is essential to consumer adoption. Early studies done by Anegawa (2010) suggest that the availability of chargers help overcome range anxiety even if the chargers themselves are not being used extensively.[5]

In terms of the number of chargers or Electric Vehicle Supply Equipment (EVSE's) installed worldwide, recent numbers are more difficult to gather than EV sales. As of December 2012, around 50,000 non-residential charging points were deployed in the U.S., Europe, Japan, and China. As of March 2013, 5,678 public charging stations existed across the United States, with 16,256 public charging points, of which 3,990 were located in California. As of November 2012, about 15,000 charging stations had been installed in Europe of which 2500 were located in the Netherlands (IEA 2013).[6]

Type 1 level charging (110 V) is most dominant in the US for home and workplace charging, while Type 2 (220–240 V) is standard for public street charging both in the US and the EU and home charging in the EU. Globally, Type 3 (AC/DC 400–440 V) is small by numbers; in terms of percentage of all EVSE's it is more dominant in Japan and the Netherlands than the US: as of September 2013, there were 3,073 quick chargers deployed around the world, with 1,858 in Japan, 897 in Europe (The Netherlands, currently 50 and 300 in 2014) and 306 in the United States (IEA 2013).

As to where EVSE's are located, California has more home and workplace charging than the Netherlands and Norway. One reason for this is the higher percentage of home owners having a private driveway in California and typically larger parking lots at the workplace, allowing for easier EVSE installation. Also, specific workplace charging incentives and targeted events like "Drive the Dream" have stimulated EVSE deployment in California. In contrast, public charging is more prevalent in the Netherlands which was spurred in large part by the e-Laad Initiative (2009–2012) supported by all major utility companies, more than 2500 EVSE's were put in the public domain for all EV owners to be used at no cost. In addition to the early push by e-Laad to install chargers, many more local, regional, and national initiatives have led to an estimated 18.000 privately owned EVSE as of March 2014.

[5]No studies were found yet to suggest that extended range for PEV's like the Tesla Model S has changes charging behavior. This would however be another option to deal with the number of EVSE's needed.

[6]Type 1 or 2 or SLOW CHARGING is the most common type of charging used which provides alternating current to the vehicle's battery from an external charger. Charging times can range from 4 to 12 h for a full charge. Type 3 or FAST CHARGING, or also known as "DC quick charging," provide a direct current of electricity to the vehicle's battery from an external charger. Charging times can range from 0.5 to 2 h for a full charge.

EVSE deployment in the Netherlands has grown rapidly and is now spreading in Europe and elsewhere (Hauser 2013). The reason for this is that governments and utility companies have early on demanded that EVSE's meet Open Data standards. Two aspects were very important for the Dutch Government: (1) Driver roaming and (2) Open protocol for network operators, OCPP = Open Charge Point Protocol. Driver roaming is all about making sure that the cost for roaming is not pushed down to the driver in the form of roaming fees (like local cell phone roaming fees that used to take place when roaming from, e.g., AT&T to Verizon). OCPP is all about having an open standard between charge station and network back office which is different from roaming interoperability[7] for the driver. Both have been important for market growth in the Netherlands.

The recent passing of Senate Bill SB454 in California is supposed to bring more harmony into the market. In addition, the Bill is assumed to bring more competition into the market which is now de facto being dominated by one single company. SB 454 has a simple goal—to allow consumers with plug-in cars the same access to charging stations that gas stations provide traditional gas cars. SB454 requires that all public charging stations which require payment accept a simple credit card transaction or provide access with a phone call. The bill also requires pricing transparency so that drivers know costs associated with particular charging stations (Plug-in America 2013)

4 EVSE Policy Comparisons

NGO's and leading charger/charging service industry, both in the Netherlands and California, claim that current public charging infrastructure is not sufficient. Already EV-drivers are not able to charge their vehicles in the public domain. And, the expected strong growth of PEV sales in the coming years will make this even worse. Although differences are seen between regions in both states, particularly urban regions seem to be at risk: Metropolitan Area of Rotterdam, Metropolitan Area of Amsterdam, San Francisco Bay Area, and Greater Los Angeles. The industry is suggesting that finance and leasing options, with great success being used by the solar panel industry in the US (Sungevity, Solar City) and by Business to Business PEV leasing in Europe (Athlon Car Lease), could help to overcome the high upfront cost and spread payments. An international comparative analysis of PEV/EVSE policies shows that implemented policies focus mainly on PEV adoption (Bakker 2013; Bakker and Trip 2013), with policies focusing on charging infrastructure lagging behind (Van der Steen et al. 2014). The question then

[7]"Interoperability" stands for the ability of PEV drivers to charge their vehicle at any given charger independent from the subscription that they may have as well as for an open standard between charge station and network back office. Both allow for easy user access and fair market competition.

becomes what should be the role of government? And, if any should it take incremental steps or the "Big Bazooka" method or a hybrid approach?

4.1 Big Bazooka Approach

Currently, arguments are being made in the European Parliament (http://www.europarl.europa.eu/news/en/newsroom/content/20131125IPR26108/html/Alternative-fuel-stations-Transport-MEPs-back-draft-law-to-expand-networks/) in favor of a so-called "Big Bazooka" deployment of EVSE's, where a clear pattern of EVSE's is agreed upon, procured, purchased, and installed. In fact, at a state level that is what has been done in Estonia, relying on a national network of fast chargers (Forbes 2013). The advantage being market-scale driving down the costs, uniformity, and clear signal to potential buyers. Disadvantages being that it presumes in-depth knowledge of user behavior, while early on PEV user numbers are small and usage is dynamic; in addition, unused EVSE's with typical added privileges like free electricity and free parking in dense urban areas may cause public resentment against PEV introduction.

In fact, over the past few years the European Parliament has been looking at a draft law to expand refueling networks in the EU resulting into a minimum of 429.000 EVSE's in 2020. Looking at the projected number of EV's in 2020, being 5.9 million (IEA 2013), about 1 EVSE would be available for 10–15 EV's. In March 2014, the European Parliament announced a much higher target of 8 million EVSE's in the EU by 2020; however, no law has been passed (Greencar 2014)

With the signing of AB32 (2006) and AB118 (2007), a unique boost[8] was given to sustainable transportation in California. While being "fuel neutral" and not picking winners and losers, these policies have resulted into PEV's taking an initial clear lead in terms of number of vehicles on the road and actual usage (Witt et al. 2012). Recently, six major new bills were approved in 2013 and signed into law further supporting major zero-emission transportation investments, some until 2023.

4.2 Hybrid Approach

At the Federal level in the US, no cohesive strategy or targets for EV infrastructure have been developed thus far. However, a large number of state and federal incentives (Plugincars 2014) are available throughout the US to support individual buyers who want to own and drive an EV. At the Federal level, until the end of 2013, EV buyers were able to claim a credit of 30 % of the purchase and costs of

[8]At least 100 million dollars per year have been made available through CEC grants (California Energy Commission).

the charging equipment, up to $1,000 for individuals and $30,000 for businesses. At the state level, e.g., California, charging equipment incentives were in place for some time (Bay Area, Greater Los Angeles) or are just starting (Central Valley).

4.3 Incremental Steps

Arguments for more incremental steps are the inclusion of user behavior, lower upfront cost, and lowering the risk of unused EVSE's. Disadvantages may be that EVSE standardization—important for the car industry—may be more difficult to achieve; another disadvantage may be that the upfront cost is higher but market development—possibly causing production cost to drop—may give governments procurement advantages.

Study by UCDavis on user behavior (e.g., Woodjack et al. 2012) suggests that incremental deployment is favored since the market is still at its early stages. Recently, the Netherlands School for Public Policy (Van der Steen et al. 2014) suggested that when the EV market is passing a certain threshold value, the "Big Bazooka" would make sense in urban areas to accommodate further acceleration of PEV sales. Looking at the introduction of the Toyota Prius and the rapid acceleration a decade ago and the assumption that many of the early Prius buyers in urban regions will switch to more electrification of their cars, a rapid acceleration in the next few years would be very likely. A recent study by Turrentine et al. (2013) demonstrates that PEV growth from 1 year to the next in California (2011: less than 1 %, 2012: 2.2 %, 2013: 4.0 %; see also Fig. 2) is outpacing the Prius sales in the late 1990's.

5 Public Charging Infrastructure: Challenges, Opportunities, and Investment Examples

So, what should be the role of government? And, if any should it take incremental steps, the "Big Bazooka" or a more hybrid approach? Before these questions can be answered, various challenges, opportunities, and investment examples of public EVSE need to be reviewed. As far as investing into public charging, a number of challenges are identified:

(a) Public charging can become quite expensive after initial support: preparation, site selection, procurement, construction/installment, and maintenance, but largely depends on civil engineering aspects like closeness to electrical sources and difficulty of trenching.
(b) Under-utilized public infrastructure creates an "image" problem for government. In fact, the question becomes: how much is needed and how fast?.

(c) What charging behaviors may become dominant: home charging, workplace charging, public street/parking garage charging? (Woodjack et al. 2012) And if you don't know that, what are you investing into?
(d) Several studies indicate that the business case for public charging is not yet strong enough (Goebel 2012)

As far as investing into public charging various opportunities are identified:

(a) Guidelines for charging infrastructure deployment are available and ready to be used (Rubin 2013)
(b) Older existing chargers may easily be retrofitted (e.g., work by ClipperCreek in the late 1990s)
(c) Investment into charging infrastructure creates new "green" jobs (e.g., Scott 1995).
(d) Visible infrastructure supports and helps consumers to be more aware of other options and more likely to switch to cleaner transportation (Geller 2013).
(e) More (fast charging) infrastructure reduces range anxiety (Anegawa 2010).

Over the past decade, various examples have been initiated worldwide to invest into public infrastructure, either by government, industry, utilities, NGOs or a combination:

- Denmark—DONG, the national electricity company teamed up with government to curtail a problem in Denmark: too much wind power generation causing on average hundreds of wind turbines to be left in idle position; DONG invested in EVSE's to stimulate the introduction of PEV's which could be using the surplus energy on the grid.
- Japan—TEPCO used government subsidies to build a large network of thousands of EVSE's (Level 2 and 3) to stimulate lesser use of carbon based fuels.
- Nissan—the first automaker to deploy a DC fast charging (ChaDeMo) network in several countries deployment at low or no cost, e.g., in the US a rebate was available for up to $15.000 until April, 2014. The reason for Nissan doing this was to obtain an early market position for its LEAF.
- Germany—Several efforts are being supported for regional deployment of EVSE's as well as a national hydrogen highway project.
- City of Amsterdam: Since 2009, the Amsterdam has invested heavily in getting EVSE's on the street by taking care of preparation, site selection, procurement, installment, and maintenance. In addition, free parking and charging were offered through 2012. Result: close to 1000 EVSE's in Amsterdam. Similar numbers apply to the City of Rotterdam, as of April 2014 about 800 EVSE's.
- The Netherlands: From 2009 through 2012, the e-Laad Foundation backed by funds from the utility companies, executed exactly what the City of Amsterdam did resulting into more than 2500 EVSE's. E-Laad was permitted by the National Government to put chargers for free in the public domain for 3 years after which the market disturbance was no longer aloud. The discontinuation of the program resulted into various regions like the Amsterdam Metropolitan Region, to look for other ways to get the necessary EVSE's in the public domain.

- Amsterdam Metropolitan Region Electric (MRA-e): In 2011 a regional cooperation between two provinces, the city of Amsterdam, Rijkswaterstaat ("Caltrans") and a few regional transportations agencies started to promote PEV's in the MRA. Since e-Laad had to stop its model of providing free EVSE's, MRA-e developed a model in 2012 to create co-funding for 1000 EVSE's backed by municipalities, provinces, car companies, and utilities. Due to smart procurement and market development, the integral price for an EVSE dropped by 50 %. MRA-e expects a breakeven point in 2017/2018 (Linnenkamp 2013); see Fig. 3.
- TESLA: As part of their ambition to be world leader for premium PEV's, Tesla has started to build a network of so-called "Superchargers." While at no cost for their clients, Tesla has been able to set up an "EV-Highway" in California and other leading states as of late 2012. Currently, Tesla is building the network in Europe allowing drivers to travel from Portugal to Norway on its SuperCharger network. Tesla has done this without any outside support. According to their plans, by the end of 2014 a complete network will be available allowing their EV's to drive throughout and across the US.
- California: In March 2012, Governor Edmund G. Brown Jr. and the California Public Utilities Commission announced a $120 million dollar settlement with NRG Energy Inc. to fund the construction of a statewide network of charging stations for zero-emission vehicles (ZEVs), including at least 200 public fast charging stations and another 10,000 plug-in units at 1,000 locations across the state. The settlement stems from California's energy crisis. Furthermore, ChargePoint and Key Equipment Finance launched a $100 million lease-to-own program for EV charging stations in October 2013, which will give small- and medium-sized companies and municipalities the opportunity to install EV chargers at no upfront cost. The program that allows for pay-as-you-go financing of charging stations, installation costs, operational services, and warranty.
- Transatlantic cooperation between leading states is found to be useful and relevant (Rubin and Van der Steen 2013). As a result Mr. Rodriquez, Secretary of the California Environmental Protection Agency and Ms. Mansveld, Secretary of the Dutch Ministry of Infrastructure and the Environment signed an agreement on e-Mobility and knowledge exchange regarding infrastructure deployment late 2013.

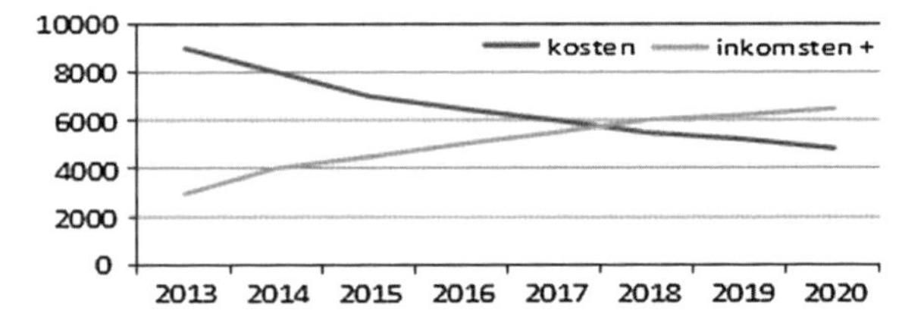

Fig. 3 Development of average total cost (="kosten") and revenues (="inkomsten") of EVSE from 2013–2020

- At the end of 2013, the California Plug-In Electric Vehicle Collaborative (PEVC) and Coast to Coast e-Mobility (C2C) signed a multiyear agreement to cooperate on knowledge transfer and business development including EV infrastructure. PEVC and C2C are large public–private partnerships, which aim to promote knowledge and innovation exchange. PEVC combines all major e-Mobility players in California. C2C connects government policies, inspires academic research, and creates business development opportunities between California and the Netherlands specifically and more in general between the West coasts of the US and Europe.

6 Investing into Public Charging Infrastructure: Six Different Models

Based on these above-mentioned challenges, opportunities, and international examples, six different models for possible investment into EVS's are proposed:

GOV100 = National or State government pays entirely for new infrastructure. The advantage being that a region continues to lead the way and that a well-defined and standardized grid of EVSE's will be initiated upon which the industry can develop their products and services. The disadvantage is that all cost would be carried by the government and that it would imply "picking winners and losers."

IND100 = Companies like Tesla and Nissan paying entirely on their own to capture first mover advantage. The advantage being that costs for government are very limited and that it would be a truly market-driven and business case-based initiative. Disadvantages are that many "solutions" will appear causing unwanted side-effects where the main public may end up paying the burden later down the line. Also, a "wait-and-see" game may start causing the introduction to stall and not contribute to the necessary reduction of greenhouse gas emissions. The question then becomes how to stimulate the industry which will most likely incur governmental costs.

Hybrid-I = State government funds match local initiatives. Advantage of state support for local initiatives which typically involve local communities, will do exactly what is needed: stimulate bottom-up change for sustainable transportation. In addition, some state regulation can be applied to ensure uniformity and overall GreenHouseGas emission reduction. Disadvantage is that some communities may have trouble getting in good proposals because of size and location. And, maybe more importantly, it is still a total government investment and seen as such by the public.

Hybrid-II = Local government funds match local initiatives. Advantage would be that the State government has no real investment in terms of people, funds, and programs, but the disadvantage is that a different form of "wait-and-see" game may be played out by local governments: "why should we (Community X) pay for the commuters from the cities Y and Z who are causing the problem to begin with." At one point of time, the state would have to step in.

Hybrid -III = Public–private partnerships like the MRA-e model (35 % state, 35 % local, 20 % industry, 10 % regional organizations). Advantage is that every stakeholder is part of the solution and cost can be spread out. In addition, combined procurement will result in lower cost and better acceptance among stakeholders to make next steps. A disadvantage could be that at present only limited experience is available with such innovative models and clear leadership is needed.

Financing = a private funding program to more quickly deploy charging stations. Objectives for such a program could include directly subsidizing customers to decrease the cost of installation or financing; recapturing or recycling funds; potentially attracting third-party providers and leveraging funds to access larger pools of capital.

7 Models Applied: California as an Example

In order to demonstrate these models, one state (California, US) was selected and for this; the authors applied these models to determine best available options and possible cost associated with model choices. The demonstration serves as an example for states and countries worldwide to determine which model(s) could best serve their interest against which costs. Although these models have not been applied, these cases exemplify the various regional differences within states or countries with strong PEV adoption.

Given the fact that California has many different regions with different levels of PEV introduction, community commitment, and industry involvement (Scott 1995), it is argued that "one size fits all" may not be the best approach (also taking other global regions into account). Rather, it would be more efficient to look at the above-mentioned models with their advantages/disadvantages, and determine if which models are most suitable for each region.

Also, given the argument that EV introduction is still at its early stages (Woodjack et al. 2012) and that paradigm shifts benefit from "innovative, open path deployment" (Van Deventer et al. 2011), it is argued that timing and local/regional sensitivity need to be taken into consideration (Tran et al. 2012). In addition, market growth can be very dynamic (e.g.. PEV sales in the Netherlands were 25 % of all cars sold in December 2013, in large part due to incentive changes) and caution has to be made in terms of expected EV growth; it is advised to make predictions for a 3-year period and evaluate/adjust annually.

Looking at the various regions and their current "PEV readiness," the following distinction would then be made for public charging: (1) Leaders: Bay Area (incl. Monterrey Bay), Los Angeles County, San Diego; (2) Followers: Most other urban areas, and (3) Start-ups: typically rural area.

The models can then be used to make an actual estimate of the total investment needed based upon a number of variables. See Table 1.

Table 1 Key assumptions and parameters of the three distinguished regions

	Leaders	Followers	Start-ups
Preferred model	Hybrid-III → Financing	Hybrid-I → II	GOV100 → Hybrid-I
% of CAL inhabitants	40 %	40 %	20 %
PEV sales % in 2013	2.5 %	1 %	0.5 %
PEV sales % in 2016	20 %	8 %	4 %
Total EV's in 2016	600,000	240,000	125,000
EV/EVSE's ratio in 2016	1 in 15	1 in 15	1 in 15
No. of EVSE's in 2016	40,000	16,000	8,000
Total cost	$200 million	$80 million	$40 million
Governmental cost	$70 million	$40 million	$30 million
Timing (focus)	2014/2015	2015/2016	2016/2017

Explanation

The six identified models will not be suitable for every type of region. Moreover, it is assumed that a region will evolve further and therefore a different model would apply better after a few years; e.g., in the table above, it is assumed that a "Leader" region could evolve from a "Hybrid-III" to a "Financing" model. In addition, various assumptions (see below) will have to be made to calculate the number of chargers needed; e.g., the PEV sales % in 2016 is not certain but assumed. Also, the necessary EV/EVSE ratio is assumed 1:15; in the Netherlands the ratio until end of 2013 was 1:1.7 for all chargers (home, work, and public), nearly ten times more. However, by the end of 2013 some 30.000 PEV's were sharing 4.000 public chargers or 1:7.5. The assumed 1:15 ratio for California is a bare-bone minimum, but this would have to be evaluated in the years to come

Local assumptions

• The spreading of PEV penetration has followed a certain pattern in most global regions; typically starting in cities, spreading to urban regions onto suburban regions and finally more rural areas
• The number of newly sold PEV's in Leader Regions will double each year for the next 3 years (using the earlier Prius distribution models)
• Follower Regions are 1 year behind Leader Regions and will see similar penetration rate in Year 2 and then follow the growth path of the Leader Region
• Start-up regions are 2 years behind and will see growth numbers similar to the Follower Region in Year 3
• The number of new car sales in California is set at 1.6 million per year compared to 0.5 million in the Netherlands
• The number of PEV's in California is expected to be 60,000 as of 12/31/13 compared to 30,000 in the Netherlands
• The number of PEV's in each region is calculated as %CAL Inhabitants X %PEV sales X 1,600,000
• Ideally, the minimum number of EVSE's is based on 0.5 mile square for Leader Regions, 1 mile square for Follower Regions, and 5 mile square for Start-up Regions

• The amount of EVSE's in each region needs to be able to meet the expected PEV growth in the next 3 years
• The growth rate of leading states like California, the Netherlands, and Norway are expected to be ahead of the curve compared to the federal (US/EU—though this is more complex in terms of actual politics and is used as a shorthand here) ambitions. For the purpose of this demonstration, it is assumed that these states will be able to achieve federal goals at least 3–4 years earlier
• As far as number of EVSE's in each CA region, a ratio is assumed of one public charger per 15 PEV's matching EU target
• The cost for Type 2 preparation, site selection, procurement, construction/installment, and maintenance is assumed to be at similar as found in large-scale procurements like MRA-e being about $5,000 per EVSE
• The number of public chargers as part of the NRG settlement are included in the totals

8 Conclusions and Recommendations

This paper has presented the next steps forward for large-scale deployment of public charging infrastructure after the first round of infrastructure was mainly financed by government agencies over the last 3–5 years. In order to create a sustainable and market-driven public charging network, governments are increasingly looking for strategies to support the next generation of public charging infrastructure with creative financing mechanisms and limited public funding.

The main goal of this paper was to research the different methods that are currently being applied in early adopter markets such as Norway, the Netherlands, and California. Based on early learnings, the various challenges and opportunities in these early markets have been discriminated and six practical models for investing into public charging infrastructure were identified. By overlaying them on local context for one US state (California), the models were applied to determine best available options and possible cost associated with model choices.

It is recommended that regions or countries worldwide wanting to invest strategically as well as practically into public charging infrastructure, consider one or more of the six models depending on the their local context. Furthermore, it is strongly recommended that these models be tested in real-life conditions, set up as pilots and be monitored for performance and sustainability. As part of the Interreg NSR E-mobility network (concluding at the end of September 2014) and the Coast-to-Coast e-Mobility partnership (at least running until the end of 2016), further information will be gathered, so that the models can be refined over time and subsequently published independently.

Acknowledgments The authors like to thank the various regional, state and national governments, academic institutions, and private organizations for providing information, discussion, and input. Although many have contributed, the efforts by the Governor's Office and the Office of Planning and Research (State of California), the PowerTeam of Ministry of Economic Affairs (The Netherlands), UCDavis/Institute of Transportation Studies, Plug-in Hybrid and Electric Vehicle Research Center (California), MRA-e (The Netherlands), and NSR e-Mobility Team of The Province of Noord-Holland (The Netherlands), are especially appreciated.

References

Ahman, A. (2006). Government policy and the development of electric vehicles in Japan. *Energy Policy, 34*(2006), 433–443.

Anegawa, T. (2010). Needs of Public Charging Infrastructure and Strategy of Deployment. Presentation at TEPCO, December 2010: http://www.ev-charging-infrastructure.com/media/downloads/inline/takafumi-anegawa-tepco-9-10.1290788342.pdf

Bakker, S. (2013). *Standardization of EV recharging infrastructure* (NSR E-mobility project report). The Netherlands: Technical University of Delft. http://e-mobility-nsr.eu/fileadmin/user_upload/downloads/info-pool/4.4_E-MobilityNSR_Recharging_infrastructure_standardization.pdf.

Bakker, S., Trip, J. J. (2013). *Stakeholder strategies regarding the realization of an electric vehicle charging infrastructure* (NSR E-mobility project report). The Netherlands: Technical University of Delft. http://e-mobility-nsr.eu/fileadmin/user_upload/downloads/info-pool/3.2_E-MobilityNSR_International_stakeholder_analysis.pdf.

Boonen, A. (2012). *Faciliteren of Stimuleren? Onderzoek naar de effectiviteit van gemeentelijk beleid om de vraag naar elektrisch vervoer te vergroten*. The Netherlands: VU University Amsterdam. http://www.appm.nl/downloads/289_Onderzoek%20effectiviteit%20gemeentelijke%20EV-beleidsmaatregelen.pdf.

Debe, M. K. (2012). Electrocatalyst approaches and challenges for automotive fuel cells. *Nature* 486, 43–51.

Diamond, D. (2008). The impact of government incentives for hybrid-electric vehicles: evidence from US states. *Energy Policy, 37*(2009), 972–983.

Hübner, Y., Blythe, P. T., Hill, G. A., Neaimeh, M., & Higgins C. (2012). ITS for electric vehicles—an electromobility roadmap. *Proceedings Road Traffic Information and Control*. London: The IET, September 2012.

Goebel, C. (2012). On the business value of ICT controlled plug-in electric vehicle charging in California. *Energy Policy, 53*(2013), 1–10.

http://www.autoblog.nl/nieuws/deze-autos-werden-het-meest-verkocht-in-europa-top-10-64553.

http://www.autoweek.nl/nieuws/27668/dringen-om-parkeerplek-door-elektrische-auto.

http://www.bizjournals.com/sacramento/blog/kelly-johnson/2013/01/california-auto-sales-surge-in-2012.html?page=all.

http://coast2coastev.org/?p=390#more-390.

http://coast2coastev.org/?p=387#more-387.

http://cleantechnica.com/2014/03/16/norway-ev-market-share-vs-france-ev-market-share-chart/.

http://e-mobility-nsr.eu/fileadmin/user_upload/events/2013_Haarlem/Julia_Williams_-_The_Dutch_approach_to_electromobility.pdf.

http://ens-newswire.com/2013/09/17/california-governor-business-leaders-pledge-support-for-evs/.

http://www.europarl.europa.eu/news/en/news-room/content/20131125IPR26108/html/Alternative-fuel-stations-Transport-MEPs-back-draft-law-to-expand-networks/.

http://evfleetworld.co.uk/news/2014/Mar/European-parliament-targets-eight-million-charging-points-by-2020/0438013536.

http://www.evcollaborative.org/.

http://evobsession.com/norway-amazing-electric-car-sales-month-march-update/.

http://www.forbes.com/sites/justingerdes/2013/02/26/estonia-launches-nationwide-electric-vehicle-fast-charging-network/.

http://gov.ca.gov/news.php?id=17463.

http://www.iea.org/topics/transport/electricvehiclesinitiative/EVI_GEO2013_FullReport.PDF.

http://insideevs.com/monthly-plug-in-sales-scorecard/.

http://insideevs.com/seldom-used-public-charging-stations-rile-up-the-locals/.

http://www.pluginamerica.org/yes-on-sb454.

http://www.plugincars.com/federal-and-local-incentives-plug-hybrids-and-electric-cars.html.

http://www.raivereniging.nl/markt%20informatie/statistieken.aspx.

http://rationalpolicy.com/articles/domestic/hydrogen-fuel-the-chicken-and-egg-problem/.

http://www.telegraaf.nl/tv/autovisie/22218466/__Te_weinig_oplaadpunten_elektrische_auto__.html.

http://theforeigner.no/pages/news/norway-tops-world-electric-vehicle-sales/.

http://www.thegreencarwebsite.co.uk/blog/index.php/2014/03/24/europe-fails-to-set-targets-for-electric-car-infrastructure/.

http://www.theicct.org/blogs/staff/if-subsidies-are-no-panacea-how-incentivize-electric-vehicles-china.

ICCT (2014). European Vehicle Market Statistics Pocketbook 2014. In P. Mock (Ed.), International Council on Clean Transportation (p. 121).

International Energy Agency. (2013). Clean energy ministerial, and electric vehicles initiative:. Global EV outlook 2013—understanding the electric vehicle landscape to 2020. *International Energy Agency*, 14–15.

Kley, F., Lerch, C., & Dallinger, D. (2011). New business models for electric cars. *Energy Policy, 39*, 3392–3402.

Kotter, R. (2013). The developing landscape of electric vehicles and smart grids: a smart future? *International Journal of Environmental Studies, 70*(5), 719–732.

Linnenkamp, M. (2013). *Business case for the recharging infrastructure*. Haarlem/Amsterdam, The Netherlands: Amsterdam Metropolitan Region, MRA-E.

Linnenkamp, M., & Horck, P. (2013). *Government tendering for a public recharging infrastructure*. Haarlem/Amsterdam, The Netherlands: Amsterdam Metropolitan Region, MRA-E.

Luis, J., Parks, R., Santos, G. (2013). Decarbonising the road transport sector: breakeven point and consequent potential consumers' behaviour for the US case. *International Journal of Sustainable Transportation, 8*. doi:10.1080/15568318.2012.749962).

Morrow, A., Karner, B., & Francfort, C. (2008). *Plug-in hybrid electric vehicle charging infrastructure review*. Washington DC, USA: US Department of Energy.

Myklebust, B. (2013). *An electric success story in Norway, through powerful, but waning incentives*. EU Interreg e-Mobility Conference "Policy, Practice and Profitability". Haarlem, The Netherlands, October 10, 2013. http://e-mobility-nsr.eu/fileadmin/user_upload/events/2013_Haarlem/Benjamin_Myklebust_-_An_electric_success_story_in_Norway__through_powerful__but_waning_incentives.pdf.

PRWeb. (2013). *California now has the most electric car charging stations of any state*. http://www.prweb.com/releases/2013/10/prweb11215744.htm.

Quandt, C. (1995). Manufacturing the electric vehicle—a window of technological opportunity for Southern California. *Environment and Planning, 27*(6), 835–862. International Bibliography of the Social Sciences (IBSS).

Rogers, E. M. (2003). *Diffusion of innovations* (5th ed.). New York, NY: Free Press.

Rubin, B. D. (2013). *ZEV readiness guidebook*. Sacramento, CA: Governor's Office of Planning and Research.

Rubin, B.D., & Van der Steen, M. (2013). The silence of success: a transatlantic comparison on government strategies to grow the EV market. *Practicing Planner, 11*(4). AICP. http://www.planning.org/practicingplanner/2013/win/.

Schot, J., Hoogma, R., & Elzen, B. (1994). Strategies for shifting technological systems—the case of the Automobile system. *Futures, 26*(10), 1060–1076.

Scott, A. J. (1995). The electric vehicle industry and local economic-development—prospects and policies for Southern California. *Environment and Planning, 27*(6), 863–875.

Schiffer, M. B. (2010). *Taking charge: the electric automobile in America*. Washington, DC: Smithsonian Books.

Snyder, J., Erstad, A., Lin, E., Rice, A. F., Goh, A. T., & Tsao, A. A. (2012). *Financial viability of non-residential electric vehicle charging stations*. Los Angeles, CA: UCLA.

Sperling, D., & Gordon, K. (2009). *Two billion cars: driving toward sustainability*. New York: Oxford University Press.

Steinhilber, S., Wells, P., & Thankappan, S. (2013). Socio-technical inertia: understanding the barriers to electric vehicles. *Energy Policy, 60*(2013), 531–539.

Tran, M., Bannister, D., Bishop, J. D. K., & McCullogh, M. D. (2012). Realizing the electric vehicle revolution. *Nature Climate Change, 2*, 328–332.

Turrentine, T., Kurani, K., Gras, D., & Tal, G. (2013). *PEV market: Glass half empty or half full?* Plug-in Electric Vehicle Collaborative, Conference, University of California, Davis, November 2013. http://phev.ucdavis.edu/ieee-event-11-7-13-presentations/IEEE_Turrentine_UCDavis.pdf.

U.S. Department of Energy. (2013). *Alternative fueling station counts by State*. Washington, DC: Alternative Fuels Data Center (AFDC).

Van der Steen, M., van Schelven, R., & Van Deventer, A.P. (2014). *Emergent strategies for an emerging technology: Government steering in the introduction of e-Mobility*. Presentation at the Interreg NSR e-Mobility Conference, London Metropolitan University, London, April 11, 2014. http://e-mobility-nsr.eu/fileadmin/user_upload/events/2014_Final_Conference/Presentations/08_vanderSteen_vanSchelven_EMOB_London2014.pdf.

Van Deventer, A. P., Van der Steen, M., Van Twist, M., De Bruijn, H., Ten Heuvelhof, E., & Haynes, K. (2011). *Governing the transition to e-mobility: small steps towards a giant leap*. The Hague: Dutch School for Public Administration (NSOB). http://www.nsob.nl/EN/wp-content/uploads/2010/06/e-mobility-webversie.pdf.

Williams, J. (2013). *The Dutch approach to eletromobility*. Presentation to NSR E-mobility conference in Haarlem. The Netherlands. http://e-mobility-nsr.eu/fileadmin/user_upload/events/2013_Haarlem/Julia_Williams_-_The_Dutch_approach_to_electromobility.pdf.

Witt, M., Bomberg, M., Lipman, T., & Williams, B. (2012). Plug-in vehicles in California review of current policies, related emissions reductions for 2020, and policy outlook. *Transportation Research Record, 2287*, 155–162. Transportation Research Board of the National Academies, Washington, D.C. http://luskin.ucla.edu/sites/default/files/WittEtAl2012TRR1stPage.pdf.

Woodjack, J., Garas, D., Lentz, A., & Turrentine, T. (2012). Consumer perceptions and use of driving distance of electric vehicles: changes over time through lifestyle learning process. *Transportation Research Record, 2287*, 1–8. Transportation Research Board of the National Academies, Washington, D.C.

Part II
Regional and City Case Studies on E-Mobility Development

Rolling Out E-Mobility in the MRA-Electric Region

Christine van 't Hull and Maarten Linnenkamp

Abstract 20,000 electric vehicles (EVs) on the road by 2015 and 200,000 EVs by 2020 … . This was regarded as an ambitious goal when it was declared in 2011, and yet the growth in the use of EVs in the Netherlands has seen rapid advancements. With more than 47,000 EVs on Dutch roads already in March 2015 the Netherlands is well on its way. Concerted efforts and initiatives were required to achieve this, among others, by the city of Amsterdam, which has been a frontrunner since 2009. Because e-mobility does not end at Amsterdam's city limits, the project MRA-Electric (Amsterdam Metropolitan Area Electric, MRA-E) was initiated by the local authorities in the Amsterdam Metropolitan Area (MRA) to stimulate, advise and assist in rolling out e-mobility in the region around Amsterdam. Essential to advancing e-mobility, and the use of e-cars, in particular, is a robust charging infrastructure, preferably powered by sustainable energy because the arguments for the environmental benefits of e-mobility rest largely on the source of the energy used to charge the batteries (http://www.rvo.nl/onderwerpen/duurzaam-ondernemen/energie-en-milieuinnovaties/elektrisch-rijden/stand-van-zaken/milieuvoordeel). Since 2009 the MRA-E region has seen the rollout of 1000 public charging points. This achievement was brought about by ratifying favourable policies at national, provincial and city levels; providing the right financial incentives; ensuring that grid operators and energy distributors are fully on-board; and obtaining a commitment from other market parties, such as lease companies, to co-finance the charging poles. Issuing calls for tenders by the province of Noord-Holland that contain unambiguous provisions has also proven highly successful in the MRA-E region. Because they are still more expensive to purchase than their petrol or diesel burning counterparts, encouraging the purchase and use of EVs also has to be stimulated with fiscal incentives. In the Netherlands, this happens at national and city levels in the form of subsidies or tax breaks, a substantial portion of which is made available to taxi and

C. van 't Hull (✉) · M. Linnenkamp
MRA-Electric, Valkenburgerstraat 218, Amsterdam, The Netherlands
e-mail: CHull@pmb.amsterdam.nl

M. Linnenkamp
e-mail: MLinnenkamp@pmb.amsterdam.nl

W. Leal Filho and R. Kotter (eds.), *E-Mobility in Europe*,
Green Energy and Technology, DOI 10.1007/978-3-319-13194-8_7

delivery vehicles (RVO.nl; http://www.rvo.nl/onderwerpen/duurzaam-ondernemen/energie-en-milieu-innovaties/elektrisch-rijden/aan-de-slag/financiele-ondersteuning?gclid=COr_-JKTs78CFUTItAod90kAbw). Cities are also leading by example: many cities in the Netherlands have already added EVs to their municipal fleets and some have incorporated hybrid and full electric buses into public transportation as well as installed public charging points at public buildings. An electric car-sharing scheme introduced by Amsterdam in 2011 has proven popular and is invaluable in raising the profile of e-mobility. With more and more e-cars on the roads, it was clear that electric driving was becoming a real alternative. This chapter provides an overview of the significant growth of e-mobility in the MRA-E region before examining in more depth a fundamental aspect that underpins this achievement, namely the rolling out of a charging infrastructure.

Keywords E-mobility • Electric vehicles (EVs) • Charging infrastructure • Subsidies • Financing • Public-private initiatives • Charging poles

1 Introduction

In response to concerns about air quality and to meet European targets set for 2015, several cities and regions in the Netherlands commenced with developing an e-mobility programme for their regions from 2009 onwards. (https://www.rvo.nl/sites/default/files/bijlagen/Plan%20van%20aanpak%20-elektrisch%20rijden%20in%20de%20versnelling-.pdf). The measures taken by these cities and regions inspired the Dutch Ministry of Economic Affairs to publish its forecasts in its *Electric Mobility Gets Up To Speed 2011–2015 Action Plan* in 2011. One of its targets related to the number of EVs on Dutch roads: 20,000 by 2015, increasing to 200,000 by 2020 (http://www.rijksoverheid.nl/onderwerpen/auto/elektrisch-rijden). Many cities, regions and companies have been working hard to achieve this, with excellent results: figures for March 2015 show that there are already over 47,000 electric vehicles (EVs) in use in the Netherlands, far exceeding expectations; moreover, on average approximately 1100 new FEVs (Full Electric Vehicles) and PHEVs (Plug-in Hybrid Electric Vehicles) passenger cars were registered in the country each month between January and April 2015 (RVO.NL). Concerted efforts and initiatives were required to achieve this, among others, by the city of Amsterdam, which has been a frontrunner since 2009. Because e-mobility does not end at Amsterdam's city limits, the project MRA-Electric (Amsterdam Metropolitan Area Electric, MRA-E) was initiated in 2011 by the local authorities in the Amsterdam Metropolitan Area (MRA) to stimulate, advise and assist in rolling out e-mobility in the region. Rather than an administrative body, the MRA is mainly a platform for regional cooperation and coordination between cities, provinces and municipalities in and around Amsterdam (Janssen-Jansen and Hutton 2011).

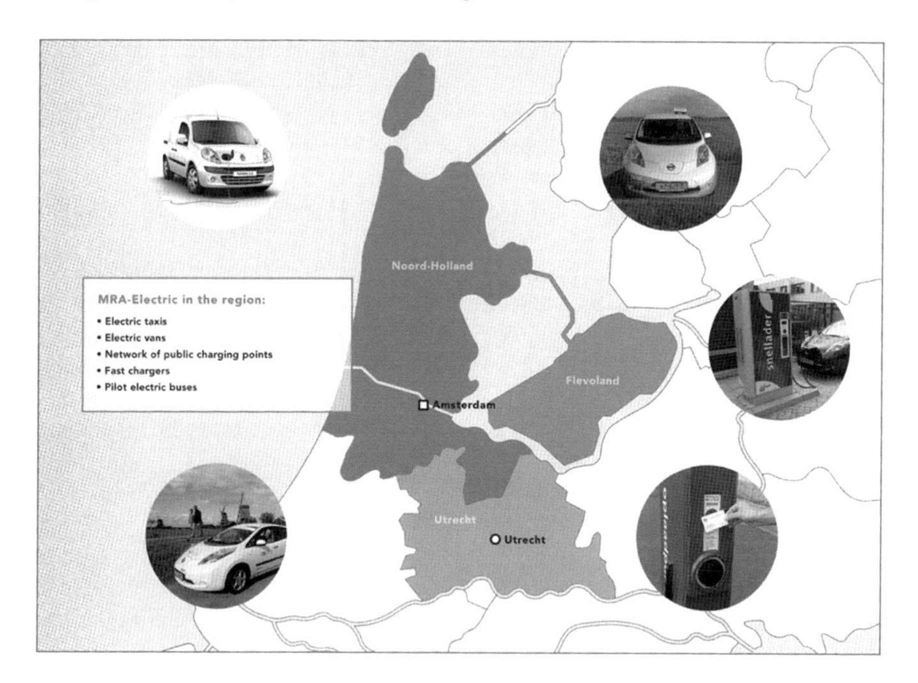

Fig. 1 Map of the MRA-E region (*Source* MRA-Electric)

Encompassing a much greater area than the MRA, the MRA-Electric works for the cities in the region which now covers three provinces, excluding the cities of Amsterdam and Utrecht itself. The region is recognised as a globally inspiring living lab and a world leader in the field of electric mobility: measuring 5464 km^2 and home to a population of 4.3 million people, it already has a network of 1000 charging points that are used by some 20,000 drivers (see Fig. 1).

2 From the Ground Up

The city of Amsterdam launched its e-mobility programme at the event "Amsterdam Electric" in March 2009. E-car drivers from all over Europe attended with their e-cars: a surprise because at that time there were hardly any e-cars available in the market. "The Think", a Norwegian plastic 100 % e-car, priced at more than €38,000, was the only 100 % e-car available in the Netherlands at that time. Just over a month later, on 1 April 2009, Amsterdam's city council ratified the *Plan van Aanpak Elektrisch Vervoer* ("Plan for Electric Mobility"). Outlining measures to stimulate the use of e-cars in the city, it set a target for the city of 10,000 EVs by 2015, equivalent to 5 % of the kilometres travelled in the city. Obviously, all these vehicles must have charging points, and as many EV users do not have direct access to the necessary facilities, this has to be addressed by rolling out a charging infrastructure on both public and private land. In Amsterdam public land receives more attention because of the lack of space in this densely populated city: most people reside in apartment blocks and do not have access to private parking facilities, so they have to park their vehicles in public space.

As a follow-up to the Amsterdam initiative, in 2010 the other three largest cities in the Netherlands—Rotterdam, Utrecht and The Hague—also began installing charging poles at their own initiative.

Amsterdam was the first city to organise a call for tenders for charging poles in public space (in the spring of 2009). Electricity supplier NUON was awarded the contract and, together with network provider Alliander, installed 100 charging poles, each containing two charging points, in Amsterdam that supplied renewable energy. The network was activated on 6 November 2009. Users could park and charge their vehicles for free until 31 March 2012—the city of Amsterdam covered the costs.

In the autumn of 2009, the city of Amsterdam undertook agreements with Nissan/Renault to ship the first 1000 e-cars to Amsterdam; the city agreed to arrange publicly accessible charging points in return. In 2010, the city of Amsterdam allocated a subsidy for the purchase of EVs. Within a year, about €3 million in subsidies had been divided over a substantial number of companies that pooled it with €8 million of their own funds to purchase 260 EVs. Also in 2010, the second tender, for another 1000 charging points, was awarded to two energy companies: Nuon/Heijmans and Essent. Other large cities in the Netherlands followed suit and organised their own calls for tenders for public charging infrastructures in their areas: Utrecht (Utrecht Elektrisch 2010) and Rotterdam (Rotterdam Elektrisch 2011) and The Hague (2012). Amsterdam was recognised as a global e-visionary city by the World Electric Vehicle Association (WEVA) in 2010 for having the best future vision with regard to sustainable mobility (compare also with Rienstra and Nijkamp 1998).

From 2009 onwards public charging points in smaller cities were fully financed and constructed by a combination of cooperative network providers through the E-laad Foundation, which was set up by the Dutch electricity network (grid) companies. By collaborating through the foundation, the network providers bundled knowledge and investments and investigated the effect that charging EV batteries had on the network.

At the end of 2012 the Minister of Economic Affairs instructed the E-laad Foundation to stop installing new charging stations, so that the market could take up that role. Local authorities were expected to organise charging points themselves, but the majority of local councils did not have the know-how, or the financial or official resources to realise this.

The project MRA-Electric was initiated in October 2011 by local authorities in the Amsterdam Metropolitan Area (MRA) to stimulate, advise and assist in rolling out e-mobility in the region. The scope of MRA-E's activities was increased in 2013 and now the project works for more than 80 municipalities in the provinces of North Holland, Flevoland and Utrecht. The cities of Amsterdam and Utrecht have their own e-mobility programmes and function independently, although they coordinate their activities with MRA-E. MRA-E focuses on knowledge sharing, arranging the financing of the charging infrastructure, cooperating with the market, and raising the profile of e-mobility, mainly by getting more EVs on the roads.

3 The Realisation of Public Charging Points in the MRA-E Region

As of March 2015 there are over 47,000 EVs (FEVs and PHEVs) in the Netherlands and that number increases daily. As a result, municipalities receive growing numbers of requests for charging poles. For the majority of local councils realising a charging infrastructure themselves is not an option. It is a relatively new technology, and the finances required to purchase the charging poles are rarely included in council budgets. However, local councils remain responsible for public space and do want to prevent the unchecked proliferation of charging poles. In January 2013, MRA-E proposed a solution for introducing more public charging points in cities: join forces and issue calls for tenders for public charging poles for all the local councils on whose behalf it operates.

A call for tenders allows the market to operate freely and produces the economically most favourable bids. Councils are allocated charging poles from the concession according to their requirements (see below for the basis of the tenders) and become the owners of the charging poles. They select the location for the charging poles and endorse its installation with a Traffic Ordinance. The role of local councils ends with organising the parking places and the installation of the charging pole: the supplier retains responsibility for their management, maintenance and eventual removal.

There are currently around 1000 charging points in the MRA-E region. Stichting E-laad installed 840 charging points during its operational period, to which another 200 have been added by the combination Nuon/Heijmans since July 2013. In the summer of 2015 a charging point operator will be selected to place another 400 charging points.

3.1 Financing the Charging Poles

Recouping the costs of installing and operating a charging infrastructure from the revenues from battery charging is currently not feasible because there are still too few EVs on the road. To ensure the rollout of the infrastructure, MRA-E collaborates with the government as well as the market. The provinces of North Holland, Flevoland, Utrecht, the city of Amsterdam, the national government and market parties, such as Leaseplan (a major lease company in the Netherlands), Mister Green (a lease company focusing on electric driving) and Nissan, contribute to financing this effort.

Car dealerships and car leasing companies that supply their customers with electric vehicles obviously have a vested interest in the provision of charging points. If someone cannot charge an electric car at home or at work, it is highly unlikely they will use one. In the Netherlands the amount of electric car owners that cannot charge their electric vehicle on their own property, and thus need a public charging point either at work or at home, is 50 %. Car dealers usually include home chargers or chargers that can be used at places of employment with the electric vehicles they supply. If there is no charging point at home or work and the client is advised to use a charging point in the public space, then MRA-E asks the car dealer/ leaser to make a contribution to a public charging point.

The cost of a public charging pole has already decreased by 50 % over the past 3 years, and will continue to do so, especially with rising demand due to the increasing volume of EVs. Local councils are required to cover the cost of organising the parking places and to make a financial contribution of €1000 per charging pole. This amount will undoubtedly fall with the decreasing prices of charging poles. For local councils, this not only simplifies the process but also makes it affordable for them to install charging poles, thereby contributing to the growth in the use of EVs. The income from a charging pole is also affected by energy prices. Until such time as the charging infrastructure proves profitable, public–private funding initiatives will be required to ensure that it can be constructed and maintained.

3.2 Call for Tenders

In the call for tenders, a commission or concession to install poles is rewarded for half a year. The management and operation of poles is awarded to the same party for 3 years. That contract could be extended for another 3 years. A government-backed call for tenders to which market parties can respond with competitive bids fosters a system where each party's role is clearly defined, ensuring that parties can fully contribute to the further rollout of the charging infrastructure in accordance with the responsibilities each of them has assumed.

The basis of the call for tenders entails the following elements:

1. The network company/the government and private companies finance the unprofitable aspects of the installation and exploitation of the poles.
2. The provincial authority calls for tenders or grants concessions.
3. The entire package is included in the tender: the supply, installation, connection, management, maintenance, power supply and exploitation of the charging points.
4. There are two charging points per pole.
5. The poles have Open Charge Point Protocol (OCCP) and are inter-operable (i.e. accept charging passes from different providers).
6. Safety measures will be identical to the E-laad charging poles and those in the four largest municipalities (Amsterdam, Rotterdam, The Hague, Utrecht).
7. The maximum charge at MRA-E poles is €0.30 per kWh.

3.3 Location of Public Charging Poles

Local councils do not anticipate the market by placing charging poles in arbitrary locations but instead react to a request from a resident who has purchased an EV and needs a charging pole close to home. A charging station should preferably be no further than 200 m from the applicant's house, in a place that does not hinder other residents. Charging poles have two charging points (cost efficiency), so two adjacent parking places have to be reserved for e-cars. In residential neighbourhoods, locating charging poles at intersections simplifies law enforcement and also raises the profile of e-mobility.

Not only do residents in the neighbourhood have to be advised of the change of status to the parking places, but the charging station also has to be clearly identified with a street tile with a distinctive e-logo, for example, and traffic signs with a special signboard (see Fig. 2). These signboards indicate that the parking places are reserved for e-cars only, hopefully discouraging drivers of other types of vehicles from using them, most likely resulting in them being fined or having their vehicle towed away.

3.4 Charging Costs

E-mobility and charging technology will continue to evolve over the coming years. Tariffs that EV users have to pay to charge their EV batteries must fit the phase that e-mobility is in, namely a period in which electric driving competes with fossil fuels.

Fig. 2 A street tile with the e-logo and a signpost indicating two parking places at a charging pole (*Photo* MRA-E Electric)

The call for tenders fixed the charging tariff at MRA-E public charging poles for the following 3 years. Local councils in the MRA-E region want to continue actively promoting e-mobility, so a low charging tariff has been guaranteed to ensure that it is more economical to drive using electric power rather than petrol. The maximum charge at MRA-E poles is €0.30 per kWh. Unlike other service providers, MRA-E does not charge a fee to start the charging-session (a so-called 'starting-tariff').

3.5 Fast Chargers

The MRA-E region is an ideal location to test fast chargers because there are relatively many FEV-drivers in the region who are at times dependent on fast chargers to be able to operate. With a fast charge the batteries of an electric vehicle can be charged to 80 % capacity in about 20 min. MRA-E started a fast charger project with the ANWB (Royal Dutch Touring Club), the province of North Holland, and Rotterdam and Utrecht municipalities. The European Union has provided a subsidy from the Life+ programme that is currently being used for the installation of 12 fast chargers and 24 normal charging points in and around the municipalities of the MRA-E region and Rotterdam. The ANWB and the provincial authorities select the locations of the charging poles.

3.6 Smart Charging

In the Netherlands, the tariff for charging an EV is the same during peak and off-peak hours. Consequently, many EV users charge their cars during peak hours, often immediately on arriving home from work between 1700 and 1900 h. This is also the time of day that households use a great deal of electricity, so charging EVs during this period could result in the power distribution network being overloaded. MRA-E and the market are investigating smart charging, allowing for variations in the price of electricity at different times of the day. Suggestions include an energy card that specifies the times the EV can be charged and at what tariff; and an app that displays the price of electricity at any time of the day, enabling EV users to choose when they would like to charge their EV batteries. This system already exists for the Tesla S.

4 E-Vehicles

It was already clear in early 2014 that the targeted number of EVs on Dutch roads by 2015, as expressed by the Ministry of Economic Affairs in its 2011 publication, *Electric Mobility Gets Up To Speed 2011–2015 Action Plan*, had been far exceeded. By March 2015, over 47,000 EVs were registered on Dutch roads—the original target was to have between 15,000 and 20,000 EVs on Dutch roads by 2015. While early innovators have already made the transition, reaching a much broader public is key to increasing the number of EV users. MRA-E regards promoting the market for EVs of primary importance and encourages the installation of charging points in parking lots at businesses (with an emphasis on taxi and delivery companies) and city councils. It is important to ensure that interested parties are fully informed of the fiscal benefits and subsidies and that they understand the "total cost of ownership" model. Figure 3 and Table 1 provide some of the statistics.

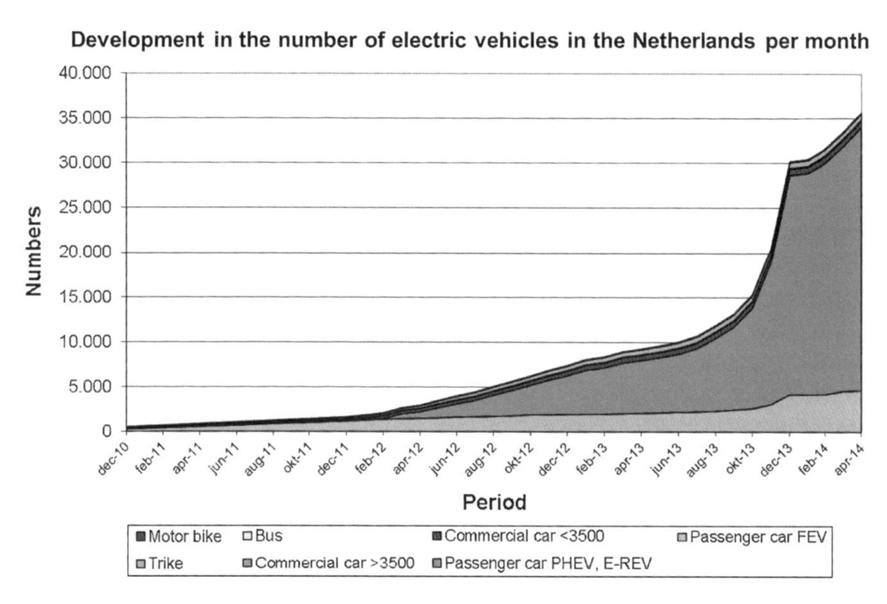

Fig. 3 Growth in registered EVs in the Netherlands (*Source* RVO.nl 2015a)

Table 1 The growth in the number of registered EVs in the Netherlands

Type of car by year	31-12-2013	31-12-2014	31-01-2015	28-02-2015	31-03-2015
Private car (FEV)	4161	6825	7152	7246	7749
Private car (E-REV[a] and PHEV)	24,512	36,937	38,978	40,255	41,802
Commercial vehicle <3500 kg load capacity	669	1,258	1,267	1,273	1,349
Commercial vehicle >3500 kg load capacity	39	46	46	50	51
Bus (incl. trolley buses and several hybrid buses)	73	80	80	80	85

Source RVO.nl (2015a)
[a]Extended Range Electric Vehicle

4.1 Fiscal Incentives

Fiscal incentives and government subsidies play an important role in advancing the growth of e-mobility. Tax incentives have dramatically increased the numbers of EVs in the Netherlands. The package was refined at the beginning of 2014—prior to the end of 2011 there were no central government subsidies or tax incentives—and now distinguishes between FEVs and PHEVs.

- Exemption from BPM (vehicle purchase) tax until 2018
- Exemption from MRB (road tax) until 2015
- Income tax addition for the private use of company cars of 4 % for FEVs and 7 % for PHEVs (1–50 g CO_2 emissions), compared to an internal combustion (ICE) car for which the income tax addition is, for most cars, 25 % of the as new-value of the car (belastingdienst.nl);
- MIA environmental investment rebate (up to 36 % of a €50,000 maximum).

The Ministry of Infrastructure and Environment has a national subsidy of €3000 for the purchase of electric low-emission taxis and delivery vans. For much of the MRA-E region, Rotterdam, The Hague, Arnhem/Nijmegen and the surrounding communities, where the most effective e-mobility measures are taken to address air quality, the subsidy provided by the Ministry was €5000 per vehicle until the 1st of January 2015. Several cities also have their own subsidies, on top of the national subsidies, to promote e-mobility. For example, the city of Amsterdam offers a €5000 subsidy on top of the national government subsidy for local businesses and taxi companies (the most polluting drivers measured by the amount of kms driven in the city) when they purchase an electric vehicle. Technological advances and increasing demand should result in the prices of EVs falling over the next few years, so the amounts of these local subsidies are adjusted annually by city councils.

4.2 Greener Taxis

Taxis travel many kilometres, often over short distances, with highly polluting diesel engines. A diesel-powered passenger car emits twice as much NO_2 as a petrol-driven car. On average, a taxi contributes 35 times more NO_2 than a personal car (*Schone lucht voor Amsterdam*, Gemeente Amsterdam (DIVV), June 2011). Hence, taxis are seen as one of the most promising market segments for e-mobility and their replacement with EVs can have a significant effect on air quality. During the week, approximately 2500 taxis are active in Amsterdam, with the number rising at weekends.

Electric taxi company Taxi Electric started in Amsterdam in 2010. It has a fleet of 20 EVs and its own charging station. Taxi Electric drove its one-millionth kilometre in Amsterdam in 2013. Among the many opportunities for greener taxis in its area of activity, MRA-E has identified Schiphol Airport as an important zone for improvement (see also Silvester et al. 2013).

Around 2100 taxis travel between Amsterdam and Schiphol Airport every day (a distance of 16 km), amounting to 80 % of the total number of taxis departing from the airport. At the request of the Netherlands Enterprise Agency (RVO.nl), the Formula E-team (a collaboration of Dutch trade and industry, institutes and administration) and MRA-E, Decisio and APPM Management Consultants provided Schiphol with the necessary knowledge that resulted in the airport's recent tender for taxis incorporating provisions for clean transport. As a result, in autumn 2014, the taxi companies Zorgvervoer Centrale Nederland BV (ZCN), Willemsen-de Koning BV and Bergisch, Boekhoff & Frissen BV (BBF) were scheduled to introduce more than 150 electric taxis at Schiphol Airport. Besides these, Schiphol has also introduced 30 electric buses, another major contribution to the greening of passenger transport to and from the airport, improved air quality and, moreover, a stimulus to Dutch industries for green growth.

4.3 Electric Car Sharing

E-mobility became much more accessible to a broader public with Daimler's introduction of its car-share system car2go in Amsterdam in November 2011. Three hundred electric smart fortwo cars are available for spontaneous rental, without the need to return them to a specific place or at a specific time (https://www.car2go.com/en/amsterdam/#pid=43913). They have a range of 135 km and can be parked for free in public parking places for as long as is necessary. Costs to users are calculated by the minute or by the day. car2go users have driven more than 4 million kilometres in 2 years; it currently has 20,000 members, who rent cars 10,000 times a week, also a valuable contribution to increasing the visibility of e-mobility. This 100 % EV car-share scheme is an ideal supplement to public transport. MRA-E is investigating the possibilities of expanding the car2go network to encompass Schiphol Airport as well.

5 Conclusion

The MRA-E region is an ideal test bed for e-mobility: it is densely populated and the distances between the cities within it across a relatively flat topography lend themselves perfectly to the use of EVs. More than 200 companies in the Netherlands operate in the field of electric transport, and many of these are based in this region. MRA-E has been engaged in a vigorous e-mobility programme, the results of which have shown that the best way for electric transport to gain a more solid market position is when market players, knowledge institutes and governments collaborate efficiently, also when it comes to financing. This has to occur on an international level as well, with emphasis on knowledge sharing as empowerment. What is evident from the rollout of the charging infrastructure in the MRA-E region is that national initiatives are often triggered by local or regional efforts, and that decision-making is most effective at a local level. As MRA-E's approach has demonstrated, anticipating possible obstacles to electric transport—in particular the rollout of the charging infrastructure—can avert delays in its overall implementation.

First of all this means that local councils should try to combine forces when calling for tenders for charging poles not only because they then have a stronger negotiating position, but also because they do not have to individually deal with the negotiations, administration and agreements. Coordinating tenders through a central project-based approach has worked well and has also greatly appealed to national and international parties, because it has simplified the process for them too. Furthermore, the locations of charging poles have to be coordinated between local authorities and market parties to ensure they are installed in the most suitable places. Their visibility prompts people to think about their choices.

Second, we are in a period of major transition in the types of fuels we use. The ultimate goal is to reduce dependency on fossil fuels, which is possible with full electric vehicles. MRA-E's ambitions include implementing a full transition to green energy whereby EV batteries are charged with power provided by solar panels that are installed on or in close proximity to the premises of EV users. Cooperating on the rollout of e-mobility as part of the measures to realise a new and clean economy means that the government and market parties can be considerably more efficient and reduce costs.

Third, promoting e-mobility should also underscore the fact that by driving electric, users contribute to improving the quality of their own lives, especially through decreasing air pollution. The field of e-mobility is advancing at a rapid rate and in the next few years should become a mature market in which EVs are regarded as a normal (and essential) part of the transport infrastructure. This increase in the number of charging point users should be anticipated, and the tariffs should be such that they encourage market parties to invest in e-mobility. The interests of EV users and improving air quality should always be paramount.

Because the field of e-mobility is advancing at a very rapid pace now, it is vital to keep abreast of technological advances—aided by knowledge-sharing at national and international levels—to ensure that strategies are not devised that incorporate outmoded systems and equipment. The highly successful rolling out of the charging infrastructure in the MRA-E region has proven that, with the right commitment, the "chicken and egg" problem—EVs and a charging infrastructure being prerequisites for each other—can be overcome.

References

Air Quality Programme Department. (2009). *Electric mobility in Amsterdam*. Amsterdam: Amsterdam City Council.

Air Quality Programme Department. (2010a). *Amsterdam Elektrisch—Het Plan*. Amsterdam: Amsterdam City Council.

Air Quality Programme Department. (2010b). *Werkprogramme Luchtkwaliteit 2010*. Amsterdam: Amsterdam City Council.

Air Quality Programme Department/Project Amsterdam Electric. (2010c). *TNO-Rapport—Elektrisch Vervoer in Amsterdam*. Amsterdam: Amsterdam City Council.

Amsterdam Development Corporation. (2005). *Ground lease in Amsterdam*. Amsterdam: Amsterdam City Council.

Amsterdam City Council, department traffic and mobility, *Schone lucht voor Amsterdam*. June 2011. Amsterdam.

Den Haag Elektrisch: http://www.denhaag.nl/home/bewoners/to/Elektrisch-vervoer-in-Den-Haag.htm

Janssen-Jansen, L. B., & Hutton, T. A. (2011). Rethinking the metropolis: Reconfiguring the governance structures of the twenty-first-century city-region. *International Planning Studies, 16*(3), 201–215.

MRA Electric. (2012). *Metropoolregio Amsterdam Elektrisch*. Amsterdam: MRA-E.

MRA Electric. (2013a). *Werkprogramma 2013*. Amsterdam: MRA-E.

MRA Electric. (2013b). *Opladen in 5 stappen*. Amsterdam: MRA-E.

MRA Electric. (2014a). *Werkprogramma 2014*. Amsterdam: MRA-E.

MRA Electric. (2014b). *Elektrisch rijden in 5 stappen van 0 naar 100%*. MRA-E: Amsterdam.

NL Agency. (2013). *We Are Holland, Ready to Market E-mobility*. The Hague: NL Agency.

Netherlands Enterprise Agency. (2011). *Electric Mobility Gets Up To Speed 2011–2015 Action Plan*. The Hague: The Ministry of Economic Affairs.

Organisation for Economic Cooperation and Development/International Energy Agency et al. (2012). *EV City Casebook*. http://www.iea.org/publications/freepublications/publication/EVCityCasebook.pdf.

Rienstra, S. A., & Nijkamp, P. (1998). The Role of Electric Cars in Amsterdam's Transport System in the Year 2015; A Scenario Approach, Serie Research Memoranda 0028, Vrie University of Amsterdam, Faculty of Economics, Business Administration and Econometrics.

Rotterdam Elektrisch: http://www.rotterdam.nl/rotterdamelektrischprojecten

RVO—Rijksdienst voor Ondernemend Nederland. (2015a, April). *Cijfers Elektrisch Rijden*.

RVO—Rijksdienst voor Ondernemend Nederland. (2015b, May). *Cijfers Elektrisch Rijden: Focusgebied Metropool Regio Amsterdam (excl. gemeente Amsterdam)*.

Schiphol Amsterdam Airport. (2014). Traffic Review 2013.

Silvester, S., Bella, S. K., van Timmeren, A., Bauer, P., Quist, J., & van Dijk, S. (2013). Exploring design scenarios for large-scale implementation of electric vehicles; the Amsterdam Airport Schiphol Case. *Journal of Cleaner Production, 48*, 211–219.

Van Woerkom, A., & Hoekstra, A. (2012). *Keuzegids Elektrisch autorijden*. The Hague: PixelPerfect Publications.

Web links

Amsterdam Metropolitan Area. Retrieved May 28th, 2014, from http://www.metropoolregioamsterdam.nl/.

Amsterdam Metropolitan Area Electric—Think Global, Charge Local. Retrieved May 28th, 2014, from http://e-mobility-nsr.eu/fileadmin/user_upload/downloads/EMIC_Amsterdam/MRA-ECOmobiel.pdf.

Retrieved May 28th, 2014, from http://martinprosperity.org/global-cities/Global-Cities_Amsterdam.pdf.
Bureau Onderzoek & Statistieken (Research and Statistics), City of Amsterdam: http://www.os.amsterdam.nl/english/.
Car2go Amsterdam: https://www.car2go.com/en/amsterdam/#pid=43913.
City of Amsterdam (n.d. as constantly updated, 2009 in original). *The City of Amsterdam's Plan van Aanpak Elektrisch Vervoer* ("Plan for Electric Mobility"). Retrieved May 25th, 2014, from http://www.amsterdam.nl/gemeente/organisatie-diensten/dmb/doet-dmb/wat/advies-beleid/milieuadvies/projecten/luchtkwaliteit/actieplan/overzicht-acties/acties/elektrisch-vervoer/@478200/pagina/.
https://www.rvo.nl/sites/default/files/bijlagen/Plan%20van%20aanpak%20-elektrisch%20rijden%20in%20de%20versnelling-.pdf.
http://www.rijksoverheid.nl/onderwerpen/auto/elektrisch-rijden.
http://www.rvo.nl/onderwerpen/duurzaam-ondernemen/energie-en-milieu-innovaties/elektrisch-rijden/aan-de-slag/financiele-ondersteuning?gclid=COr_-JKTs78CFUTItAod90kAbw.
http://www.rvo.nl/onderwerpen/duurzaam-ondernemen/energie-en-milieu-innovaties/elektrisch-rijden/stand-van-zaken/milieuvoordeel.
Map of Operational AC fast chargers in the Netherlands and Belgium: www.ac-lader.nl.
Nederland Elektrisch. Retrieved May 2014, from http://www.nederlandelektrisch.nl/english/.
Population statistics. Retrieved June 21st, 2014, from http://home.kpn.nl/pagklein/provincies.html.
Utrecht Elektrisch: http://www.utrecht.nl/utrecht-elektrisch/.

Electrifying London: Connecting with Mainstream Markets

Stephen Shaw and Louise Bunce

Abstract London's Mayor Johnson has given a high priority to Electric Vehicles (EVs) as the most appropriate road transport technology to reduce CO_2 emissions and improve air quality. This chapter outlines the policy rationale that underpins the Mayor's strategy to stimulate the early market for EVs, and reviews its implementation over the period 2009–2014. The Mayor of London's commitment has been demonstrated through initiatives that include the development of the diesel-electric hybrid 'New Bus for London', experimental hydrogen fuel cell powered buses and taxis, by collaboration with commercial operators to pilot electrification of freight transport, and by plans to create 'the world's first Ultra Low Emission Zone' by 2020. Another key objective has been to create an extensive infrastructure for recharging electric scooters, motorcycles, cars, vans and light trucks by 2015. The authors consider these developments with reference to recent research on early market adaptation to electric driving, and the prospects for converting mainstream private drivers and firms. This leads to a discussion of some important challenges that suggest the need for further research to help decision-makers improve the effectiveness of interventions, especially at the user interface: locating and designing EV infrastructure, supporting longer distance electric driving and informing current and potential EV users.

Keywords Electric vehicles • Policy interventions • Psychology • Mass marketization • Acceptability

S. Shaw (✉)
Faculty of Business and Law, London Metropolitan University, Calcutta House, Old Castle Street, London E1 7NT, UK
e-mail: s.shaw@londonmet.ac.uk

L. Bunce
Department of Psychology, University of Winchester, Herbert Jarman Building, King Alfred Campus, Sparkford Road, Winchester SO22 4NR, UK
e-mail: louise.bunce@winchester.ac.uk

W. Leal Filho and R. Kotter (eds.), *E-Mobility in Europe*,
Green Energy and Technology, DOI 10.1007/978-3-319-13194-8_8

1 Introduction

In the UK, as in other European countries discussed in this volume, central government has established a supportive strategy to nurture the development of electric vehicles (EVs) and other ultra low emission vehicles (ULEVs) (Kotter and Shaw 2013). In tandem with decarbonisation of power generation, the public benefits include significant reductions in CO_2 emissions to mitigate climate change, better energy security and improved air quality, along with the anticipation of badly needed economic growth as EVs replace the 'old technology' of internal combustion engine (ICE) vehicles. Bearing in mind that, unlike some of the countries discussed in this volume, the UK does not have regional government, local authorities in partnership with other agencies in London and other city regions are championing the electrification of road transport to deliver the desired environmental and economic outcomes for their respective areas (Hodson 2013; Hickman and Banister 2014).

Since his election in 2008, Mayor Boris Johnson has given a high priority to stimulating the early market and preparing the way for mass marketization for EVs in Greater London: the administrative area for the UK capital with a population of 8.17 million (ONS 2012) that comprises 32 Second-tier London Boroughs and the City of London. This chapter opens with a brief synopsis of the UK Government's policies to support the switch to EVs and other ULEVs, and how these have been interpreted and implemented over the last 5 years in Greater London. It considers a range of incentives and initiatives that include the piloting, development and demonstration of commercial passenger and freight transport operations, reviews progress with plans create an extensive infrastructure for recharging electric scooters, motorcycles, cars, vans and light trucks across the metropolitan area by 2015, and highlights the significance of conversion to EVs for air quality and public health.

In the section that follows, the authors reflect on the findings of recent research on the factors that have motivated or deterred the early market for EVs, and the challenge of convincing the more cautious and sometimes sceptical mainstream private drivers and fleet managers of the personal and business advantages of electric driving, as well as the wider benefits to the environment and economy. This leads to a discussion of some key issues that have emerged and which suggest the need for further investigation and analysis to guide decision-makers in the critical years ahead on how the high-level policy aspirations for Greater London and other city regions can be implemented more effectively.

2 National Policy Context

In a policy statement published by the Office for Low Emission Vehicles (OLEV)—the UK's special purpose intergovernmental agency for promoting zero and ultra low emission vehicles—the incoming Conservative-Liberal Democrat Coalition

Government confirmed its commitment to reducing the UK's greenhouse gas emissions by 50 % in 2027 and by 80 % by 2050, its belief that '[p]lug in vehicles will make a substantial contribution to meeting these targets', and that it wanted to grow 'the market for plug-in vehicles in the UK because of the contribution they… can make across our economic and environmental priorities' (OLEV 2011, p. 19). Re-emphasising its support for electrification of road transport, the UK Government acknowledges the inevitability of a 'once in a lifetime technology change' in the global automotive industry (OLEV 2013, p. 6). The emergence of market-ready ULEVs as a real option for consumers has begun, and the Government is supporting the nascent industry and its associated supply chain with £400 m funding through to 2015 that focuses on the following interventions (OLEV 2013, pp. 8–9):

- *Helping to support the purchase of ULEVs*, especially through the Plug-in Car and the Plug-in Van grant to help reduce the purchase price differential compared with conventional vehicles, as well as through incentives through the tax system, together with local incentives such as exemption from the London Congestion Charge (see below);
- *Facilitating the provision of recharging infrastructure*, including the Plugged-in Places (P-IP) scheme that matched funded 'Source London' (see below) and seven other regionally based public–private consortia that by June 2013 had in total installed over 5500 charge points (with an estimated 5000 additional points delivered by non-P-IP commercial companies (OLEV 2013, p. 8);
- *Preparing for hydrogen fuel cell electric vehicles*, working with companies to develop a business case for the roll-out of vehicles and associated infrastructure from 2015, including a demonstration fleet of five hydrogen fuel taxis in London during the 2012 Olympics (see below);
- *Encouraging and investing in research and development*, with Central Government funding for a programme to support the new generation of ULEVs and to help build the necessary skills and knowledge, including support through the Low Carbon Innovation Platform; and
- *Lowering emissions from other vehicles*, including support through the UK Department of Transport's Green Bus Fund which has supported more than 1200 new low carbon vehicles in England, including 350 in London, while the Low Carbon HGV Technology Task Force and Low Carbon Truck and Infrastructure Trail support demonstrations and other initiatives for logistics and freight (see FDT et al. 2013, pp. 153–175; and FEVUE, see http://frevue.eu/london/).

The suite of interventions above is coordinated by OLEV to create the 'long-term and stable framework' that is needed to stimulate investment in the fledgling sector and nurture its growth (OLEV 2013, p. 6). The UK Government emphasises that [i]f the UK wants to benefit from the employment and economic opportunities, as well as cleaner and quieter towns and cheaper motoring, which these new vehicles can bring 'it is vital that this country is in the vanguard of this change' (ibid, p. 4). To secure the anticipated benefits, appropriate strategies must be implemented through collaborative local initiatives in Greater London and other areas. These should be considered in the wider context of the broader aspirations of environmental and economic policy. The following section outlines the Mayor of London's (2009a, b) strategic vision for electrifying road transport in the broader context of public policy, and considers how the principles have been put into practice through to 2014.

3 The Mayor's EV Delivery Plan and EV Infrastructure Strategy

The Mayor of London is the elected leader of the Greater London Authority (GLA): the strategic metropolitan authority with powers that include transport and economic development. He is also responsible for Transport for London (TfL), the agency that implements the Mayor's Transport Strategy and manages transport services, along with major highways in the UK capital. Further, the Mayor's strategic policy frameworks for transport and land use planning across Greater London provide the context for local implementation by the second-tier authorities. It is therefore significant that Mayor Johnson has given his personal support to electrification of road transport and stimulation of the early market for EVs. The drive to decarbonise road transport and improve air quality in Greater London is enshrined in public policy, principally in: the Mayor's Transport Strategy (GLA 2010); the Air Quality Strategy, (Mayor of London 2010a), the London Plan, (GLA 2011); and the Climate Change Mitigation and Energy Strategy (Mayor of London 2010b).

Comparing developments in London with Berlin, a report by LSE Cities, ICLEI and Global Green Growth Institute (2013, p. 98) observe that although E-Mobility policy remains in its infancy, some differences of approach are emerging. Renewable energy integration is a central component of Berlin's strategy and programme, which 'aims to establish Berlin as a centre for development and production of smart grid components, renewable energy systems, vehicle-to-grid technology and mobility services'. Although this may remain an aspiration, '[s]uch a strategy does not exist in London and the integration of renewable energy with electric mobility is a much lower priority' (ibid, p. 98). Berlin has also focused more on promoting systemic change of urban transport rather than providing subsidies to individual car owners (ibid, p. 99).

Nevertheless, an ambitious vision for electric mobility was introduced in Mayor of London (2009a, p. 5) *An Electric Vehicle Delivery Plan for London*. The companion document Mayor of London (2009b, p. 41) *Turning London Electric: London's Electric Vehicle Infrastructure Strategy* anticipated that the significant public benefits of converting to electric mobility included reduction in carbon emissions so that, 'over the period to 2020, emissions of CO_2 for cars will decrease by approximately 40 % from the current EU average of 153.5 g/km for new vehicles to 95 g/km'. This should be seen in the broader context of a 60 % reduction for CO_2 emissions by 2025 applied across *all* sectors on a 1990 base (GLA 2007): a challenging target to achieve without compromising economic and quality of life goals (Hickman and Bannister 2014, pp. 104–106). With respect to air quality, it is anticipated that by 2020 EVs 'may reduce annual NOx emissions by up to 100 tonnes across London and lower the emissions of PM10 by several tonnes per year'.

In addition to the national incentives above, the Mayor of London (2009a, p. 32) confirmed that he would guarantee an important personal concession for EV users introduced by his predecessor Mayor Livingstone: exemption from payment of the central London Congestion Charge (ibid, p. 28), which [in 2014] represents a saving of around £2000 p.a. for regular users. The Delivery Plan also emphasised that the GLA would 'lead by example and set challenging targets for the procurement of EVs for its own fleet' (ibid, p. 24). Further, it would work with other agencies, e.g. the Public Carriage Office to trial low carbon taxis (ibid, p. 26), and large fleet operators, including express parcel carriers, and retail delivery fleets (ibid, p. 28). As yet, pure battery powered buses would not be able 'to meet the arduous operational requirements' of working 18 h a day in London's heavy traffic. Nevertheless, from 2012 onwards all new buses entering service in London would be hybrids, and TfL would 'work with bus manufacturers and other cities… to develop technical solutions' (ibid, p. 27).

The EV Infrastructure Strategy acknowledged that electrification of road transport in Greater London presents particular challenges. Whereas nationally, the 'norm' is to charge EV cars overnight at home, this arrangement is impossible for two-thirds of London households who do not have off-street parking Mayor of London (2009b, pp. 16–17). Provision of non-domestic charging points will therefore be essential, and throughout Greater London 'an extensive charging infrastructure network' will be installed (Mayor of London 2009b, p. 6). The Delivery Plan (Mayor of London 2009a, p. 33) envisaged a simple and consistent 'London brand' for marketing and communications: 'strong and easily recognisable' to make it 'clearer for users and help to encourage potential users to purchase EVs'. A number of websites relevant to electric mobility in London had been set up. Now, there is an opportunity to deliver 'a significant in the quality of information provided' by creating a single portal (ibid, p. 33).

4 Operationalizing the Vision in Greater London

4.1 Passenger and Freight Carriers Go Electric

The testing and development of mainstream passenger and freight undertakings in Greater London has been an important signifier of progress. The 'New Bus for London', designed as a 21st century version of the historic 'Routemaster', uses diesel-electric hybrid technology (Fig. 1). It is driven by an electric motor and powered by batteries that are recharged by a diesel generator (that only runs when

Fig. 1 The New Bus for London is powered by diesel-electric hybrid technology (*photo* S. Shaw)

the batteries need recharging), and by regenerative braking. TfL commissioned prototype models of the new vehicle, following a high-profile design competition. The bus has capacity for 87 passengers, with three doors and two staircases for speedy boarding and exit. The first prototype entered regular service in February 2012, and in September 2012, TfL announced its intention that over 600 vehicles will be introduced by 2016: the largest order for hybrid buses ever made in Europe. By April 2014 they were operating on six of London's busiest bus routes.

London is also paving the way for full market commercialization of Fuel Cell Hydrogen powered (FCH) buses. By 2013, TfL bus route RV1—a route that links many landmark visitor attractions—was fully converted to FCH buses. Further, the Hydrogen Transport for European Cities Project (HyTEC) is endowed with a demonstration fleet of five hydrogen fuel cell powered taxis, supported by a hydrogen fuelling station at Heathrow airport. The first phase saw the pioneering fleet of fuel cell electric London Taxis transport visiting dignitaries and the VIP guests of the Greater London Authority during the 2012 Olympic and Paralympic Games (HyTEC 2012). Based partly in the 'Tech City' centre in London's East End, Qualcomm in partnership with Renault have pioneered the UK's first wide-spread trial of wireless technology to enable EV users to charge their battery by driving over an electric pad (Kotter and Shaw 2013, p. 59).

'Electric 20' is a working group of twenty leading companies that use electric commercial vehicles in their fleets and have agreed to work with the Mayor to share information about their experiences and help increase the uptake of commercial EVs in London. The London Electric Vehicle Partnership, chaired by the Mayor's Environment Advisor, brings together key stakeholders and decision-makers from within the vehicle manufacturing industry, London Boroughs, the Greater London Authority, energy and infrastructure providers and EV users to accelerate the delivery of new EV technology and increase the level of support for EV drivers in London. Through these initiatives to support electrification of road transport, both for passenger and freight movement, the Mayor envisages a future scenario in which London becomes the 'epicentre of electric driving in Europe' (Source London 2014).

4.2 Wiring up the Infrastructure: Source London

As leader of the first-tier authority for Greater London (above), the Mayor has a strategic role in coordinating and promoting the infrastructure for recharging EVs across the capital. This required the integration of eleven earlier local schemes established by the second-tier London Boroughs. In 2010, 50 % match funding from central government was secured under the P-IP initiative (above): a public–private consortium led by TfL (Mayor of London 2010c). In May 2011, the Mayor launched 'Source London': the unified brand anticipated in the EV Delivery Plan (above), and by December 2012 a publicly accessible network of over 200 points had been installed (Source London 2014).

The EV Infrastructure Strategy (Mayor of London 2009b, p. 15), anticipates that 25,000 charging points will be installed by 2015, 90 % (22,500) of which will be at workplaces and homes, where vehicles are 'typically parked for seven or more hours in a relatively secure location'. These will facilitate charging of company-owned fleet cars/vans, and provide top-up charging for staff and visitors' EVs. The other 10 % (2500) will be in publicly accessible locations, including 2 % (500) in dedicated on-street parking bays. The remaining 8 % (2000) will be in parking spaces managed by landlords such as supermarkets, the second-tier London Boroughs, airports, Underground, and Train Operating Companies. Source London posts with fast charge sockets (7KW 32A) are double headed with a standard socket that can be used by all EVs. A further development has been the installation of 16 'rapid' points (20–50KW 33A+) by March 2014 (Source London 2014).

The subscription-based scheme with a smart card offers EV owners in London unlimited use of the public network. In 2012, annual membership was reduced from £100 to just £10 (2012) giving users access to the Source London points across the capital at nominal rate, and by March 2014 over 1400 charge points were available. In line with the Mayor's vision to enable commercialisation of the public charging infrastructure, TfL has announced that the French organisation IER has been selected to take over the management and operation of Source London from summer 2014 onwards, and that IER will be looking to expand the number of Source London points to 6000 to meet the growing needs of EV drivers by 2018 (Source London 2014).

4.3 Addressing Air Quality and Public Health

As Hanley and Colin Buchanan (2011, p. xvi) observe, assessment of air quality in the UK focuses on concentrations of specific elements in the atmosphere that are monitored against statutory thresholds set out in EU Directive 2008/50/EC on ambient air quality and cleaner air for Europe: a regulatory framework that was made law in England through the Air Quality Standards Regulations 2010. Unfortunately, despite these legal imperatives and the aspirations of Mayor of London to reduce NO_x emissions (above), these thresholds have all too often been exceeded in 'pollution hotspots' across Greater London. Vidal (2013) commented in *The Guardian* newspaper on a rising political concern by governments and environmental groups over the 'public health crisis' associated with poor air quality and in particular with the transport sector; Client Earth—an organisation of activist environmental lawyers—initiated legal action against the UK government in the Supreme Court to remedy the alleged breach of European law.

Although the growing popularity of diesel cars and vans has made a positive contribution to reducing CO_2 emissions (Hickman and Bannister 2014, p. 121), its effect on air quality and thus on public health appears to have been highly detrimental. The Mayor's senior environment advisor Matthew Pencharz commented "the dieselisation of the car fleet that we've seen in recent years has greatly exacerbated the air quality problem in London" (Gibbs 2014 M3). Local initiatives may seem marginal compared to 'top-down' regulation by the EU (especially the European emission standards for particulates that define acceptable exhaust emissions of new vehicles), the technologies developed by manufacturers, and the preferences of mainstream buyers. Nevertheless, Hanley and Colin Buchanan (2011, p. xvi) note examples that demonstrate commitment, e.g. adoption by the London Borough of Camden of a hierarchy of vehicle fuels and technologies for procurement of new vehicles that takes into account particulate matter, nitrogen oxides and carbon emissions. Further, the UK's first electric car club is establishing schemes in East London [http://www.e-carclub.org/your-community/poplar-east-london/].

Mayor Livingston had previously used his powers to establish by 2008 a Low Emission Zone (LEZ) based on European emission standards (above); non-compliant diesel-powered commercial vehicles—such as lorries, buses and coaches—are charged and thus deterred from entering the LEZ which covers most of Greater London. In 2013 Mayor Johnson announced his intention that the world's first Ultra Low Emission Zone (ULEZ) would be created in central London by 2020: "My vision is a central zone where almost all the vehicles running during working hours are either zero or low emission". This will "deliver incredible benefits in air quality and stimulate the delivery and mass use of low emission technology" (GLA 2013). TfL's MD Planning (Dix 2013, p. 12) confirmed that the scheme would support 'the Mayor's ambition for electric, plug-in hybrid and alternative fuel vehicles to be common place on London's streets'. Given the range of incentives and support now available in London, consideration must be given to the attitudes and perceptions of potential converts to E-Mobility.

5 Acceptability to Mainstream Drivers and Firms

Global market leaders in the car industry have made rapid advances in EV technology, meaning that a number of high performance, well-branded EVs are currently available on the consumer market. However, the extent to which these are viewed as an acceptable alternative to driving an ICE depends largely on the perceptions and attitudes of would-be consumers (Schuitema et al. 2013; Lane 2011). Unfortunately, it seems that a sizable portion of the UK public hold stereotypes of EVs that are based on outdated associations with golf buggies and milk floats, implying an interpretation of EVs as having low speed and low range capabilities (Burgess et al. 2013; Graham-Rowe et al. 2012).

In contrast, experimental evidence from potential consumers who have experienced driving an all electric vehicle on a regular basis has revealed that these old fashioned stereotypes largely do not apply to current EV models (Carroll et al. 2013; Franke and Krems 2013; Switch 2013; Graham-Rowe et al. 2012; Vilimek et al. 2012; Everett et al. 2011; Franke et al. 2012; Turrentine et al. 2011; Carroll 2010). For example, after driving an EV for 1 year, 77 % of participants in the 'Switch EV' trial in the North East of England (Switch 2013) thought that the performance of their EV, specifically the acceleration, was as good as or even faster than their traditional ICE. When drivers were describing their experience of driving an EV they made comments such as: *It can outperform any other car that I've driven* (ibid, p. 6). However, perceptions of EV range, requirements for recharging, and the visibility of the public charging infrastructure remain important concerns among both current and potential EV drivers.

5.1 *Range Acceptability*

Fully electric vehicles have a range of approximately 100 km, which is significantly lower than that of an ICE. This difference in travel distance between a full charge and a full tank has led to the phenomenon of 'range anxiety'. Range anxiety is a psychological phenomenon caused by a fear of running out of charge before reaching your destination (Franke et al. 2012). The term 'psychological' is significant because the range of an EV has been found to be more than sufficient for the majority of journeys commuted on a daily basis. In the UK's Ultra Low Carbon Vehicle Demonstrator Program (ULCVP) (Carroll et al. 2013; Everett et al. 2011) over 300 EVs were tested on everyday journeys by 'real-life' users who were interviewed about their pre-EV use expectations and 3-month post-driving experiences. At the start of the trial, both private and fleet drivers indicated that they would be more concerned about reaching their destination in their EV compared to their 'normal' ICE car. After 3 months, however, the number of drivers who expressed this same concern dropped by 35 %. The drop in range anxiety was possibly due to an increased understanding of the vehicle capabilities, driving techniques and journey planning. Despite this, most drivers agreed that a visible public charging infrastructure would be required to support the emerging EV market.

However, range anxiety has also recently been described as something of a 'range paradox' (Franke and Krems 2013). This is because an approximate 100 km range of an EV has been found to be sufficient based on analyses of people's actual travel behaviour (e.g. 61 % of Europeans drive less than 100 km/day and 24 % even less than 20 km, see Franke and Krems (2013), for a summary of this research. On average, 84 % of car trips made by Londoners are less than 20 km and 95 % of London motorists travel less than a total of 75 km/day (Mayor of London 2009a, p. 9). However, this is discrepant with, i.e. falls short of, drivers' stated range preferences. Drivers do not seem to want to compromise on range or settle for a vehicle that appears to achieve less than an ICE (Chéron and Zins 1997).

Despite this range paradox, negative perceptions of range still need to be overcome. Although technological advances in battery capacity may be one way to overcome range anxiety, this is costly in terms of time and money. The results of the ULCVDP imply another strategy based on changing people's attitudes and behaviours when it comes to driving an EV (Burgess et al. submitted). There was a subset of drivers (about one quarter) who successfully overcame range anxiety by deliberately testing the range of the vehicle, or after an unexpected incident in which range threatened to run out accidently. Nonetheless, these drivers still managed to get to their destination successfully, and in the process they learnt how route selection, regenerative breaking, and driving style interconnected to maximise the range. Thus taking drivers on a test drive in which they are challenged and taught to maximise the range of the vehicle may be an effective strategy for overcoming range anxiety.

5.2 Recharging Acceptability

The location of charging points and the length of time taken to recharge an EV are two fundamental concerns of potential EV users (Carroll et al. 2013). Charging is often perceived as 'inconvenient' because drivers foresee spending hours waiting for the vehicle to recharge as opposed to refuelling at a petrol station in under 10 min (Bunce et al. 2014). In addition, drivers want the reassurance of a public charging infrastructure to enable them to recharge their EV on longer journeys. In practice, however, recharging the EV is usually done at home overnight, similarly to a mobile phone, while the driver does not need to use it (Carroll et al. 2013; Turrentine et al. 2011). This represents a particular challenge in London since, as noted above, two-thirds of households do not have off-street parking, meaning that recharging their EV from home is difficult, if not impossible. Despite this practical barrier, which will be overcome according to the Delivery Plan of the Mayor of

London, experimental trials of EV drivers have revealed that the charging process itself does not present a barrier to mainstream adoption.

Drivers in the ULCVDP found the charging process easy and convenient: *I found charging very easy: plug it into the mains... that's it... wake up in the morning and it's all done... and away I go.* (Everett 2011, p. 9). Similarly in Graham-Rowe et al. (2012) , drivers actually thought that recharging, as opposed to refuelling, was convenient and *time-saving* because it meant that you never needed to stop en route to refuel. One driver explained: *It was a delight during the week not having to go to a petrol station* (Graham-Rowe et al. 2012, p. 146). The speed of charge was originally thought to be unsuitable for requirements in over half of the drivers in the ULCVDP, however, after 3 months experience of driving an EV saw these figures drop significantly to just a quarter of drivers (Carroll et al. 2013). However, other drivers in Graham-Rowe trial which took place in the South of England, UK, bemoaned the lack of public charging infrastructure, which meant that they sometimes had to use an ICE car to make a journey. Despite this, over time, drivers generally become more relaxed about recharging and find that they do not need to recharge every day, sometimes only charging every few days (Bunce et al. 2014; Switch 2013; Turrentine et al. 2011 on California). These findings have important implications for the requirements of the future public charging infrastructure, support for longer distance electric driving, and communications strategies to inform current and potential users in London and other city regions.

6 Emerging Issues: Interventions and the User Interface

The positive narrative of adjustment to vehicle range and recharging routines, diminishing concerns and increased satisfaction provides a powerful message that can be emphasised in communication strategies to promote the personal and business advantages of electric driving. Nevertheless, in Greater London as in other areas, public policy accepts the need for interventions to overcome the barriers that still inhibit mass-market acceptance of EVs at the user interface. Work carried out by the authors through the E-Mobility NSR project partnership suggests the need for further research to help decision-makers overcome current shortcomings:

6.1 Locating and Designing EV Infrastructure

As in other countries and city regions discussed in this volume, the provision of an effective network of Public Charging Infrastructure (PCI) has been subject to something of 'chicken-and-egg' debate. Should a comprehensive network of charge points be provided at the outset to reassure drivers and thus stimulate demand, with public subsidy where necessary? Or should this be left to market forces with commercial charge point providers responding to the demand pattern of EV drivers, as and when it develops? The UK Government has made it clear (OLEV 2011, p. 42) that they 'do not want to see a charge point on every corner; this is an unnecessary and expensive approach to allaying range anxiety'.

The policy aspiration (Mayor of London 2009b, p. 24) is that by 2015 'no Londoner [will be] more than 1 mile from a public charging point by 2015'. Nevertheless, with limited public funding, provision has been targetted at the 'hostspots' where higher demand is anticipated. To this end, Source London has used the 'Mosaic' public sector data analysis tool (ibid, p. 27) to identify the socio-economic characteristics of current EV and hybrid owners and map their location by neighbourhood. In principle, siting and design should accommodate travel patterns and user convenience, e.g. proximity to attractors, anticipated dwell time, visibility, personal and vehicle security. Figure 2 illustrates an on-street EV parking bay that has proved popular among staff, students and visitors to a nearby to a university in the London Borough of Camden and is highly visible to the many passers-by.

In practice, however, 'rational' matching of charge points to anticipated demand may be compromised at 'street level' (Kotter and Shaw 2013). Some points may be conspicuously underused, as illustrated in Fig. 3. Shaw and Evatt (2013) outline a methodology to support practitioners working to improve point location from the user's perspective at the micro level through negotiation between key stakeholders that typically include:

1. PCI network providers, e.g. Source London;
2. Distribution Network Operators (DNOs) from whom permission must be obtained to secure connection with the grid for installation of charge points;
3. Local Authorities responsible for land use and transport planning, highways and pedestrian environments;
4. Landlords, especially owners of car parks, e.g. at shopping centres, supermarkets, railway stations;
5. Adjacent owners of neighbouring properties.

Fig. 2 Well-sited, well-used points can attract favourable attention from passers-by (*photo* S. Shaw)

Fig. 3 Conversely, those that are conspicuously underused do little to promote EVs (*photo* S. Shaw)

The authors highlight the need for knowledge exchange between researchers and practitioners to establish how the location and design of EV charge points could be improved to optimise usage and enhance user convenience.

6.2 Supporting Longer Distance Electric Driving

Although Source London now provides a relatively high density of charge points, Londoners who want to drive EVs beyond their 'home' network must plan their itinerary to ensure the availability of points en route. Figure 4 illustrates the potential deterrent of gaps between 'refuelling' points along a major route between North East England and London, and a cross-border route from the UK to Norway that would require considerable planning compared with an equivalent ICE vehicle. Limited integration between different networks presents a challenge that may not be acceptable to mainstream drivers and fleet managers. Thus the House of Commons Transport Committee (2012, p. 25) recommended that the UK's Department for Transport should give priority to 'making sure that vehicle owners can access charge points across the UK'.

As noted above, the Source London smart card has given members unlimited use of their 'home' network for a nominal annual subscription. Initially, however, it did not give access to other networks, as its back-office system communicated only with its own members at the interface of charge points. The challenge of interoperability was addressed by the East of England P-IP (EValu8), which negotiated a Memorandum of Understanding with Source London and other schemes to enable mutual 'roaming' with one smart card from 2012. Further, the Charge Your Car network (see chapter by Herron et al. in this volume), which began in North East England, has expanded geographically and provides interoperability for users with a pay-as-you-go radio-frequency identification (RFID) card.

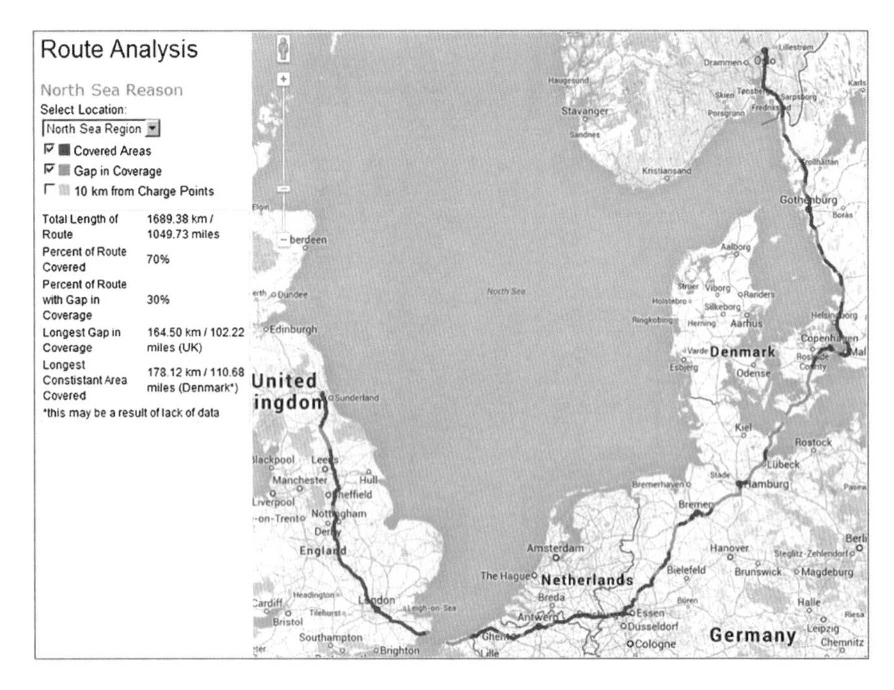

Fig. 4 Planning longer distance routes for electric driving remains a challenge. (*Source* E-Mobility NSR Work Package 4) ©2013 Google Maps (http://maps.citiesinstitutesurveys.org/UKEmobility.html (last accessed 18.03.14))

As Bakker (2013, p. 16) observes, however, in general across the UK 'separate systems have been developed and roaming is by and large impossible'. The Netherlands provides a useful model whereby drivers can access around 2500 points nationwide with a pay-as-you-go RDIF card (ibid, p. 17). However, this is exceptional as most EU countries have at least two providers with separate schemes. Although the European Commission has encouraged standardisation of the Type 2 plug (Mennekes), the 'real problem (both within and between countries) is much more in the different identification systems that are used to grant access to chargers and to arrange payment'. (ibid, p. 16). Bakker (ibid, p. 22) concludes that given the unlikeliness of a 'single European system for roaming between the networks, it would be best if charging station operators would be able to include ad hoc and payment systems in their chargers (using (short text message) SMS or credit card payments for instance)'.

The authors suggest that empirical evidence, such as travel diaries, is needed to inform policy-makers and PCI providers how EV drivers plan and complete longer itineraries—including cross-border trips—beyond the range of their vehicles, and to establish what support can be provided.

6.3 Informing Current and Potential EV Users

The findings of EV trials (above) confirm that the pioneers and early adopters seem to have adapted to electric driving with relative ease and some enthusiasm. Nevertheless, a recent survey by Rexel (2013) of mainstream drivers in the UK (n = 1188) revealed that whereas 41 % of respondents would consider purchasing an EV in the foreseeable future, 50 % were put off because they did not know where they could plug-in. Over 60 % perceived EVs to be impractical because of the insufficient number of charge points, and over 70 % had never seen a charge point. OLEV (2013, p. 11) acknowledges that, as yet, 'most people have little, if any knowledge' of these new vehicles and that 'insufficient or inaccurate information can put off potential buyers' (ibid, p. 10). Hutchins et al. (2013, p. 7) concur and recommend that key players should 'develop a united voice to educate and inform the public'.

In January 2014, the UK Government—in collaboration with Vauxhall (GM), Renault, Toyota, Nissan and BMW—launched a national campaign 'Go Ultra Low' (http://www.goultralow.com). This includes a 'one stop' portal for existing and potential users of EVs and other ULEVs. The Go Ultra Low website fits well with the recommendations in E-Mobility NSR Report Work Package 6 (Hoeje-Taastrup Commune 2013) that the information provided by Electric Mobility Information Centres (be they physical or virtual) should include unbiased content, especially concerning: driving and charging an EV; environmental benefits; market-ready options; and economic benefits for the user. The portal also provides access to a 'Zap Map' to inform EV drivers 'on the move' where public charge points are available nationwide. However, obtaining cross-border information on point location remains a challenge (Lilley et al. 2013).

An extensive survey of recent converts to electric driving showed that respondents had absorbed information from a range of media sources, e.g. newspapers, motoring magazines, television programmes, and that 'the two sources deemed most useful in supporting decision-making were internet forums and test drives' (Hutchins et al. 2013, p. 29). They conclude that deliberative research could further explore the underlying lack of knowledge and awareness; based on the outcomes of such research, especially which barriers could be broken down by information provision, 'it may be possible to determine which areas should be focused on to improve the uptake of EVs' (ibid, p. 96).

The authors conclude that campaigns such as Go Ultra Low can help improve communications to potential mainstream converts, but suggest that further research is needed to assess the complementary role of portals such as Source London that set electric driving in the local context, and outreach initiatives that offer hand-on experience.

7 Conclusions

The public policy rationale for electrifying road transport in the UK capital echoes the broader national and transnational aspirations for ‘green growth’ in city regions across the North Sea Region, with a particular emphasis on the imperative to reduce CO_2 emissions and improve air quality. The Mayor of London (2009a, b) set out its strategy to stimulate the early market for EVs and thus achieve the desired economic as well as environmental benefits across Greater London, in anticipation of full commercialization when mainstream markets adopt the new vehicle technology. In this context, demonstration projects are showcasing the benefits of converting high profile commercial transport undertakings, notably development and roll-out of the custom-designed diesel-electric hybrid ‘New Bus for London’, the piloting of hydrogen fuel cell powered buses and taxis, and collaboration with leading logistics and freight delivery operators.

A significant incentive to both commercial and private EV users is their exemption from paying the Central London Congestion Charge that applies to most other vehicles: a saving worth around £2000 per annum for regular users in 2014. Match funding from the UK Government has also supported expansion of Source London, launched by the Mayor in 2011 to provide an extensive network of points for recharging electric scooters, motorcycles, cars, vans and light trucks across Greater London. Although the Mayor envisaged that 90 % of the projected 25,000 points to be installed by 2015 would be in workplace car parks, the provision of a Public Charging Infrastructure (PCI) is particularly important given the high proportion of residents who lack off-street parking, and thus the ability to recharge at home.

Empirical evidence from trials in London and elsewhere in the UK highlights the relative ease with which the early market for EVs has adapted to new routines for journey planning, driving and recharging. Analysis suggests that, for most people, the initial psychological deterrents do not translate into significant real-world barriers. Nevertheless, both the UK Government and the Mayor of London acknowledge the continuing need to pump prime and support the transition to mass-market acceptance of EVs and other ULEVs. The E-Mobility NSR project partnership has highlighted a number of issues that are likely to deter mainstream private drivers and firms. These challenges suggest the need for action-based research to help decision-makers in Greater London and other city regions improve the effectiveness of policy interventions at the user interface, in particular:

Locating and Designing EV Infrastructure Source London has developed a network of 1400 publicly accessible charge points (data from March 2014). And, as demand increases, IER is expected to expand the network to around 6000 points by 2018. Nevertheless, creating a PCI ahead of market demand presents a considerable challenge. An optimum number of points must be installed in areas where the new EV owners are expected to park their vehicles. Further, they must be conveniently positioned to meet user requirements at street level, subject to negotiation of siting and design with key stakeholders. The authors highlight the need for knowledge exchange between researchers and practitioners to establish how the location and design of EV charge points could be improved to optimise usage and enhance user convenience.

Supporting Longer Distance Electric Driving Some progress has been made to improve interoperability between network providers, a notable example being the agreement between Source East and Source London that enables mutual roaming. Across the UK, however, fragmentation of Public Charging Infrastructure (PCI) creates a challenge remains for longer distance journeys with respect to access, payment and information on the location of charge points. Mapping of PCI provision on transnational routes across the North Sea Region highlights the degree of planning and preparation required: a deterrent to mainstream drivers to which solutions must be found. Empirical evidence such as travel diaries is needed to inform policy-makers and PCI providers how EV drivers plan and complete longer itineraries beyond the range of their vehicles, and to establish what support might be provided.

Informing Current and Potential EV Users: Recent surveys suggest that public awareness of user benefits of EVs remains low, negative stereotypes persist, and few know-how or where EVs can be recharged. In 2014, the UK Government launched a national campaign in collaboration with EV manufacturers: 'Go Ultra Low'. This features a 'one stop' portal for existing and potential users that includes digital map-based information on where public charge points are available. The authors conclude that such campaigns can help improve communications to potential mainstream converts, but suggest that further research is needed to assess the complementary role of portals such as Source London that set electric driving in the local context, and outreach initiatives that offer hand-on experience.

References

Bakker, S. (2013). *Standardization of EV recharging infrastructures*, E-Mobility NSR, Report. Retrieved March 18, 2014, from http://e-mobility-nsr.eu/fileadmin/user_upload/downloads/info-pool/3.2_E-MobilityNSR_International_stakeholder_analysis.pdf.

Bunce, L., Harris, M., & Burgess, M. (2014). Charge up then charge out? Drivers' perceptions and experiences of electric vehicles in the UK. *Transportation Research Part A, 59*, 278–287.

Burgess, M., King, N., Harris, M., & Lewis, E. (2013). Electric vehicle drivers' reported interactions with the public: Driving stereotype change? *Transportation Research Part F, 17*, 33–44.

Burgess, M., Mansbridge, S., Harris, M., Carroll, S., & Walsh, C. (submitted). Initial mind-set and challenging experiences influence drivers' adaptation to their electric vehicle. *Transportation Research Part A*.

Carroll, S. (2010). *The smart move trial: Description and initial results*. London: Technology Strategy Board.

Carroll, S., Walsh, C., Burgess, M., Harris, M., Mansbridge, S., King, N., & Bunce, L. (2013). *Assessing the viability of EVs in daily life*. Doc TSB ULCVD-12-036 v1.0. Retrieved March 18, 2014, from http://www.innovateuk.org/ulcv-demonstrator-report.

Chéron, E., & Zins, M. (1997). Electric vehicle purchasing intentions: The concern over battery charge duration. *Transportation Research A, 31*, 235–243.

Dix, M. (2013). Ultra low emission zone, TfL stakeholder event slides, 14th November. Retrieved March 06, 2014, from http://cleanairinlondon.org/wp-content/uploads/CAL-264-TfL-FINAL-ULEZ-Stakeholder-event-MD-slides-STATIC-141113.pdf.

Everett, A., Burgess, M., Harris, M., Mansbridge, S., Lewis, E., King, N., et al. (2011). *Initial findings from the ultra low carbon vehicle demonstrator program. How quickly did users adapt?*. Retrieved March 18, 2014, from http://www.innovateuk.org/_assets/pdf/press-releases/ulcv_reportaug11.pdf.

Franke, T., & Krems, J. F. (2013). What drives range preferences in electric vehicle users? *Transport Policy, 30*, 56–62.

Franke, T., Neumann, I., Bühler, F., Cocron, P., & Krems, J. F. (2012). Experiencing range in an electric vehicle: understanding psychological barriers. *Applied Psychology, 61*(3), 368–391.

FDT, TU Delft, ZERO, LSP. (2013). *Comparative analysis of European examples of schemes for freight electric vehicles—compilation report*. NSR E-mobility project report. http://e-mobility-nsr.eu/fileadmin/user_upload/downloads/info-pool/E-Mobility_-_Final_report_7.3.pdf.

Gibbs, N. (2014). London Mayor's office urging government to increase tax on Diesels, In *The Daily Telegraph*, Motoring section, Saturday 24th May.

Graham-Rowe, E., Gardner, B., Abraham, C., Skippon, S., Dittmar, H., Hutchins, R., et al. (2012). Mainstream consumers driving plug-in battery-electric and plug-in hybrid electric cars: A qualitative analysis of responses and evaluations. *Transportation Research Part A, 46*, 140–153.

Greater London Authority. (2007). *Climate change action plan*, London: GLA.

Greater London Authority. (2010). *Mayor's transport strategy*. London: GLA. Retrieved March 18, 2014, from http://www.london.gov.uk/sites/default/files/MTS_Executive_Summary.pdf.

Greater London Authority. (2011). *The Mayor's London Plan*. London: GLA. Retrieved March 18, 2014, from http://www.london.gov.uk/shaping-london/london-plan/.

Greater London Authority. (2013). *Mayor announces air quality game changer*, London: GLA, 13 February. Retrieved March 06, 2014, from https://www.london.gov.uk/media/mayor-press-releases/2013/02/mayor-of-london-announces-game-changer-for-air-quality-in-the.

Hanley, C., & Colin Buchanan, S. K. M. (2011). *How local authorities can encourage the take-up of lower-carbon vehicles*. London: RAC Foundation.

Hickman, R., & Banister, D. (2014). *Transport, climate change and the city*. Abingdon: Routledge. Esp. chap. 3: Ambitions towards sustainable mobility (London).

Hodson, M. (2013). *Low carbon nation?* Abingdon: Routledge.

Hoeje-Taastrup Commune. (2013). *Mapping of public and private E-Mobility awareness needs*. E-Mobility NSR, report. Copenhagen capital region. Retrieved March 18, 2014, from http://e-mobility-nsr.eu/fileadmin/user_upload/downloads/info-pool/E-Mobility_NSR_Activity_6.6_Report.pdf.

House of Commons, Transport Committee. (2012). *Transport committee-fourth report plug-in vehicles, plugged in policy*. 12th September. London. Retrieved March 18, 2014, from http://www.publications.parliament.uk/pa/cm201213/cmselect/cmtran/239/23902.htm.

Hutchins, R., Delmonte, E., Stannard, J., Evans, L., & Bushell, S. (2013). *PPR668*: *Assessing the role of the plugged-in car grant and plugged-in places scheme in electric vehicle take-up*. London: DfT, Transport Research Laboratory. Retrieved March 18, 2014, from https://www.gov.uk/government/uploads/system/uploads/attachment_data/file/236748/research-report.pdf.

Hydrogen Transport for European Cities Project. (2012). Retrieved March 18, 2014, from http://hy-tec.eu/2012/10/hydrogen-fuel-cell-electric-taxis-drive-olympic-vips-2500-miles/.

HyTEC. (2012). Retrieved March 18, 2014, from http://hy-tec.eu/2012/10/hydrogen-fuel-cell-electric-taxis-drive-olympic-vips-2500-miles/.

Kotter, R., & Shaw, S. (2013). *Micro to macro investigation*. E-Mobility NSR. Report. Newcastle upon Tyne/London. Retrieved March 18, 2014, from http://e-mobility-nsr.eu/fileadmin/user_upload/downloads/info-pool/E-mobility_3.6._Main_Report_April_2013.pdf.

Lane, B. (2011). *Market delivery of ultra-low carbon vehicles in the UK: An evidence review for the RAC foundation*. Cambridge: Ecolane Transport Consultancy. Retrieved March 18, 2014, from http://design.open.ac.uk/documents/Market_delivery_of_ULCVs_in_the_UK-Ecolane.pdf.

Lilley, S., Kotter, R., & Evatt, N. (2013). *A review of electric vehicle charge point map websites in the NSR region. E-Mobility NSR project report*. Newcastle upon Tyne/London. Retrieved March 18, 2014, from http://e-mobility-nsr.eu/fileadmin/user_upload/downloads/info-pool/Review_of_NSR_charge_point_maps_interim_report_June_2013_final.pdf.

LSE Cities, ICLEI Local Governments for Sustainability and Global Green Growth Institute. (2013). *Going Green: How cities are leading the next economy*. London: The London School of Economics and Political Science. Retrieved June 03, 2014, from http://www.lse.ac.uk/businessAndConsultancy/LSEConsulting/pdf/LSECities-Green.pdf.

Mayor of London. (2009a). *An electric vehicle delivery plan for London*. London: GLA. Retrieved March 18, 2014, from http://www.london.gov.uk/archive/mayor/publications/2009/docs/electric-vehicles-plan.pdf.

Mayor of London. (2009b). *Electric vehicle infrastructure strategy: turning London electric: London's electric vehicle infrastructure strategy*. London: GLA. Retrieved March 18, 2014, from https://www.sourcelondon.net/sites/default/files/draft%20Electric%20Vehicle%20Infrastructure%20Strategy.pdf.

Mayor of London. (2010a). *Clearing the air. The Mayor's air quality strategy*. Retrieved March 18, 2014, from http://www.london.gov.uk/publication/mayors-air-quality-strategy.

Mayor of London. (2010b). *London climate change mitigation and energy strategy*. Retrieved March 18, 2014, from http://www.london.gov.uk/sites/default/files/CCMES_public_consultation_draft_Oct%202010.pdf.

Mayor of London. (2010c). *London secures £17 million funding for UK's largest electric vehicle charge point network*. Press release. Retrieved March 18, 2014, from http://www.london.gov.uk/media/press_releases_mayoral/london-secures-%C2%A317-million-funding-uk%E2%80%99s-largest-electric-vehicle-charge.

Office for Low Emission Vehicles. (2011). *Making the connection: The plug-in vehicle infrastructure strategy*. London: OLEV. Retrieved March 18, 2014, from http://assets.dft.gov.uk/publications/making-the-connection-the-plug-in-vehicle-infrastructure-strategy/plug-in-vehicle-infrastructure-strategy.pdf.

Office for Low Emission Vehicles. (2013). *Driving the future today: A strategy for ultra low emission vehicles in the UK*. London: OLEV. Retrieved March 18, 2014, from https://www.gov.uk/government/uploads/system/uploads/attachment_data/file/239317/ultra-low-emission-vehicle-strategy.pdf.

Office for National Statistics. (2012). *Census result shows increase in population of London as it tops 8 million*. London: ONS. Retrieved June 03, 2014, from http://www.ons.gov.uk/ons/rel/mro/news-release/census-result-shows-increase-in-population-of-london-as-it-tops-8-million/censuslondonnr0712.html.

Rexel, C. (2013). *Cited in next Greencar*. Retrieved March 18, 2014, from http://www.nextgreencar.com/news/6335/40-UK-drivers-would-consider-buying-EV.

Schuitema, G., Anable, J., Skippon, S., & Kinnear, N. (2013). The role of instrumental, hedonic and symbolic attributes in the intention to adopt electric vehicles. *Transportation Research Part A: Policy and Practice, 48*, 39–49.

Shaw, S., & Evatt, N. (2013) *Methodologies for mutual learning*. Supplementary Paper to Main Report: Micro to Macro Investigation. E-Mobility NSR project report. London. Retrieved March 18, 2014, from http://e-mobility-nsr.eu/fileadmin/user_upload/downloads/info-pool/Emobility_3.6._Supplementary_Paper_to_Main_Report_April_2013.pdf.

Source London. (2014). Retrieved March 18, 2014, from http://www.sourcelondon.net.

Switch, E. V. (2013). *Switch EV final report*. Newcastle University. Retrieved March 18, 2014, from http://vehicletrial.switchev.co.uk/the-project.aspx.

Turrentine, T., Garas, D., Lentz, A., & Woodjack, J. (2011). *The UC Davis MINI E consumer study*. Institute of Transportation Studies, University of California, Davis, Research Report UCD-ITS-RR-11-05. Retrieved March 18, 2014, from http://pubs.its.ucdavis.edu/publication_detail.php?id=1470.

Vidal, J. (2013) 'UK air pollution: why are we only now waking up to this public health crisis?' in *The Guardian* Tuesday 19th. March. Retrieved June 03, 2014, from http://www.theguardian.com/environment/2013/mar/19/uk-air-pollution-health-crisis.

Vilimek, R., Keinath, A., & Schwalm, M. (2012). The MINI E field study—similarities and differences in international everyday EV driving. In N. A. Stanton (Ed.), *Advances in human aspects of road and rail transport* (pp. 363–372). Boca Raton: CRC Press.

Technology Trajectory and Lessons Learned from the Commercial Introduction of Electric Vehicles in North East England

Colin Herron and Josey Wardle

Abstract North East England (NE) is at the forefront of low carbon vehicle development with Nissan manufacturing both the Nissan LEAF and lithium-ion batteries at its Sunderland plant from 2013. Since 2010, the region has installed a comprehensive recharging infrastructure, run major electric vehicle (EV) trials and awareness raising campaigns, and has consequently seen a fast increase in EV adoption. The NE has now become a major hub for vehicle, battery and energy research and development, as well as a UK centre for manufacturing and training facilities throughout the EV supply chain. This paper reflects on the experience gained over the last 4 years of EV and infrastructure roll-out and the lessons learned which may be of use to other regions considering large-scale adoption of low carbon vehicles and recharging infrastructure.

Keywords Electric vehicle infrastructure · Training · Integration · Strategy · Technology

1 Background to North East England

North East England (NE) covers 8,600 km^2 and is the second smallest region in the UK. Over 80 % of its 2.6 million population (2011 Census data) lives in urban areas which span a small area covering only 27 % of the region, clustered around the three main rivers. Although gross value added (GVA) per head is only £16,000 (2012 Office of National Statistics (ONS) estimates), the lowest among the nine

---

C. Herron (✉) · J. Wardle
Zero Carbon Futures, Future Technology Centre, Nissan Way, Sunderland SR5 3NY, UK
e-mail: colin.herron@gateshead.ac.uk

J. Wardle
e-mail: josey.wardle@gateshead.ac.uk

W. Leal Filho and R. Kotter (eds.), *E-Mobility in Europe*,
Green Energy and Technology, DOI 10.1007/978-3-319-13194-8_9

English regions, car ownership has continued to rise steadily increasing by 7 % in the 10 years since 2001 (2011 Census data).

Over 500 leading international companies, including 40 % of Dow Jones listed businesses, operate in the NE region. Sunderland, in the heart of the region, is home to a large automotive sector including Nissan which directly employs 7,000 people (Nissan UK, May 2014). In 2011, the Nissan Sunderland plant was chosen as the European manufacturing base for the Nissan LEAF, making it one of only four Nissan plants worldwide to invest in EV manufacture. LEAF and lithium-ion battery production represents a £420 million investment, maintaining 2,250 jobs at Nissan and in the UK supply chain (Nissan UK, May 2014). Nissan has been the UK's largest car producer for more than 15 years, producing 510,272 cars in 2012 (Nissan UK, March 2013). As well as Nissan, the region is home to Smith Electric Vehicles—an international manufacturer of commercial EVs (though production has just stopped in the NE), AVID, Hyperdrive as well as Sevcon, who are a world leader in the design and manufacture of electric motor control systems for EVs—including components for the Renault Twizy.

NE England faces a significant challenge in reaching national CO_2 reduction targets. In 2009 all of its 12 local authorities recognised that these targets could not be met under existing policies. By 2050 the NE's emissions would be nearly 39 million tonnes of CO_2 equivalent and this must be reduced to 7.7 million tonnes in order to meet the Climate Change Act Obligations (AEA Energy & Environment 2008).

NE England was the first region in the UK and Europe to put itself firmly behind the commitment to reduce CO_2 emissions. In 2008, all 12 local authorities in the region signed up to the European Covenant of Mayors an ambitious initiative aiming to tackle climate change and reduce carbon emissions by 20 % by 2020 (The Convenant of Mayors, n.d.). Since that date a number of e-mobility initiatives have been introduced into the region to support this.

The challenges of emission reduction, coupled with the urban concentration of the NE population and the importance of Nissan to the regional economy, made the adoption of Electric Vehicles (EVs) an ideal opportunity for the region (Robinson et al. 2013).

2 Background to Electric Vehicle Adoption in the UK

With the emergence of EVs came the obvious challenge of developing a new fuelling (re- charging) infrastructure. Car manufacturers have traditionally had little involvement in the development of such infrastructure, and thus this challenge by default fell to national governments to develop.

In 2011 the UK government, through its newly established Office for Low Emission Vehicles (OLEV), issued its first low emission vehicle strategy entitled 'Making the Connection' (Office for Low Emission Vehicles 2011) which set out a framework for creating a recharging infrastructure for plug-in vehicles. The plugged-in places programme (PIP) (Office for Low Emission Vehicles 2010) was created to address this challenge, alongside a range of vehicle incentives. These include the plug-in car (PiCG) (Office for Low Emission Vehicles 2011b) and Van Grants (PiVG) (Office for Low Emission Vehicles 2012), the plugged-in fleets initiative (Energy Savings Trust 2013) and the low carbon vehicle public procurement programme (LCVPP) (Office for Low Emission Vehicles 2011a), which are all designed to encourage the uptake of low carbon vehicles in the UK. In September 2013, the UK Government released its latest strategy document for ultra-low emission vehicles (ULEV) entitled 'Driving the Future Today' (Office for Low Emission Vehicles 2013a). This represents the latest phase in the UK government's activity aimed at cutting transport emissions and making the UK highly attractive for ULEV inward investment. The ULEV strategy states that "a portfolio of solutions will be required to decarbonise road transport" and also that "It is not Government's role to identify and support specific technologies at this early stage". Therefore, the strategy includes a variety of low carbon technology solutions for transport (i.e. is expressly technology neutral), but it also recognises that electrification is likely to be central to many of these.

The results to date from the PIP programme were used to inform the ULEV strategy, including the fundamental requirements for creating a plug-in vehicle recharging scheme and the importance of public recharging facilities to EV buyers. The PIP programme has created (central government) subsidised EV recharging networks in eight regions of the UK with the specific objectives of investigating different recharging strategies, evaluating recharging technologies, investigating user behaviour and advancing common standards. Regional stakeholder funding was aided by match funding from OLEV, delivered through the eight regional PIP projects (Hodson and Marvin 2013; Lumsden 2012; Kotter and Shaw 2013).

The first report on the effectiveness of the UK's low emission vehicle strategy to date and its suite of incentives concerning low emission vehicles and infrastructure was published alongside the ULEV strategy in September 2013. This contained early analysis of the charge point usage data resulting from the UK's plugged-in places trials (Office for Low Emission Vehicles 2013) and an independent report on the impact of the government's EV incentives to date (Hutchins et al. 2013). The impact report highlights the continuing frustrations of EV users resulting from high EV costs and incompatibility between recharging schemes, and reinforces the requirement for public and domestic infrastructure with rapid chargers required to extend EV range. The PIP report also cautions that the early data is heavily influenced by the early mix of recharging equipment types with very few rapid chargers and relatively few home chargers in place. A report on the lessons learned (Office for Low Emission Vehicles 2013b) from the eight regional plugged-in places infrastructure trials was also released in parallel, to provide guidance to any organisation considering installing plug-in vehicle recharging facilities.

The EU's Clean Power for Transport policy (European Commission 2013) seeks to break Europe's dependence on oil for Transport, and therefore sets out a package of measures to facilitate the development of a single market for alternative fuels for transport in Europe. A proposal for a Directive on the Deployment of Alternative

Fuels Infrastructure (European Commission 2013a) was released in January 2013, which if adopted would require Member States to adopt national policy frameworks for the market development of alternative fuels and their infrastructure, including the provision of specific quantities of recharging points as defined by the EU. This is currently being commented upon by all Member States and the outcome will have a major effect on the volume and technologies deployed in future recharging infrastructure provision.

The EU's Directive on the Promotion of Clean and Energy Efficient Road Transport Vehicles (European Commission 2013b) aims at a broad market introduction of environmentally friendly vehicles. It requires that energy and environmental impacts linked to the operation of vehicles over their whole lifetime are taken into account in all purchases of road transport vehicles. The EU's Clean Vehicles Task force activities are aimed at enhancing the competitiveness of the automotive sector by encouraging investment in clean technologies, in order to increase their global market share. Its scope covers the lack of infrastructure and standards, long-term legislation and training requirements for low carbon powertrains (European Commission 2012).

3 An Overview of EV and Infrastructure Activity in the North East

The North East of England continues to be one of the most advanced UK regions for the demonstration of electric vehicles and associated technology. Over 1100 charge points (public, workplace and domestic) were installed in the region between 2010 and 2013 through the Government funded Plugged-in Places (PIP) regional trial scheme. In parallel, more than 50,000 electric vehicle journeys were studied by Newcastle University's Transport Operations Research Group (TORG) through the Switch EV trial (Hübner et al. 2013), one of the Technology Strategy Board (TSB)'s Low Carbon Vehicle Demonstrator projects, amongst other EV trials.

The region continues to take part in recharging infrastructure programmes across the UK, with a focus on EV rapid charger roll-out. One key example is the Rapid Charge Network project ('The Rapid Charge Network project'), which is creating a multi-standard rapid charger network, including the CHAdeMO, CCS and AC standards, to link the UK with the Irish e-Cars programme, funded by Nissan, BMW, Renault, VW and ESB, and match funded by the EU's TEN-T programme. Additionally, the region was instrumental in creating the UK's first open access, national, pay-as-you-go recharging network operator, Charge Your Car Limited in 2012.

In parallel with vehicle and recharging activity, the region continues to take part in leading energy research programmes, many of which are funded by the UK non-ministerial government department and independent national regulator Office of Gas and Electricity markets (Ofgem)'s Low Carbon Networks Fund ('Low Carbon

Networks Fund'). The Customer-Led Network Revolution, the UK's biggest smart grid project, is helping customers find ways to reduce their energy costs and carbon emissions in the years to come (www.networkrevolution.co.uk); 14,000 homes and businesses, mostly in the North East and Yorkshire, are involved in this innovative £54 million project, led by Northern Power Grid, the Distribution Network Operator (DNO) for the region. The Durham Energy Institute (DEI), part of Durham University, provides academic input to the trial design, hypotheses and analysis of this important project. In the 'My Electric Avenue' Project (www.myelectricavenue.info) led by Scottish and Southern Energy (SSE), EA Technology is developing an EV charge control system to balance out the charging cycles of EVs at times of network stress. Half of the trial's participants, most of whom are private individuals located in domestic clusters, are located in the NE.

One North East, the Regional Development Agency (RDA) for NE England initially led on the development of a regional Low Carbon Vehicle strategy from 2008 and funded relevant projects including NE PIP and Switch EV. However, in 2010 the UK Government announced that the English RDAs, which had been set up by the then Labour Government from 1998, were to be dissolved at the end of March 2012. The North East Local Enterprise Partnership (NE LEP) was then established from 2011 onwards to work on an agenda for growth, looking at where support for communities and businesses should be focused, in order to grow the area's economy and create more and better jobs (NE LEP, n.d.).

The region's academic and development institutions continue to be active in many European projects and there is a thriving culture of research and development by local Small and Medium-sized Enterprises (SMEs), many of which have received match funding support to develop new products and create or safeguard skilled jobs. Over the last 6 years more than £250 million has been invested in developing world-leading expertise in emerging technology and the region has gained an international reputation. The NE region is now considering more than just the car; it is working towards the integration of EVs and associated technologies into society, e-mobility.

Zero Carbon Futures was created in 2011 as a subsidiary of Gateshead College to focus on Low Carbon Vehicle (LCV) development, technology and associated skills. In parallel, Gateshead College also took on the staff and remaining obligations of the NE PIP project. Gateshead College and Nissan entered a Zero Emission agreement in 2012, and Zero Carbon Futures began its existing relationship with Nissan in Europe and Japan acting as a technology partner. These actions complement the college's existing role in supporting the development of the NE as a low carbon vehicle demonstrator region by providing the development and training facilities for key personnel throughout the LCV chain. Funded in part by the Government's Regional Growth Fund (www.gov.uk/understanding-the-regional-growth-fund), a number of initiatives have been put in place. This includes the NE's Collaborative Projects Fund (www.collaborativefund.org.uk), a grant programme supporting SMEs and academia to work together to develop products and services, and the development of The Future Technology Centre, a dedicated business centre for those working in the LCV sector.

4 Recharging Infrastructure Roll-Out in the North East

NE England was one of the first three UK regions to be awarded OLEV funds in March 2010, to establish a regional recharging infrastructure, in order to seed the uptake of low carbon vehicles. The aims of the pilot regional PIP projects were to feedback experience gained from creating and operating EV recharging infrastructure in diverse situations, into future policy decisions at both regional and national levels. This included the development of standards, evaluation of technologies, the harmonisation of local incentives, understanding users' behaviour and its impact upon the infrastructure. A further five regional PIP projects were subsequently added in 2011 and a national infrastructure funding scheme was then launched by OLEV in 2013, using the experience gained from the eight pilot projects.

The NE PIP project created a regionwide integrated network of over 700 publically accessible EV charge points between April 2010 and June 2013, spanning a region of 8,600 km^2, enabling EV journeys to become feasible across neighbouring regions in the UK, Scotland and Europe. As one of the UK's EV industry pioneers, the NE was the first area to create a regional network of 50 kW DC rapid charge points which enable EVs to be recharged to 80 % in just 30 min. 12 rapid chargers were installed by the NE PIP project at key staging points across the region. The public estate now combines 3, 7 and 22 kW AC and 50 kW DC rated charge points, with single and double outlets, located in public and workplace areas.

Potential hosts were attracted to have charge points installed on their property by various levels of grant incentives covering equipment and installation costs. More than 65 partners became charge point hosts, ranging from local authorities, academia and the NHS through large retail and business parks, and small businesses. Charge points have consequently been installed in public places, workplaces and domestic locations in accordance with demand from interested hosts. In exchange for this grant funding, each host provided free electricity and parking to EV drivers during the 3 year trial period, which ended in June 2013. The charge point hosts now own the NE's EV recharging infrastructure which forms the NE recharging estate, and have an obligation to maintain them. All publically accessible charge points were operated by a single network operator, Charge Your Car (CYC), which was funded to provide customer service and charge point information via a live availability map on a dedicated website. EV drivers joined the project's CYC membership scheme at a cost of £100 per year or £10 per month in order to receive access to this website and their own recharging records, as well as free electricity and free parking in public and workplaces. In addition to this public and workplace infrastructure, the project also installed over 400 domestic chargers with captive cables for EV drivers in the region to use in their own home environment.

Having addressed consumers' EV concerns by providing more charge points than anywhere in the UK outside London, by the end of 2013 the NE had the highest number of EVs in use per head of population in the UK. The visibility and public availability of charge points was seen by both local and national government as being key to increasing the number of EVs sold, and the Switch EV trial supported by awareness campaigns in the region helped to reassure potential EV customers. The Switch EV trial showed that, for 90 % of the time, North East EV drivers were no more than 5 km from a charge point and that 93 % of all trial participants' journeys were less than 40 km (Hübner et al. 2013). A core strength of the NE infrastructure is the number of motorists who now rely on it, and the frequency with which the charge points are used. By December 2013, NE public charge points had been used over 43,000 times, delivering 286,000 kWh of electricity and reducing EV drivers' fuel bills by £142,998 over a total of 1,119,664 miles. This equates to a saving of over 232 million grams of CO_2 versus comparable C class Internal Combustion Engine (ICE) vehicle (NE PIP Project 2013).

The NE PIP project was also one of the leaders in addressing many of the technical and process issues that installing EV infrastructure can bring. In the absence of specifications and standards, the NE team was amongst the first to deal with many standardisation issues that are now established for other schemes to follow. The project was instrumental in working with bodies such as the Institution of Engineering and Technology (IET) to set current standards for charge point installation. The experience of the NE PIP project was used to inform the IET Standards' Code of Practice for Electric Vehicle charging equipment installation (Institution of Engineering and Technology 2012). The NE's experience was also used as a case study within a report written by Future Transport Systems, also a NE based e-mobility business, and published by the IET to give local authorities advice on how to successfully implement EV infrastructure (Lumsden 2012).

The NE PIP project also promoted electric vehicle ownership to both the general public and the business community in the NE, through a range of marketing channels and events. This work led to the region being named as the place most likely to adopt EVs in the UK by the Green Car website in 2012. During 2012, the region played host to the first GreenFleet North East event which brought together over 100 regional businesses to test drive electric vehicles on the regional test track. Delegates heard from a range of speakers about the benefits of incorporating EVs into their business. The regional awareness campaign '*Why Not Electric?*' was promoted through a range of marketing channels, including advertisements, bus backs, online, direct mail and PR. The website www.whynotelectric.com contained a wealth of information about EVs including an annual fuel calculator, to encourage people to think about the savings possible by driving electric. A competition to win a Renault Zoe for a year also encouraged people to take test drives and over 150 test drives were taken during the campaign at regional dealerships offering Mitsubishi, Nissan, Renault, Peugeot and Vauxhall EVs.

Connecting drivers to the recharging network was also an important achievement for the project with the launch of the UK's first back office system in 2010, allowing NE drivers to sign up for an annual membership which provided unlimited use of all NE public charge points. The first generation back office provided a wealth of data which was analysed regularly, helping to understand usage patterns, driver behaviour and informing partners and future partners on future provision.

It was essential that the public network should remain sustainable after the project's funding ceased in 2013, so the project took the bold step during 2012 to procure a joint venture partner and Elektromotive, Europe's leading provider of electric vehicle charge points, was appointed following a detailed procurement exercise. The resulting commercial venture, Charge Your Car Limited (CYC) (www.chargeyourcar.org.uk), created a second generation solution featuring the UK's largest pay-as-you-go, open source network of public access charge points, which has since expanded to become a successful national network operator. EV drivers can now gain ready access to charge points across the UK, without the need to pay a monthly subscription. Using the CYC tools including mobile device Apps, interactive voice response (IVR) services and radio-frequency identification (RFID) cards, EV drivers can locate, start and pay for their charge. The second generation system offers charge point owners the opportunity to generate revenue and manage and promote their charge points to drivers across the UK. The first step in this roll-out began in August 2012 when Transport for Scotland announced that all 500 of its EV charge points would be operated by the CYC network and subsequently Manchester, Northumberland, York, Bristol and Sussex have taken up CYC's services, making it the largest UK operator of EV charge points, 1600 and growing daily. CYC Ltd is also working with BMW to support the Charge Now Card which offers BMW i3 drivers a total E-Car solution. The data provided by CYC's second generation system is now being used by charge point owners across the UK to further support and develop CP installation and use strategy.

The NE PIP project is now known internationally for its expertise and the outcomes and achievements have been disseminated widely at industry events and in publications. The team has also advised many other UK regions on the learning outcomes of the project and have also hosted visits from delegations from countries such as India, Malaysia, Australia, China and Estonia in order to share its experience. In 2013 the NE PIP project was awarded the *Low Carbon Road Transport Initiative of the Year* at the LowCVP Champions Awards. The awards celebrate outstanding and innovative practice in accelerating the shift to lower carbon vehicles and fuels and reducing road transport emissions.

In addition, key suppliers such as Newcastle City Council Technical Services division have extended their customer base beyond the regional boundaries, now providing installation and maintenance services for EV infrastructure across the north of England and in Scotland.

5 EV Trials in the North East

The second key element of the North East's electric vehicle activity was centred on 44 EVs trialled under the TSB's Ultra-low carbon vehicle demonstrator (ULCVD) programme. The Switch EV trial, led by North East-based company, Future Transport Systems, brought together a consortium of vehicle manufacturers, data collection experts and project managers to deliver 44 new and innovative full-electric production vehicles onto the roads of the NE. The NE's electric vehicle trials were unique in that drivers were allowed a comprehensive choice of recharging infrastructure (public, work and domestic) provided by the NE PIP project. This allowed behaviour to be monitored through quantitative data on vehicle and charging equipment use.

The Switch EV trial ran from November 2010 until May 2013 and saw 44 full EVs cover over 400,000 miles across the NE, which accounted for over 90,000 journeys and over 19,000 recharging events (Hübner et al. 2013).

The vehicles were fitted with data loggers that provide a range of driving and vehicle performance data as well as GPS and a time stamp. These data points were collected and analysed by Newcastle University's Transport Operations Research (TORG) group. In parallel, driver attitudes towards EVs were gathered through questionnaires and focus groups. These two sets of data were then correlated to explore trends, changes in driving and recharging behaviour and attitudes to electric vehicles, charging and key issues such as cost and 'range anxiety'. Most of the Switch EV drivers were also members of the CYC scheme and used the recharging infrastructure created by the NE PIP project.

The vehicles used in the Switch EV trial were mostly commercially available vehicles, including Nissan LEAF, Peugeot iOn, Avid Cue-V, Liberty electric cars eRange, and the Smith Electric Vehicle Edison Minibus. Trial participants were a mixture of companies and local authorities who used the vehicles as part of their fleet as a pool vehicle or for the sole use for one individual. A small number of cars were also leased to private individuals.

This structure of EV, and infrastructure trials running in parallel, allowed behaviour to be monitored through quantitative data on vehicle and recharging equipment use. The data collected and analysed show clear patterns of behaviour (Hübner et al. 2013; also see Robinson et al. 2013). However, these patterns were in part impacted by external influences, such as free parking with the CYC membership, and further analysis of EV driving and recharging behaviour continues to be performed now that the CYC membership and payment system are changing as subsidies are removed.

6 Lessons Learnt from the Introduction of an EV Infrastructure

A number of valuable lessons were learnt in the process of delivering such a ground breaking project, and these are summarised below.

The need to start deployment as quickly as possible in 2010 in support of the Nissan LEAF launch and the Switch EV project, as well as the UK government pressure to spend, meant that the project had limited suppliers and products to call upon. The UK EV charging market was in a very early stage of development at this time and this limited the success of the public procurement exercises carried out to create frameworks for supply. Models, specifications and prices have moved on substantially since 2010 but the project was unable to take advantage of these developments due to timing and resource constraints. There is therefore a need to be able to keep procurement frameworks agile, with the ability to add new products and take advantage of technical enhancements as markets evolve. The public procurement exercises, although lengthy and resource intensive, were detailed and therefore provided some degree of control over product specifications, costs and delivery timing. However, supplier constraints, in light of the boost the PIP programme provided to charge point manufacturers, also caused problems in supply. Therefore, the need for a physical demonstration of capability has been suggested as a valuable addition to the process used in the NE. The frameworks were set up so that other PIP regions and public bodies could use them, but this had limited benefit until 2013 when large numbers of local authorities began procuring using new OLEV funding.

The nature of the NE PIP project consortium was unique and complex. In order for the project to succeed, the project team needed to engage with over 100 different partners. Many site operators did not own the land in which the charge points were to be installed, so legal agreements had to be put in place with the land owners. The legal support required in developing these formal agreements to engage with site hosts was an essential but expensive resource requirement. The aim was to establish a 'one size fits all" solution and also provide a sound basis for other PIP projects to start from. However, the effort expended did not prevent individual site hosts from demanding their own conditions, resulting in many lengthy negotiations and some exceptions being adopted on individual bases. In some cases, these negotiations could not even be completed within the 3 years of the project, for example where large national firms were involved with little regional interest or control. The time taken and effort required in stakeholder engagement proved to be one of the biggest barriers to the project and should not be underestimated by future providers. Interest, motivation and engagement differed greatly between stakeholders. Finding a good contact person with responsibility, influence and understanding with which to engage was a very difficult task, even with significant regional knowledge and contacts in place.

The project used a large regional installation supplier with relevant experience of public infrastructure installation, maintenance and strategic importance. However, the supplier found the administrative burden of such large-scale delivery to be a challenge, since it was essential to track every stage from survey to commissioning across a large number of sites, and report frequently. The additional time taken to engage site hosts also had a major impact on the installer's resources as installation timeframes were squeezed continually. In future, detailed Service Level Agreements would be considered from the outset to protect all parties.

Both the installer and the project team had good relationships with the regional Distribution Network Operator (DNO). However, this did not result in any significant benefits in the time taken for new electrical connections to be established. Indeed, the availability of electrical capacity in a suitable location for a charge point became one of the major restrictions to the project, particularly for rapid charger sites.

The project benefitted from key staff secondments from interested regional organisations such as Nissan and the National Renewable Energy Centre, Narec (Narec, no date). The knowledge and experience gained by all parties as a result of their involvement in the NE PIP project continues to provide benefit going forwards.

Due to the start time of the NE PIP project, there was minimal demonstration available of long-term charge point capability in the field, and of certain desirable features such as access methods, communications and payment methods. As a result of the early technology procured, the project has experienced continuing reliability problems, and a lack of integration capability still affects the NE network today. The project found that, the more immature the technology, the harder it was to integrate it into a long-term, sustainable future system. Some equipment had to be upgraded to meet evolving standards, placing additional pressures on both budget and time, but some cannot be upgraded. Unfortunately, budget eligibility rules prevented the project from replacing old equipment with new, higher specification units. The project's public funders were unwilling to allow the evolving standard to be procured from the outset of PIP, in advance of standards being formally set by the appropriate bodies (Institution of Engineering and Technology, IET). As with many new technology projects, Intellectual Property issues also caused concerns with some suppliers.

Technology readiness also proved a challenge in some novel areas of the project, such as the proposed induction recharging trial. Suppliers were unwilling to quote for real world trials in the region in 2011.

Continuing changes in standards have also caused some challenges for the North East infrastructure. The emergence of the current type 2 connector standard for EV charging came from the demands of the EV manufacturers and was therefore championed by OLEV through the PIP projects from 2011. This led to a large-scale programme of updates to older 3 pin plug connector charge points that had been installed at the start of the programme. However, many of these manufacturers continue to supply EVs into the UK market without this standard deployed. This causes uncertainty for EV dealers and potential purchasers, and frustration and additional cost for EV drivers.

The situation has recently been further complicated by different rapid charging protocols which have been introduced by international car manufacturers. At the beginning of the project, all EVs available on the market with rapid charging capability utilised the CHAdeMO protocol and connector for rapid charging, and this was therefore the standard adopted by the NE PIP. However, this changed in 2013 with the introduction of EVs mainly by BMW and VW using the Combined Charging Station (CCS) protocol. This introduction has meant that the existing rapid charge points installed as part of the project are not suitable for new EV models emerging from those manufacturers.

The NE PIP's technical involvement with the IET standards bodies regarding the development of EV charging standards proved beneficial, for the PIP programme nationally and for charge point manufacturers. However, this cost the NE PIP project a significant amount in staff costs for highly experienced staff, and no funding was available from OLEV to contribute to the national benefit that was a direct result. It is suggested that a technical representative from OLEV, or a nominated body, joins such essential bodies in future.

Inevitably circumstances change in evolving technology projects. However, the entire environment in which the NE PIP project was created changed in 2010, with the UK Government's announcement that the Regional Development Agencies (RDAs) were to be dissolved at the end of March 2012. This presented a major challenge for the continuation and completion of the NE PIP project which was operated by One North East, the RDA for NE England. Gateshead College, with its interests in training for new technology topics, particularly in the automotive sector, stepped forward to take on the staff and remaining obligations of the project in December 2011.

Whilst the OLEV public funding was essential to the creation of volume EV infrastructure within the region, this funding only contributed towards capital costs and was based on the quantity of charge points installed. This funding did not cover the true resource costs of creating, and then operating, such an innovative network of infrastructure going forwards. Nor does it provide proper recognition and recompense for essential research and development activities. Unfortunately, this has resulted in a lack of recognition for the wider achievements made against the original aims of the programme by politicians, leading to fears for future policy decisions.

The NE's early-to-market EV recharging estate was created and operated under public subsidy, in order to seed the marketplace for further EV and recharging equipment adoption. Until June 2013 the NE PIP project paid all system operating costs and charge point owners were therefore shielded from the true costs of operation. The result is that NE charge point owners have adopted recharging infrastructure without being fully aware of the costs of operation and maintenance.

The charge point owner is now at the centre of a system of infrastructure which is seen as essential by many stakeholders, such as Government, local authorities, environmental bodies and EV drivers, and they are expected to at least maintain the current level of supply. However, the expectations and assumptions based upon which NE hosts made the decision to adopt charge points have not materialised. The actual uptake of EVs and therefore demand for recharging equipment has been both lower and slower than predictions made in 2010. The early forecasts of likely EV sales have not been achieved to date, which affects the early business models envisaged for on-going operation and increasing provision. The consequence is that NE charge point owners are continuing to financially support the costs of operation of recharging infrastructure and cannot foresee an acceptable conventional business model.

7 The Future for EV Recharging in the UK

One of the greatest opportunities presented to the UK Government today is the installation of a national network of rapid charge points. NE was a leader in the development of a regional rapid charge network, with 12 rapid chargers placed on key spine roads across the region. This has proved key to EV drivers with the rapid charge points being used extensively. Although making up only 2 % of the estate, the rapid charge points have delivered 15 % of the transactions and 16 % of the energy. Conversely, the public standard charge points which make up 56 % of the estate only delivered 34 % of the energy (The NE PIP project, December 2013).

The challenge is now to replicate what has been achieved in the NE across the rest of the UK which will be made possible through the UK Government's continuation of their Plugged-in Places programme, as well as the European Union funded 'Rapid Charge Network' project. These networks stand to benefit from the emergence of new multi-standard rapid charge points which are available to charge any standardised electric vehicle regardless of manufacturer or rapid charging protocol. This technology was not available in the UK at the time of the NE PIP project.

Looking towards the future, the potential for induction charging is one still being investigated. This is an area that is still in its infancy and is currently subject to a lack of agreed standards, protocols and commercial viability.

8 Wider Opportunities for the NE Presented by EV Technology

The growth in this emerging industry has led to exceptional facilities for training in LCV development with a range of academic, research, Continued Professional Development (CPD) and training opportunities emerging in the region, ranging from CPD certificate to National Vocational Qualification (NVQ) level up to degree, taught postgraduate to Ph.D. doctoral level. This is providing the region with a skilled workforce able to respond to the new demand created by the growth in production and adoption of LCV and associated technologies. Gateshead College, of which Zero Carbon Futures is a subsidiary, are a European leader in this area. Since 1984, the College has worked with Nissan to provide a wide and diverse range of training to both their new and existing staff aimed at improving productivity. With the announcement of the Nissan LEAF manufacture, Gateshead College developed, with Nissan, a full curriculum on battery development, handling and maintenance. Their expertise has now led to the development of training in all areas of EV including driver training, repair and maintenance, domestic charge point installation and emergency and first responder training. In 2011, Gateshead College opened its £9.8 million Skills Academy for Sustainable Manufacturing and Innovation. Based directly opposite the Nissan plant, the Academy trains Nissan apprenticeships in the state-of-the-art environment.

The opportunities presented by EV battery life are a key focus for the region. With the arrival of advanced battery technology to power pure EV and Hybrid EV there will come a time when a battery has deteriorated to a level where it is not fit for purpose in a passenger car yet still has a value. 4R projects—Reuse, Remake, Reconfigure and Recycle—investigate the second life use of the Li-ion battery. Extending the life of the battery should reduce the overall cost of EVs and increase the environmental impact.

The NE PIP project commissioned a feasibility study conducted by Northumbria University to better understand the EV battery life cycle and the opportunities for second life EV battery use (Lacey et al. 2013a). This is connected with factors that influence the state of health of the battery (Lacey et al. 2013b).

The study was subsequently developed to explore the possibility of harnessing power in EV batteries as energy storage in a domestic or workplace environment, either for recharging vehicles, providing an emergency power source or feeding power back into the grid with smart and controlled/optimised ways of doing this bidirectionally (Lacey et al. 2013c; Jiang et al. 2013). A demonstrator unit has now been incorporated into the North East's Future Technology Centre, a national centre for low carbon vehicle technologies, and the results will be assessed to help establish the true potential of this technology.

Another area being looked into, and which was funded by CYC, was an investigation and the development of a modelling tool to investigate the effect of electrical vehicle charging on low voltage networks (Lacey et al. 2013c).

The introduction of LCV onto the UK roads is also having a direct impact on homes and cities, which is in turn leading to future research opportunities in these areas. By using an alternative fuel (electricity) to power transport in the future, it is inevitable that major countries and cities are reappraising their power generation strategies and looking to see what potential there is for the integration of electric vehicles into society on a large-scale.

Vehicle to Home (V2H) and Vehicle to Business (V2B) technologies are being heavily investigated in Japan (Nissan Motor Company 2014). Technologies exist to allow a driver to effectively run their property using the energy stored within their EV, diminishing the impact of power cuts and potentially opening up the possibility of avoiding peak energy tariffs. Whether or not these technologies will change the way cities are run in the future has yet to be seen but they will potentially offer further research and commercial opportunities for NE companies.

The NE has also seen a number of advancements in fleet uptake. Since 2008, the LCVPP (Office for Low Emission Vehicles 2011a) has been used by the region's local authorities to obtain subsidies towards the additional costs of procuring LCVs for their fleets, therefore using the public sector's purchasing power to accelerate market introduction of LCVs. For example, Newcastle City Council now has 22 EVs in their fleet, and Stockton Council is about to introduce their 18th electric vehicle in addition to their two electric pick-up trucks. After successful trials as part of Switch EV, the emergency services including North East Fire Service and Northumbria Police have also taken up government grants to introduce EVs into their fleets. The North East Ambulance NHS Foundation Trust is now beginning collaboration with Newcastle University's TORG to look at ways to save both carbon through nationally pioneering work on its fleet and moving to electric.

NE England's not-for-profit car club, Co-wheels (Co-wheels car club, n.d.), is based in Durham and now operates across the UK, working in partnership with local authorities, business and communities. Co-wheels has also purchased a number of EVs using LCVPP funding which are now available throughout the region for hire. Most recently Co-wheels has been working with Transition

Tynedale, a sustainability charity in rural Northumberland, to introduce two hybrid EVs into the community.

Buses and taxis in NE England have also made the transition to hydrid and full-EV technology, further raising the profile of electric forms of transport in the region. In 2011, a £5 million fleet of greener hybrid electric buses were introduced into Newcastle by Stagecoach North East, using £2.2 million in support funding from the Government's Green Bus Fund ('Background to the Green Bus Fund') to purchase a fleet of 26 new Enviro400H Euro5 double-decker buses. NE taxis are also embracing the benefits of EVs. Phoenix Taxis (Phoenix taxis, n.d.) began introducing EVs into their fleet in early 2013, supported by the installation of charge points at their head office using funding from the NE PIP project. Phoenix are now planning to introduce 30 electric vehicles onto the road by summer 2014, supported by a significant increase in rapid charging infrastructure in the area which will enable multiple deployment.

9 Future Opportunities

In summer 2013, an economic impact study was commissioned by the NE LEP to investigate the potential opportunity that the sector can have on the economic development agenda in the region (E4Tech 2013). The overriding question that this study set out to answer is: 'How can the North East build a basis for sustainable competitive advantage as an LCV innovation region?' The report responded by laying out the specific areas in the overall LCV sector that may constitute an opportunity for the NE, and proposed specific focus areas, outlined current short-comings and an indicative plan that may lead the region to exploit the potential.

The study concluded that while established powerhouse regions in Germany and the UK Midlands will continue to dominate the efficiency of internal combustion engines, the NE could capitalise on its emerging EV manufacturing status by relying on its skills base, automotive productivity record and ease of doing business. While Nissan provides a strong anchor for such manufacturing activity, another Original Equipment Maker (OEM) and a deeper LCV supply chain would significantly broaden the proposition.

Wider e-mobility research, demonstration and deployment may also constitute opportunities in testing and adapting solutions to the UK context: the development of business models and solutions, implementation of intelligent transport infra-structure for the regional Urban Traffic Monitoring Centre (UTMC centre), and the development and testing of hydrogen infrastructure roll-out. Indicatively, the report concluded that success in these activities could provide 4,000 jobs and £1 billion GVA/year plus many other benefits-associated broader policy objectives ranging from air quality and affordability of e-mobility to make the NE an attractive place to live and work.

However, the report acknowledged that a regional strategy based upon EV and Fuel Cell EV prospects also carries risks: despite good recent progress EV sales have yet to achieve mass market volumes; commercial models for infrastructure investment have to be created; battery costs and performance need to be improved significantly; the Midlands region may capitalise on its capabilities in hybrids and enter the EV market; and technologies such as automotive fuel cells still need a step change improvement over technology in use today.

10 Conclusions

NE England is home to the UK's largest car plant and, with the automotive industry facing its biggest transformation in many years and Nissan leading the way with the launch of the first mass-produced EV, this provided an opportunity not to be missed by the region. The region's e-mobility journey therefore began as early as 2008 with the region's ambition to become home to one of only three Nissan LEAF production sites worldwide.

The region first recognised that, in order to secure EV production, a regional network of charging infrastructure would be required, making sure the region was geared up for this transition in driving behaviour. Developing the charging infrastructure was essential for two reasons: first, to provide EV drivers with the necessary network to travel widely around the region and second, for visibility and awareness raising purposes. The lack of recharging infrastructure is often stated as one of the biggest barriers to EVs, so the visibility of charge points on street and in car parks across the region has helped to break down this barrier. A recent market research poll of the general public in North East England (Other Lines of Enquiry 2012) showed that over 60 % of people knew about the recharging network and 46 % of people said that there were charge points near them.

Establishing the recharging infrastructure was not a simple task, and a variety of challenges faced the NE PIP project during its life and some continue to cause complications today. Some of these challenges were unique to the NE region and the changing political environment. However, there are many findings which are relevant to other cities, regions and countries today. Issues including finding suitable locations and charge point hosts (see also Namdeo et al. 2013) changes in international standards and protocols, procurement and supplier constraints and ensuring a legacy strategy is put in place to keep charge points operational and maintained after subsidies end, are all areas that will continue to affect the roll-out of EV recharging infrastructure throughout the world.

Nissan LEAF production has had an enormous impact on the region. It has resulted in the creation of an EV-ready environment. Vehicle trials, education, training and awareness programmes and events, fleet events, public and private sector initiatives to encourage electric driving, taxi firms and car sharing schemes have all benefited from the environment created.

The region's vision has continued to widen to cover a number of infrastructure, R&D and capital projects in the area of e-mobility and its associated opportunities for Smart Cities. Capital investments, support for manufacturing companies large and small, collaborative projects and a programme of skills development in this area are just some of the projects that have been developed in the region. The opening of The Future Technology Centre in Sunderland in June 2014 is the next piece of the jigsaw, creating a national home for the demonstration of LCV technologies both now and in the future.

References

AEA Energy & Environment. (2008). *North East Assembly*. Cheshire: Climate North East England greenhouse gas emissions baselines and trajectories study.

Background to the green bus fund. Retrieved May 25, 2014, from https://www.gov.uk/government/collections/background-to-the-green-bus-fund.

Co-wheels car club. Retrieved May 24, 2014, from http://www.co-wheels.org.uk.

Energy Savings Trust. (2013). Plugged in fleets initiative. Retrieved from http://www.energysavingtrust.org.uk/Organisations/Transport/Products-and-services/Fleet-advice/Plugged-in-Fleets-Initiative-100.

E4Tech. (2013) *Low carbon vehicles in the North East of England*. Final report for the North East Local Enterprise Partnership, September 2013. London/Newcastle upon Tyne.

European Commission. (2012). Implementation of the 2012 industrial policy communication—state of play of the task forces. Brussels.

European Commission. (2013a, Jan 24). Proposal for a directive of the European parliament and of the council on the deployment of alternative fuels infrastructure. Brussels.

European Commission. (2013b, Feb 28). Guidelines on financial incentives for clean and energy efficient vehicles. Brussels.

European Commission. (2013). Clean power for transport. Retrieved from http://ec.europa.eu/transport/themes/urban/cpt/index_en.htm.

Hodson, M., & Marvin, S. (2013). *Low carbon nation?* London: Abingdon.

Hübner, Y., Blythe, P., Hill, G., Neaimeh, M., Austin, J., Gray, L., Herron, C., & Wardle, J. (2013). 49,999 electric car journeys and counting. Electric Vehicle Congress 27 (EVS27), Barcelona, Nov 2013.

Hutchins, R., Delmonte, E., Stannard, J., Evans, L. & Bussell, S. (2013). Assessing the role of the plug-in car grant and plugged-in places scheme in electric vehicle take-up. In Department for Transport (DfT) (Ed), Queen's Printer and Controller of Her Majesty's Stationary Office.

Institution of Engineering and Technology. (2012). IET code of practice for electric vehicle charging equipment installation. London: Developed by the ISL Committee 1.1 EV Charging Equipment Installation.

Jiang, J., Putrus, G., Gao, Z., Conti, C., McDonald, S., & Lacey, G. (2013). Development of a decentralized smart charge controller for electric vehicles. *Electrical Power and Energy Systems 61*(2014), 355–370.

Kotter, R., & Shaw, S. (2013). *Micro to Macro Investigation*. Project Report for NSR E-mobility Network. Retrieved April 2013, from http://e-mobility-nsr.eu/fileadmin/user_upload/downloads/info-pool/E-mobility_3.6._Main_Report_April_2013.pdf.

Lacey, G., Putrus, G., & Salim, A. (2013a). The Use of Second Life Batteries for Grid Support. In *Proceedings of EuroCon 2013*, 1–4 July 2013, Zagreb, Croatia. Retrieved from http://ieeexplore.ieee.org/stamp/stamp.jsp?arnumber=06625141.

Lacey, G., Putrus G., Jiang, T., & Kotter, R. (2013b). The Effect of Cycling on the State of Health of the Electric Vehicle Battery, In *Proceedings of 2013 Universities Power Engineering Conference (UPEC)*. Dublin Institute of Technology: Dublin. Retrieved from http://e-mobility-nsr.eu/fileadmin/user_upload/downloads/info-pool/Northumbria2013_CyclingEffectsOnEV battery.pdf.

Lacey, G., Putrus G., Bentley, E., Johnston, D., Walker, S., & Jiang, T. (2013c). A modelling tool to investigate the effect of electric vehicle charging on low voltage networks. In *Proceedings of Electric Vehicle Congress 27 (EVS27)*, Nov 2013. Barcelona, Spain. Retrieved from http://e-mobility-nsr.eu/fileadmin/user_upload/downloads/info-pool/Barcelona_EVS27_GL_Conference_Proceedings_EVS27.pdf.

Low Carbon Networks Fund. Retrieved May 20, 2014, from https://www.ofgem.gov.uk/electricity/distribution-networks/network-innovation/low-carbon-networks-fund.

Lumsden, M. (2012). *Successfully implementing a plug-in electric vehicle infrastructure; a technical roadmap for local authorities and their strategic partners*. Institution of Engineering and Technology: London.

Namdeo, A., Tiwary, A., & Dziurla, R. (2013) Spatial planning of public charging points using multi-dimensional analysis of early adopters of electric vehicles for a city region. *Technological Forecasting & Social Change,* in press. Retrieved from http://www.sciencedirect.com/science/article/pii/S0040162513002175.

Narec (The National renewable energy centre). Retrieved May 25, 2014, from http://www.narec.co.uk/.

Nissan Motor Company, The 2nd V2X Forum, Copenhagen, May 2014.

North East Local Enterprise partnership (NELEP). Retrieved May 25, 2014, from http://www.nelep.co.uk/.

Office for Low Emission Vehicles. (2010). Plugged in places. Retrieved from https://www.gov.uk/government/publications/plugged-in-places.

Office for Low Emission Vehicles. (2011a). Low carbon vehicle public procurement. Retrieved from https://www.gov.uk/government/publications/the-low-carbon-vehicle-public-procurement-programme-support-for-low-carbon-vans.

Office for Low Emission Vehicles. (2011b). Plug-in car grant. Retrieved from https://www.gov.uk/government/publications/plug-in-car-grant.

Office for Low Emission Vehicles. (2012). Plug-in van grant. Retrieved from https://www.gov.uk/government/publications/plug-in-van-grant.

Office for Low Emission Vehicles. (2013a). Driving the future today—a strategy for ultra low emission vehicles in the UK. In Olev ed. London.

Office for Low Emission Vehicles. (2013b). Lessons learnt from the plugged-in places project. In Olev ed. DfT.

Office for Low Emission Vehicles. (Sept 2013). High level analysis of the plugged-in places chargepoint usage data. In Olev ed. London.

Office for Low Emission Vehicles. (June 2011). Making the connection, the plug-in vehicle infrastructure strategy. In Olev ed. London.

Office for National Statistics (various dates). Retrieved from http://www.ons.gov.uk/ons/index.html.

Other Lines of Enquiry. (Sept 2012). Charge your car advertising awareness & effectiveness study. Location?

Phoenix taxis. Retrieved May 25, 2014, from http://www.phoenixtaxis.net/.

Robinson, A. P., Blythe, P. T., Bell, M. C., Hübner, Y., & Hill, G. A. (2013). Analysis of electric vehicle driver recharging demand profiles and subsequent impacts on the carbon content of electric vehicle trips. *Energy Policy, 61*(2013), 337–348.

The Covenant of Mayors. Retrieved May 25, 2014, from http://www.covenantofmayors.eu/about/covenant-of-mayors_en.html.

The rapid charge network project. Retrieved May 20, 2014, from http://rapidchargenetwork.com/.

Stuttgart Region—From E-Mobility Pilot Projects to Showcase Region

Rolf Reiner and Holger Haas

Abstract In 2009, the Stuttgart Region of Baden-Württemberg in Germany was awarded funding from the Federal Ministry of Transport, Building and Urban Development (BMVBS) as one of eight "E-Mobility Pilot Regions" and started to implement several projects—from 2 wheelers to buses, including the development of full electrical vehicles (e.g. Vito E-Cell). Meanwhile, Stuttgart Region's LivingLab BW^e mobil in April 2012 became one of four national "E-Mobility showcase regions" for 3 years, and is also hosting the "Leading Edge Cluster Electric Mobility South-West". The LivingLab BW^e mobil started 40 projects in and around the Stuttgart Region where the scale and scope is even broader than in the Pilot Region activities, including field tests of e-mobility business models. The more than 100 partners from business, science and public authorities are testing electric mobility in actual practice. In their activities the projects concentrate on the Stuttgart Region (Württemberg) and the city of Karlsruhe (Baden), and promote the visibility of their work right up to the international level. The project aims to put around 2000 electric vehicles on the road by 2015, and to install more than 1000 charging points. This paper highlights three select case studies on the projects Get eReady, Car2Go full electric and Stuttgart Services.

Keywords E-mobility · Charging infrastructure · Carsharing · Mobility services

R. Reiner (✉)
Innovationhouse Deutschland GmbH, Friedrichstr. 10, 70174 Stuttgart, Germany
e-mail: rolf.reiner@innovationhouse.de

H. Haas
Wirtschaftsförderung Region Stuttgart GmbH, Friedrichstr. 10, 70174 Stuttgart, Germany
e-mail: holger.haas@region-stuttgart.de

W. Leal Filho and R. Kotter (eds.), *E-Mobility in Europe*,
Green Energy and Technology, DOI 10.1007/978-3-319-13194-8_10

1 Introduction: National Policies for E-Mobility in Germany

During the economic crisis of 2007/2008 Germany decided to establish a national support scheme for electric mobility. Based on the traditional strengths of the national carmakers, this smart specialization strategy (BMWi et al. 2009) aims at closing the gap to leading electric mobility nations and placing the German automotive industry on top of global development. Germany should not only become a lead market for electric mobility but a lead provider of electric vehicles and sustainable mobility solutions.

The German National Platform for Electric Mobility (Nationale Plattform Elektromobilität—NPE) was established in 2009 by the Federal government as a think tank and policy advisory group. It brings together representatives of industry, the research and political communities, the trade unions and civil society in Germany. Its members have agreed on a systemic, market-driven approach characterised by a readiness to deploy a variety of different technologies in order to achieve the goal of making Germany the world's leading supplier and market for electric mobility by 2020. The NPE produced a general roadmap for its systemic approach during the market ramp-up phase, which has been published in September 2013 (Heidenreich et al. 2013).

In parallel to the establishment of the NPE on 12 January 2009, half a billion Euro programme for research, technology development and innovation (RTDI) was launched (Economic Stimulus Package II), including Information and Communication Technology (ICT) for electric mobility (IKT II) and eight pilot regions of e-mobility schemes. This programme to promote electric mobility came to an end in October 2011 (Tenkhoff et al. 2011). As a next step, four showcase regions were selected in April 2012 in a national competition and will be supported with an overall Federal funding of 180 Mio Euro over 3 years to demonstrate everyday usability of electric vehicles and prepare for the market rollout of electric mobility solutions (including the LivingLab BW^{e} mobil in Baden-Württemberg). The showcase regions offer potential users and the general public in Germany the opportunity to gain first-hand experience of the system electric mobility.

In the summer of 2009 the German Federal Government published the "German Federal Government's National Electromobility Development Plan" (German Federal Government 2009). The goal of the National Development Plan for Electric Mobility is to advance research and development, market preparation for and introduction of battery powered vehicles in Germany. The measures adopted by the Federal Government in its economic stimulus package II served as first catalysts. Germany's National Development Plan for Electric Mobility is based on a strong and broad foundation. In order to advance more rapidly on the path described above, the public sector and above all industry need to step up their efforts. The BDEW (Federal Association of the Energy and Water Sectors), VDA (German Automobile Industry Association) and ZVEI (German Electrical and Electronic

Manufacturers' Association) national industry associations acknowledged their responsibilities in 2009 (BDEW et al. 2009).

As a prerequisite for the large-scale introduction of electric vehicles in the years to come, appropriate political, regulatory, technical and infrastructural frameworks have to be created (Peters et al. 2012). For example, open European standards, which will also serve as ambitious global benchmarks, are necessary to ensure interoperability, safety and acceptance (Reiner et al. 2010). In the framework of the National Development Plan for Electric Mobility the Federal Government will contribute to this process until 2020. In addition to regulatory measures to support in particular progress in the areas of battery technology, grid integration and market preparation and introduction (Reiner et al. 2010), the launch of a market incentive programme and its form are currently being reviewed by the four Federal Ministries (BMWi—economy and energy, BMVI—transport and infrastructure, BMBF—research and innovation, and BMUB—natural and build environment) jointly responsible for electric mobility in Germany.

2 The Story: A Regional Roadmap Towards Sustainable Mobility

The Stuttgart Region is home to one of the most comprehensive automotive clusters in Europe with 190,000 employees in car manufacturing and mobility industries. Car manufacturers (OEM) and suppliers dominate the economic structure and employment in Stuttgart Region. The automotive industry plays a key role in the economy of the Stuttgart Region, accounting for a sixth of all local jobs. With respect to geographic concentration, density of companies and specialisation, Stuttgart Region is the leading automotive region in Germany. Besides Daimler AG and Porsche AG approximately 400 suppliers are located in the region, with companies of different sizes ranging from small to medium-sized enterprises (SMEs) to global players like Bosch in the field of electronics, from small automotive-design offices to Bertrandt AG, a leading engineering company. In contrast to other automotive regions, many of the suppliers still are independent companies, which are not part of any corporate group. Robert Bosch GmbH, Mahle GmbH, Behr GmbH & Co. KG, Eberspächer GmbH & Co. KG and Mann+Hummel GmbH are all headquartered in the Stuttgart Region, and are listed in the group of "Top-100-Automotive-Suppliers 2010". In 2010/11, the automotive industry produces with 11.4 % of all employees in the German automotive industry a turnover of 39,452 million € (12.4 % of national turnover in the sector), of which 28,509 million € (14.3 % of Germany) are exports (Dispan et al. 2011).

The Stuttgart Region is one of the hot spots of electromobility related RDTI projects in Germany (Wintjes et al. 2013). The main funding sources are Federal programmes, followed by funding provided by the State of Baden-Württemberg as well as Stuttgart Region and the European Commission. The funding resources are deployed under the monitoring of the Stuttgart Region Economic Development Corporation (Wirtschaftsförderung Region Stuttgart GmbH), which is overseen by the regional body (Verband Region Stuttgart), where the project ideas, stakeholder groups and the necessary areas of action are matched with the appropriate constellation. The main projects in the Stuttgart Region belong to one of the three pillars (i) "leading edge cluster (basic research)", (ii) "pilot region (applied research)" and (iii) "showcase projects (demonstration)". Further research in close cooperation with these pillars complements RDTI activities in the region.[1]

3 The Basics: Charging Infrastructure

Under the lead of the stock market listed utility company EnBW AG the project ALIS (Aufbau Ladeinfrastruktur Stuttgart und Region), funded by the State of Baden-Württemberg, installed 500 AC charging points with a connected load of 22 kW each in the Stuttgart Region in 2012 and 2013.[2] This will not only expand the essential infrastructural basis for electrical mobility, but will also enable the development of sustainable business models and the analysis of user response and behaviour. Through this project, Stuttgart became the first city in Germany with a comprehensive charging infrastructure. In order to guarantee access to this infrastructure for drivers of electric vehicles from all over Germany and many European countries. The charging stations are connected via a common backend to European and national platforms allowing "e-roaming" for incoming and outgoing electric drivers. The city council of Stuttgart decided on a 3-year exemption of parking charges for electric vehicles (Battery Electric Vehicles and Plugin Hybrid Vehicles) in all public parking areas. Through the integration of the Car2Go system with e-vehicles, the Federally and State funded LivingLab BW^{e} mobil aims at finding out about barriers and problems when setting up a public charging infrastructure for e-vehicles. 500 charging points co-funded by the LivingLab serve as energy source for the 500 Smart electric drive of Car2Go, and at the same time makes e-mobility more usable for private persons.

Once charging infrastructure is installed, connected and in use, charge management becomes important to avoid local peak loads. Electric supply and demand need to be managed in smart grids (Pehnt et al. 2011). Thus, the project inFlott coordinated by the utility company EnBW AG and funded by the Federal Ministry for Economic Affairs and Energy (BMWi) aims to implement and verify smart charging solutions. An integrated fleet and charging management was developed, demonstrated and tested.[3] As a first spot to implement the system, a car park in Stuttgart was selected, which is operated by a company owned by the state of

[1]For details see http://www.e-mobilbw.de/en/ and http://nachhaltige-mobilitaet.region-stuttgart.de/.

[2]For details see http://www.livinglab-bwe.de/projekt/alis/ (in German only).

[3]For details see http://www.livinglab-bwe.de/projekt/inflott/ (in German only).

Baden-Württemberg. The charging of electric vehicles parked in this garage is managed by a server which decides when to charge which vehicle according to its requirements ("to be charged with x kWh at time y"), and is thus planning the electricity consumption of the overall system. Similar solutions are already installed in other semi-public parking areas, e.g. company car parks.[4] Based on experiences collected through these projects, future solutions for urban quarters will be developed and implemented.

4 The Ice Breaker: Fleet Integration of Electric Vehicles

Whereas private users mainly focus on the price of a car when deciding whether or not to buy an electric vehicle (Dütschke et al. 2012), decisions of fleet managers often are based on total cost of ownership (TCO) calculations. Therefore, company fleets are a core target for the market ramp-up of electric vehicles (NPE 2013). In mixed fleets of battery electric vehicles and internal combustion engine cars, the operation of EVs can be optimized by smart fleet management approaches, maximizing mileage and thus minimizing costs per km for electric vehicles (NPE 2013). The Get eReady Project led by Bosch Software Innovations GmbH and funded by the Federal Ministry for Economic Affairs and Energy (BMWi) aims at integrating e-vehicles into company fleets and surveys a software-based linking of the charging infrastructure within a high density of vehicles. Through Get eReady, in total 750 electric and hybrid vehicles will be brought on the streets of the Stuttgart Region by 2015. The project offers operators of business, public and non-commercial fleets an easy way to gain experience of e-mobility, with increased user acceptance helping to improve the prospects for mass implementation. Participating fleet operators should have fleets of at least ten vehicles—though it is in the project's interest for the vehicles involved to cover a broad range of sizes and functions as this will ensure that the results obtained are robust.

Fleet managers can book services to analyse in-depth the operation of the fleet vehicles by individual trips, resting times, daily mileage, etc. Based on this analysis, a proposal will be made as to how many vehicles of which size in the fleet can be replaced by electric vehicles. The recommendation includes cost calculations, operational recommendations and optimization of the total fleet. The fleet manager is supported in the purchasing process for vehicles and appropriate charging equipment. Get eReady furthermore offers an e-fleet management software package to operate a mixed fleet including full electric and hybrid as well as non-electric vehicles.

[4]For details see http://www.livinglab-bwe.de/meldungenarchiv/inflott-startet-pilotbetrieb/ (in German only).

5 The Enabler: Stuttgart Services

Electric mobility still is currently available for limited user groups only. Even offers open to everybody such as electric car sharing or pedelec rentals, addressing different needs of citizens, are not accessed by large groups of consumers. In order to overcome the threshold to acceptance of these systems, Stuttgart Services aims to integrate such services into everyday use cases. The consortium of 22 partners ranging from public transport providers, carsharing providers, bike and pedelec rental schemes, municipal undertakings, a bank and technology providers[5] is developing and implementing an integrated mobility services platform. Based on a general applicable identifier, e-mobility services, car sharing, pedelec rental, public transport and other services will be able to be booked with one multifunctional service card only. The electronic card is currently undergoing field testing by selected users interested in multimodal mobility and will be rolled out to more than 150,000 public transport users by the end of 2015. The core use case will be intermodal traffic. The card offers an e-ticketing function for all holders of annual tickets for the public transport provision, and seamless access to and special benefits for the solutions offered by all cooperation partners, e.g. DB Rent with its Flinkster carsharing (including electric vehicles) and its call-a-bike rental, Car2Go with its 500 Smart electric drive, and the municipal e-bike stations of the Netz-E-2-R community managed by NAMOREG (Nachhaltig Mobile Region Stuttgart) and operated by nextbike. This smart card-based solution will serve a double function as an information and a booking platform not only for electrical mobility, but also for public transport and diverse additional services, for instance for leisure time. It is a primary goal of the implementation to connect other projects of the LivingLab and facilitate easier access by the public.

Netz-E-2-R offers a sustainable and convenient e-bike-based last-mile-home mobility solution at train stations in Stuttgart Region, especially in rural areas. Booking is possible around the clock (24-7 service) via smart phones or the call centre, and the users of the system will be most welcome to keep the e-bikes overnight. The e-bikes are hosted at dedicated energy autonomous hire-stations located at the stations of the regional train and suburban rail (S-Bahn) network. The hire-stations store the electricity from solar panels in Li-ion batteries. The surplus electricity is fed into the grid, generating additional income for the municipality owning the hire-station. Up to 15 hire-stations with up to 220 e-bikes are under construction or already up and running.[6] The system will also allow the drop-off at other stations, and thus the one-way use of this service. This especially addresses touristic purposes, for instance for biking tours. In effect, the resulting high frequency use will reduce the unit operational costs.

The general idea behind Stuttgart Services is rather simple: give people who are used to travel daily without their own car the key to multimodal services, and

[5]The complete list can be found at http://www.stuttgart-services.de/projektpartner.html.

[6]For details see http://www.livinglab-bwe.de/projekt/netz-e-2-r/.

thereby increase the probability of first contacts with e-mobility. In almost all user evaluations it turned out that persons' acceptance of electric mobility significantly increased with personal experiences driving electric vehicles (Dütschke et al. 2012).

6 First Results: E-Mobility in Everyday Life

Meanwhile, there is no day where you cannot see electric vehicles on the roads in Stuttgart Region. Hybrid buses, full electric vans, electric cars, electric scooters made in Germany, France, Japan or Korea (and sometimes the USA) are all part of the urban and regional traffic ecosystem. The regional body responsible for transport planning (Verband Region Stuttgart) became authorized in February 2014 for "regional mobility management" overall, as well as specific services for electric mobility. Handbooks and guidelines for fleet managers and charging infrastructure operators are available,[7] and a lot of test ride events for the public with several types and models of electric vehicles are organized around the year by municipalities, car dealerships and event organisations (e.g. accompanying fairs or conferences). Most car dealerships are offering electric vehicles in addition to their traditional assortment of internal combustion engine cars. Altogether, it looks like the beginning of the ramp-up of the market for electric mobility both in terms of supply and demand (especially from selected company fleets).

7 Conclusion and Outlook

Therefore, priority for ongoing and future public support should turn to focus on improving the framework for users of electric vehicles, but not any more support vehicle and drivetrain technologies development. Further progress of the latter increasingly will be driven by competition in the growing market. Tax incentives or subsidies for electric and hybrid vehicles are able to speed-up the market penetration, but need to be handled carefully in order to avoid market disturbances. Crucial for the success of electric mobility will be an appropriate barrier-free and seamless charging infrastructure. At present, no noteworthy return on invest exists for charging station, nor any proven business model. Moderated by the Federal accompanying research of the e-mobility showcase region programme, a specific working group consisting of operators of charging infrastructure, municipalities and scientific experts has been established to elaborate and discuss different financing models for public charging infrastructure. A systematic development of a fast charging network will be essential for long-distance use of electric vehicles. On top

[7]See for example http://www.now-gmbh.de/de/publikationen.html.

of the "charging agenda", however, stands the comprehensive e-roaming between all charging suppliers in public and semi-public areas, offering access for charging with a single registration point for users at their preferred energy provider only.

References

BDEW, VDA, & ZVEI. (2009). Gemeinsame Position der Verbände BDEW—Bundesverband der Energie- und Wasserwirtschaft e.V., VDA—Verband der Automobilindustrie e.V., ZVEI e. V. —Zentralverband Elektrotechnik- und Elektronikindustrie e.V. zur Elektromobilität. Berlin/ Frankfurt.

BMWi, BMVBS, BMU, BMBF, & BMELV. (2009). Bericht an den Haushaltausschuss Konjunkturpaket II, Ziffer 9 Fokus "Elektromobilität", 19.03.2009, Berlin.

Dispan, J., Koch, A., Krumm, R., & Seibold, B. (2011). *Strukturbericht Region Stuttgart 2011, Entwicklung von Wirtschaft und Beschäftigung, Verband Region Stuttgart, Handwerkskammer Region Stuttgart, Industrie- und Handelskammer Region Stuttgart*. Stuttgart/Tübingen: IG Metall Region Stuttgart.

Dütschke, E., Schneider, U., Sauer, A., Wietschel, M., & Hoffmann, J. (2012). *Roadmap zur Kundenakzeptanz*. Berlin: Bundesministerium für Verkehr, Bau und Stadtentwicklung (BMVBS).

German Federal Government. (2009). German federal Government's national electromobility development plan. Retrieved August 2009, from https://www.bmwi.de/English/Redaktion/Pdf/national-electromobility-development-plan.

Heidenreich, L., Heim, R., Pasch, A., & Runge, S. (2013). *Vision and roadmap of the national electric mobility platform, national electric mobility platform (NPE), systemic approach working group*. Berlin: Chairman Dr. Rudolf Krebs.

NPE. (2013). *Elektromobilität in Deutschland—Ergebnisse aus einer Studie zu Szenarien der Marktentwicklung*. Berlin: Nationale Plattform Elektromobilität, AG 7 Rahmenbedingungen.

Pehnt, M., Helms, H., Lambrecht, U., Dallinger, D., Wietschel, M., & Heinrichs, H., et al. (2011). Elektroautos in einer von erneuerbaren Energien geprägten Energiewirtschaft. *Zeitschrift für Energiewirtschaft 35*(Nr.3), 221–234.

Peters, A., Doll, C., Kley, F., Möckel, M., Plötz, P., & Sauer, A., et al. (2012). *Konzepte der Elektromobilität und deren Bedeutung für Wirtschaft, Gesellschaft und Umwelt, Innovationsreport* (Nr. 153). Berlin: TAB-Arbeitsbericht.

Reiner, R., Cartalos, C., Evrigenis, A., & Viljaama, K. (2010). *Challenges for a European market for electric vehicles*. Brussels: European Parliament, Directorate-General for Internal Policies, Directorate—Industry, Research and Energy.

Tenkhoff, C., Braune, O., & Wilhelm, S. (2011). *Ergebnisbericht 2011 der Modellregionen Elektromobilität*. Berlin: Bundesministerium für Verkehr, Bau und Stadtentwicklung.

Wintjes, R., Turkeli, S., & Henning F. (2013). Innovation policy in metropolitan areas: Addressing societal challenges in functional regions. Regional Innovation Monitor, Thematic Paper 6, Brussels.

Launching an E-Carsharing System in the Polycentric Area of Ruhr

Timm Kannstätter and Sebastian Meerschiff

Abstract RUHRAUTOe is the first e-carsharing project in Germany that includes public transport and housing associations. The project was initiated in November 2012 and spans a period of 18 months. It is government-sponsored by the Federal Ministry of Transport and Digital Infrastructure as an undertaking of the "Modellregion Elektromobilität Rhein Ruhr". The RUHRAUTOe-consortium comprises Duisburg-Essen University, Drive CarSharing GmbH, Vivawest Wohnen GmbH and Verkehrsverbund Rhein-Ruhr. It has been closely collaborating with municipalities, local energy suppliers, public and private initiatives and foundations, as well as the private sector. The overriding goal is to establish a demonstration and test model of a multi-modal mobility system in the Ruhr area. More specifically, the project's objectives are providing people with opportunities to encounter e-mobility and future urban mobility concepts, exploring e-carsharing applications with a high customer value and promising target groups, revealing consumer acceptance amongst drivers, developing a sustainable business model, exploring ways to endorse public transport in a sensitive way, contributing to the concept of eco-friendly housing projects and gathering both subjective and objective data in order to unveil generic drive habits and technological requirements. RUHRAUTOe has been operating a traditional station-based carsharing approach. Currently, 20 PHEVs and 26 BEVs are available at 28 carefully selected and continuously monitored public and private charging stations in eight Ruhr cities. So far, a total number of 1,102 users have driven a total distance of approximately 170,000 km.

Keywords Carsharing · Electric cars · Public transport · Urban mobility · Urban mixed mode commuting

---

T. Kannstätter (✉) · S. Meerschiff
Universität Duisburg-Essen, CAR – Center for Automotive Research, Bismarckstraße 90, 47057 Duisburg, Germany
e-mail: timm.kannstaetter@uni-due.de

S. Meerschiff
e-mail: sebastian.meerschiff@uni-due.de

W. Leal Filho and R. Kotter (eds.), *E-Mobility in Europe*,
Green Energy and Technology, DOI 10.1007/978-3-319-13194-8_11

1 Introduction: Antecedents of RUHRAUTOe

The role of cars in metropolitan regions will change significantly in the next years. Mounting costs of car ownership, limited capacities of parking facilities, a growing environmental awareness and a change in values jointly contribute to this development (Loose et al. 2006). Recent findings indicate that especially among younger people a change in values regarding car usage and ownership has been taking place since the early 2000s. Simultaneously, the trend of reurbanisation has resulted in generally shorter travel distances to cover and a growing demand for intermodal urban mobility services. Moreover, the demand for a more individualized mobility is growing quickly (Deffner 2011). In combination with the current wide scale implementation of e-mobility in Europe and many other parts of the world, these developments give broad potential for implementation of innovative urban mobility systems (Canzler 2010).

The origins of RUHRAUTOe date back to early 2011, when the Center for Automotive Research (CAR) at Duisburg-Essen University conducted a series of empirical studies on the consumer acceptance of electric vehicles. Within the study setting, a total number 257 of study participants went on a test drive with a selection of electric vehicles. A t-test that was used to compare pre- and post-survey results displayed significantly higher ratings of self-reported attitudes towards e-mobility amongst the study participants. Beyond that, following the test drive, 71 % of the participants were willing to include electric vehicle in an imminent vehicle purchase decision-making process, another 23 % would consider buying an electric vehicle as their next car within the next 4 years. The findings showed that driving experiences with an electric car can be considered a very powerful means to reduce doubts and misgivings regarding e-mobility in people's minds. Thus, in particular at the beginning of the e-mobility market development, it is a sensible move to create a platform that allows interested parties to drive electric cars and discern the benefits for themselves. In order to attract a large number of users, adequate practices of electric vehicle-utilization which compensate the shortcomings (i.e. limited range) of the current EV-technology and reduce financial risks for users must be found. The rapidly emerging concept of urban carsharing, which allows spontaneous booking and temporary use of rental cars at fixed stations by means of a standardized process (Katzev 2003), has been identified as an ideal practice of electric vehicle-utilization (Müller et al. 2011; Kley et al. 2011). But even though the potential for e-carsharing was given, electric vehicles were used by very few carsharing providers in daily operations (Müller et al. 2011). On the basis of all these considerations, CAR and the newly formed project consortium jointly devised the innovative RUHRAUTOe-concept in mid-2011.

2 Public Project Sponsorship and Project Objectives

The RUHRAUTOe concept was approved for public funding in November 2011, thereby becoming one of 11 promoted projects of "Modellregion Elektromobilität Rhein-Ruhr Phase II". The "Modellregion Elektromobilität Rhein Ruhr" is part of the e-mobility promotion funding of the German Federal Ministry of Transport and Digital Infrastructure (BMVI), which is fixed in the National Development Plan for Electric Mobility (ElektroMobilität NRW 2014). The total public sponsorship for RUHRAUTOe amounted to 1.15 Mio Euros. Expenses of the partners in trade and industry are thereby eligible for a 50 % level of co-funding, and in case of the research partners a 100 % funding is guaranteed. The project was officially scheduled to start in September 2012 and to end in February 2014. Nevertheless, vehicle supply difficulties did cause a 2-month delay.

In addition to the general social, technical and economic objectives of the government funding for e-mobility (e.g. climate protection, technology improvement, location policy, development of new market segments, local air pollution improvements), the RUHRAUTOe concept has pursued the following funding policy objectives:

- a large-scale implementation and demonstration of electric mobility
- support and development of new business models in the context of electric mobility
- economic growth through new products/services and the creation of new jobs
- the linking of electric mobility and public transport
- development of intelligent and intermodal mobility concepts
- an investigation of user acceptance and customer behaviour
- the integration of local partners
- environmentally friendly processing of urban traffic flows

Drawing on the classification by Kley et al. (2011) based on Tukker (2004), RUHRAUTOe can be seen as a service-oriented, user-oriented business model within the framework of different mobility concepts. For efficiency analysis of the tested business model, a comprehensive evaluation of the economic viability and the basic requirements for a self-supporting e-carsharing system was conducted. The continuous monitoring and controlling of operating expenditures and revenues, individual vehicle utilization levels, and charging station performances provided the consortium with adequate data basis for this purpose.

3 Project Scope

Conceptual (intermodal) linkages between public transport and the concept of carsharing were investigated long before RUHRAUTOe (Nobis 2006). But with the exception of Huwer (2002), all sifted investigations were carried out in monocentric urban spatial structures. The selected agglomeration along the river Ruhr, however, forms Europe's largest polycentric urban spatial structure with 5.3 million inhabitants (Knapp et al. 2006). Thus, the wide scale spatial scope conferred an innovative nature on RUHRAUTOe: the major cities along the river Ruhr formed a uniquely widespread urban demonstration and test field.

The overriding objective of RUHRAUTOe has been to gain multiple insights into an urban intermodal passenger transport concept that combines an all-electric carsharing fleet as a key single-mode means of travel with the local public transport. The undertaking is designed to be an open platform that invites essential collaborating partners such as municipalities, local car dealerships and Original Equipment Manufacturers (OEMs), as well as local energy suppliers to participate and contribute to a successful implementation. A holistic approach has also been pursued in the case of the potential customer base, since potential commercial and private and public users have been addressed.

Within the spatially extended test field, a multi-make all-electric fleet of 46 vehicles with different drivetrain technologies forms a carsharing network with 28 fixed stations in residential and inner city areas. With Duisburg, Oberhausen, Mülheim an der Ruhr, Gelsenkirchen, Bottrop, Essen, Bochum and Dortmund, all major cities along the river Ruhr have been included. Table 1 gives an overview of the applied RUHRAUTOe-vehicle fleet.

For implementation in inner city areas, RUHRAUTOe mainly drew on existing public charging networks, whereas the utilization of private charging infrastructure merely occurred in the case of residential carsharing. Since RUHRAUTOe was designed to complement public transport by means of maximized connectivity, vehicle stations have mostly been placed nearby public transport hubs. From these hubs, public transport users make of use of e-carsharing vehicles in order to move to their final destination. Due to this strategy, the RUHRAUTOe vehicle fleet was designed to be an additional means of transportation within local mixed mode commuting, which allows a more comprehensive and flexible coverage of the last-mile, "i.e. the last link in the transport chain to the destination" (Edwards et al. 2010). The main objective of the residential e-carsharing approach is to test the acceptance of urban mixed-mode commuting enhanced with e-carsharing among local residents in selected housing areas.

In all cases, RUHRAUTOe has pursued the traditional carsharing approach, which—unlike the emerging commercial free floating-models (e.g. Car2go, DriveNow)—is based on fixed stations (Bundesverband CarSharing 2014). Thus, it is mandatory for users to pick up and return the rentals at the same charging location. This approach excludes flexible short-range one-way utilization of cars (Barth and Shaheen 2002).

Table 1 Overview of the vehicle fleet of RUHRAUTOe

Vehicle model	Number of applied vehicles	Classification	Drivetrain	Range	Locations	Source of funding (coverage of leasing costs in percent)	Period of application
Opel Ampera	20	Medium-sized 4 passengers	Plug-in hybrid electric vehicle	60 km electric; total >500 km	Essen, Mülheim, Oberhausen, Dortmund, Bottrop, Bochum	50 % public sponsorship/50 % rental earnings	11/2012 –
Smart fortwo electric drive	10	A-mini car 2 passengers	Battery electric vehicle	150 km	Essen, Mülheim, Bottrop, Bochum, Gelsenkirchen	50 % public sponsorship/50 % rental earnings	05/2013 –
Nissan Leaf	7	Medium-sized 5 passengers	Battery electric vehicle	150 km	Duisburg, Mülheim, Bottrop, Essen	100 % rental earnings	07/2013 –
Renault Zoe	2	Compact-car 5 passengers	Battery electric vehicle	180 km	Oberhausen, Bochum, Dortmund	100 % rental earnings	01/2014 –
Renault Twizy	5	A-mini car 1 +1 passengers	Battery electric vehicle	80 km	Duisburg, Bochum, Essen, Mülheim	100 % rental earnings	06/2013 –
Peugeot i-On	1	A-mini car 4 passengers	Battery electric vehicle	130 km	Dortmund	100 % rental earnings	01/2014 –

4 RUHRAUTOe-Consortium: The Project Partners and Their Duties

The RUHRAUTOe-consortium comprises CAR—the Center for Automotive Research at Duisburg-Essen University, Drive CarSharing GmbH, Vivawest Wohnen GmbH and Verkehrsverbund Rhein-Ruhr. The selection of consortium partners reflects the project scope of RUHRAUTOe.

Verkehrsverbund Rhein-Ruhr is the governing body of the public transport operators in the district of Rhine-Ruhr. It has been engaged in RUHRAUTOe to test whether an optimized networking of e-carsharing and public transport is a practical strategy to cover a growing demand for a more flexible intermodal passenger transport. Moreover, the VRR aims to enhance its service through covering existing spatial voids and temporal timetable-gaps in its transport services with e-carsharing applications. The VRR has provided the data basis for the analysis of existing infrastructures and timetables, local transport modes, public transport connections and transport hubs. Another key contribution was the extension of the newly introduced single smart card system (e-ticket) in order to give VRR-customers easy access to RUHRAUTOe-vehicles. The local VRR-subsidiaries have taken over the initial client consulting and the registration process of new customers at their customer centers in the Ruhr area.

Vivawest Wohnen GmbH is the largest housing stock holder and developer in the Ruhr area. Within the framework of RUHRAUTOe, the company has investigated the acceptance of RUHRAUTOe among their tenants to identify valuable target groups and profitable vehicle locations for residential e-carsharing applications.

Drive Carsharing GmbH has been engaged in the carsharing business since 1993. The firm has been serving as a carsharing provider that allows a variety of parties to integrate their vehicles into a thoroughly cross-linked nationwide carsharing system. RUHRAUTOe has been added to this system, thereby providing users of other carsharing organizations (e.g. DBrent-operated Flinkster) with easy access to RUHRAUTOe vehicles. Beyond that, this medium-sized firm has taken over the operative tasks of fleet and station management, as well as the whole process of customer administration. The scope of duties has further included the management of clients' accounts and their monthly settlements, vehicle and customer insurance, vehicle support and maintenance, vehicle equipment with the necessary carsharing hardware, as well as provision of the carsharing software applications (incl. mobile app) for vehicle reservations and booking. Beyond that, the company has acquired electric vehicles for RUHRAUTOe in consultation with the consortium, initiates business connections, and has been leading negotiations with associated project partners.

The professorial chairs of CAR—Center Automotive Research were in charge of the superordinate project management, associated marketing and public relation measures, the evaluation of potentials for a self-sustained business model, as well as the accompanying business and technical research. The main research emphasis has been on the user perceptions and acceptance of RUHRAUTOe. Thus, the accompanying business research investigated the customer profiles (demand structure) and customer acceptance at different levels and with different approaches. University researchers have investigated what influence RUHRAUTOe has exercised on urban mixed-mode commuting behaviours, and to what extent the processing of urban traffic flows is viable through this undertaking. Actual RUHRAUTOe users and further groups identified as potential users (e.g. local residents, institutions, and firms nearby the charging stations) were frequently questioned on their awareness, user preferences and basic requirements, perceived usage barriers, acceptance of different price structures and willingness to pay, as well as preferred incentives and other acceptance drivers (e.g. usability, vehicle attributes). In addition, qualitative interviews with users and experts were conducted. For the purpose of vehicle usage patterns analysis and profound recommendations on the dimensioning of vehicle drivetrain components, project cars were equipped with data loggers to gather multiple driving data sets.

D+S Automotive GmbH, a CAR spin-off consulting firm, has been responsible for analysis of existing infrastructures, transport services and public transport interfaces in order to identify points of contact for the e-carsharing system. The determination of adequate pitches for RUHRAUTOe vehicles has also been among the firms' responsibilities. In this process, D+S Automotive has been closely collaborating with the VRR and Drive CarSharing GmbH.

5 Impacts of RUHRAUTOe

Since the launch of RUHRAUTOe, the project has had significant influence in a variety of different areas. The impacts have affected the environmental, as well as the political and social spheres.

From an environmental perspective, the total mileage of 170.000 electrical kilometres has demonstrated the quantities of vehicle emission cuts that could be obtained through the operation of electric rental vehicles instead of conventional vehicles with internal combustion engines in carsharing operations. The annual reports on the carmakers' progress towards EU CO_2 target reveal a mean vehicle fleet emission level of 150 g/km for Germany from 2008 to 2012 (European Federation for Transport and Environment 2010, 2013). Calculations based on these numbers reveal that more than 25 tonnes of local CO_2 emissions have theoretically been saved. By incorporating the emissions of electricity generation (well-to-wheel analysis) with the 2007 power mix of Germany (Jahns 2009), the amount of saved CO_2 emission accounts for a 6-tonne reduction.

In the political and social spheres, major obstacles and barriers of urban e-mobility applications have been identified, and reported to local and federal policy makers. In the course of the implementation of the RUHRAUTOe system, a frequent exchange with decision-makers appears to have raised their awareness of specific problem areas. To cite an example, to this very day there is no legal certainty for municipalities when it comes to towing conventional vehicles from parking spaces that are signposted for "electric vehicles during loading process". Since RUHRAUTOe draws on the public charging infrastructure, this missing legal certainty turned out to be a significant blow to the projects' usability, especially in the early stages, when even local residents were unaware of the newly placed electric vehicles at unsigned parking space nearby their homes. The project consortium reported this key problem area to the responsible political entities and its project sponsor, the Federal Ministry of Transport and Digital Infrastructure (FMTDI), in the context of its regular project progress reports and Model Region e-Mobility Meetings. Recently, the FMTDI promised to address privileged parking for electric vehicles in public spaces through a legal regulation in the current legislative period (Bundesregierung 2014).

Another example for the local impact of RUHRAUTOe has been the integration of its carsharing service into the mobility concepts of local municipalities. Several departments in the cities of Oberhausen and Mülheim an der Ruhr have become frequent users of vehicles placed at charging stations nearby their facilities. Instead of augmenting their own vehicle fleet or the use of private cars to cover situations of peak demand, the municipal departments have started to make use of the carsharing system of RUHRAUTOe. The municipal policy makers have reported considerable cost savings, a positive effect on the environment, and a significant image boost for their cities through this "e-mobility on demand" approach.

It has to be emphasised that without the support of the municipal administrations, the RUHRAUTOe concept would not have been feasible. In cases of exclusive occupation of public parking spaces within the reach of public charging stations, the utilization of the public charging infrastructure itself, permissions to promote the concept in public areas and local business networking, a supportive city government are of vital importance. Thus, for implementation of RUHRAUTOe in new cities, the respective municipalities—as well as the local energy suppliers—had to be thoroughly involved. Luckily, the vast majority of local policy makers have been open-minded about RUHRAUTOe. The best practices for the integration of RUHRAUTOe in the urban mobility concepts and municipal mobility itself have been illustrated by the cities of Oberhausen and Mülheim an der Ruhr. In the fields of charging infrastructure and business networking, the city of Bochum and its affiliated local energy provider, Stadtwerke Bochum, deserve to be mentioned. Exclusively for RUHRAUTOe, Stadtwerke Bochum built three additional charging stations in the city centre, which will remain in open-access operations after the project is concluded. Altogether, 13 new open-access charging points were built through the cooperation with local authorities and businesses. The newly constructed recharging facilities range from public available household outlets to high-end charging solutions.

RUHRAUTOe has had considerable impacts on the commercial sector as well. Altogether, 14 local companies of various branches have placed RUHRAUTO e-vehicles in commercial fleet operations. Fleet managers considered e-carsharing an attractive way of electric vehicle utilization, since it allows extensive test driving without further financial obligations and helps to meet fleet emission-targets. Even though the feedback was mostly positive, the stated purchase intentions among the questioned fleet managers remained generally low. Nevertheless, e-carsharing can be considered a powerful concept in order to pave the way for a future wide scale integration of electric vehicles in commercial vehicle fleets. Furthermore, several OEMs and affiliated local dealerships have become aware of the project. They have regarded RUHRAUTOe as an opportunity to promote their electric vehicles and bring them to market. One OEM-PR-Manager described RUHRAUTOe as a "cost-efficient advertising campaign in attractive urban surroundings". The Renault–Nissan-group has contributed 12 all-electric vehicles to the vehicle fleet, which has enabled RUHRAUTOe to cover additional cities and charging stations, thereby exceeding the initial project scope.

An extensive marketing campaign, ranging from local flyer programmes in pedestrian areas nearby the vehicle locations to the hosting of major test drive events, has helped to firmly anchor RUHRAUTOe and the concepts of carsharing and e-mobility in numerous minds of the Ruhr locals. In a series of surveys which were conducted with locals living within a 500-m radius around three charging stations in the city of Essen, between 28 and 37 % of the study participants reported a personal awareness of the concept of RUHRAUTOe. Based on a cautious estimation which includes all project activities, on can guess that approximately 3,000 people have had their first ride with an electric vehicle through RUHRAUTOe over the course of the 18-month project duration. The number of people who have first-time encountered general topics like e-mobility and innovative urban mixed mode commuting is much higher, yet only dubiously to quantify.

6 Results

From the 18-month of fieldwork a range of objective and subjective data sets have been obtained. The analysis of such data sets has provided researchers of CAR with numerous viable findings on the best practices, problem areas, and efficient business operations and strategies for potential project extension. Beyond that, conclusions for an overall successful implementation and economic development of similar projects can be drawn as well. The following pages display some of the important findings of RUHRAUTOe. Since the project is still running at the time of writing in April 2014, further data analyses are scheduled to be performed in order to gain deeper insights.

6.1 Concept

When RUHRAUTOe was launched in November 2012, it was the first project that operated carsharing with electric vehicles using the public recharging infrastructure. Therefore, the project consortium faced a general internal and external lack of prior experiences in advance. For the project consortium and even the associated communities, the issue of parking privileges for electric vehicles in public places was uncharted territory. Since there are no general legal regulations on this issues provided by superior political entities, there was no certainty of the law when issues in areas regarding electric vehicle-parking privileges occurred. This uncertainty turned out to be a significant obstacle for the implementation of RUHRAUTOe, since a station-bound carsharing system relies on exclusive parking. Due to the absence of a general legal regulation for this issue, each municipality pursues its own policies regarding the provision of public parking spaces, park space signage and conventional vehicle-towing, as in Table 2.

In the case of the City of Duisburg, the local energy municipal utilities did not allow RUHRAUTOe to place the vehicles at public charging stations urging the argument that this would impede their own e-mobility customers when it comes to public recharging. Therefore, all bases in Duisburg were situated on private or university properties.

Besides the constant risk of the occupation of public parking spaces by conventional vehicles, one has to take into account that other electric vehicles are entitled to recharge at public charging stations as well. Specific traffic signs for a legally unbinding signage of parking spaces with charging facilities for electric vehicles exist. Nevertheless, an exclusive provision of public parking spaces for certain parties is excluded at all levels. Thus, the concept of RUHRAUTOe could only be implemented due to the goodwill of the associated municipalities and the relatively low total number of electric vehicles in the Ruhr area.

An additional problem area that undermines the concept of RUHRAUTOe is the heterogeneity of the available public charging infrastructure. Various energy suppliers provide various charging solutions with different conditions for access. In order to recharge an electric vehicle at any of the available charging stations in the Ruhr area,

Table 2 Municipal policies regarding the provision of parking space, signage and conventional vehicle towing

Municipality	Signage	Compensation payment for vehicle parking	Towing conventional vehicles
Essen	"Electric Vehicle Charging Station"	Yes	Yes
Bochum	"Electric Vehicle Charging Station"	No	No
Oberhausen	"Electric Vehicle Charging Station"	No	No
Bottrop	"No stopping" + "Except for Electric Vehicle Charging"	No	No
Duisburg	"Electric Vehicle Charging Station"; "No Stopping" + "Except for Electric Vehicle Charging"	–	Yes
Mülheim an der Ruhr	"Electric Vehicle Charging Station" + "RUHRAUTOe charging station signage"	No	No
Dortmund	"Electric Vehicle Charging Station"	No	No

Table 3 Charging infrastructure in the associated cities

City	Energy supplier	Plug-in device charging station	Access
Essen	RWE	IEC 62196 Type-II	ID + password
Bochum	Public Services Bochum	IEC 62196 Type-II/Schuko	RFID-card
Oberhausen	RWE/EVO	IEC 62196 Type-II	ID + password
Bottrop	RWE/ELE	IEC 62196 Type-II	ID + password
Duisburg	Public Utilities Duisburg	IEC 62196 Type-II/Schuko	RFID-Karte
Mülheim an der Ruhr	RWE	IEC 62196 Type-II	ID + password
Dortmund	RWE/DEW21	IEC 62196 Type-II	ID + password
Gelsenkirchen	RWE/ELE	IEC 62196 Type-II	ID + password

each car would have to be equipped with at least three different access cards. Under these circumstances, any planned measures regarding a more flexible concept with free station choice are very complicated to bring about. Beyond that, each performed substitution of vehicles for technical or capacity-related reasons have required tedious exchanges of the vehicle-related charging equipment. Table 3 gives an overview of the utilized public charging infrastructures in the associated cities.

Another problem caused by the heterogeneity of charging infrastructures originated in the special features of the local RWE vehicle recharging-business model. RWE is one of Europe's five leading electricity and gas companies. Several local municipalities like Essen and Mülheim an der Ruhr are still among the largest shareholders, and their municipal electric and gas utilities are joint enterprises or at least economically linked to RWE as well. Thus, even after the liberalization of the German energy market, the firm and its local affiliates dominate the power market in the Ruhr area. The RWE-division "eMobility" constructed the vast majority of public charging stations there.

Whereas the connection of the vehicle and the RWE charging station is an unproblematic process, users have had to initiate the actual charging process manually by entering a username and a password via phone or mobile app. This procedure was widely rejected and heavily criticized as "tedious", "annoying" and "frequently defective" by early RUHRAUTOe users. In order to secure a high level of usability, the project consortium acquired special "intelligent" RWE-charging cables. These cables have an additional ID box installed, which transmits the customer data and initiates the recharging process automatically after plug-in.

The glove boxes of the vehicles located in Bochum and Duisburg, however, have had to be equipped with RFID charge cards, since another concept of public vehicle recharging with a different type of charging station forms the public recharging infrastructure there. To unlock this type of charging station before connecting the charging cable, the users have to hold the charge card over a station's reader. This procedure was generally accepted among RUHRAUTOe users, and cases of system malfunction have been rarely recorded. In cases of malfunction, however, neither of the charging stations transmits a clearly discernible error message with the result that the following users found an empty vehicle at the charging stations.

6.2 Usability

The system usage consists of three key elements: one-time registration, vehicle reservation/booking procedure, and the actual vehicle usage. In an online survey with RUHRAUTOe users conducted by Fraunhofer ISI in collaboration with CAR in late 2013, users assessed the usability of the RUHRAUTOe service itself and the deployed electric vehicles. The results displayed in the figure underneath indicate that both usability indicators were predominantly classified as "easy" (Fig. 1).

6.3 Evaluation of the Applied Vehicles

The overall reliability of the deployed electric vehicles in daily use has widely been approved by RUHRAUTOe users and the project consortium. Electric car-related malfunctions, such as severe battery failures or a significant degradation of vehicle performance over time, have not occurred. The users have widely appraised the road performances of the vehicles regardless of the model, make and hybrid or all-electric drivetrain: "Impressive vehicle acceleration" and high levels of "driving dynamics" and "driving pleasure" were frequently reported. The vast majority of the users questioned assessed the different all-electric vehicle ranges as "completely sufficient" for their individual applications. However, a significant loss of range has been reported during winter.

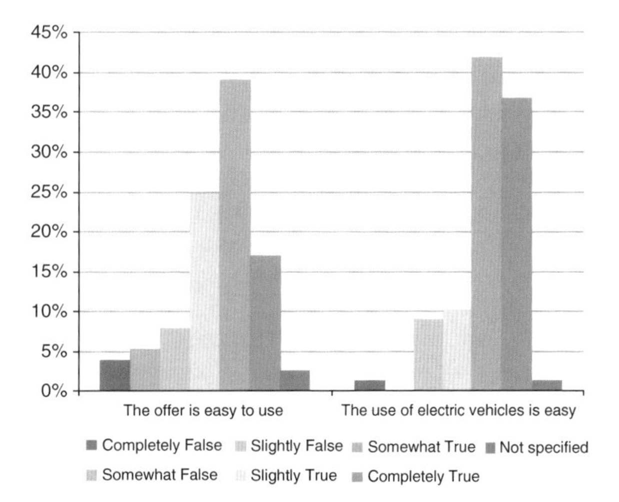

Fig. 1 Results of a RUHRAUTOe user survey regarding the approval of the overall system and vehicle usability (in percent)

6.4 Vehicle Utilization

As illustrated in Fig. 2, the use of the applied electric vehicles has risen over the project duration. After a prolonged stagnation in the initial stages which lasted until June 2013, the number of trips and the total monthly kilometres driven steadily increased to satisfactory levels in the period to late summer 2013. This degree of vehicle utilization allowed project partner Drive CarSharing to cover just the proportion of monthly leasing payments which had to be generated through rental revenues in case of the public sponsored 20 Opel Ampera and 10 smart fortwo electric drive. However, a cost-covering operation of the vehicles was still beyond reach. Another phase of stagnation during the winter months 2013/2014 followed, until a further significant increase in both the number of trips and the total monthly kilometres occurred in March 2014. The most obvious explanation for the persistently low number of monthly trips until mid-2013 is the limited spatial extent and the incomplete vehicle fleet at that time. Another explanation could be the long-reigning winter conditions, which lasted until April 2013. In general, winter times appear to curb the evolvement of RUHRAUTOe. A further explanation could be provided by the "diffusion of innovations" theory (Rogers 2003). People went through an extensive knowledge phase, because an unprecedented mobility concept like RUHRAUTO requires a lot of information to be obtained, before subsequent decision making-phases can be passed through and a positive adoption decision can be taken. Interestingly enough, two major price adjustments were implemented in March and August 2013. The first price cut coincides with the "take-off" of the total monthly number of trips and total kilometres driven per month (Fig. 2).

Figure 3 gives an overview of the utilization of the deployed vehicle models. However, preferences for specific car models cannot be validly derived from the

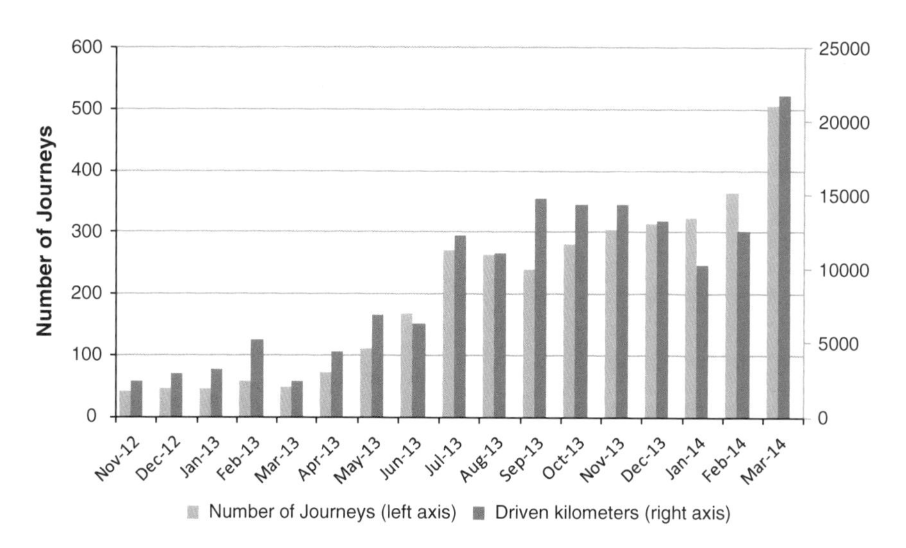

Fig. 2 Development of the numbers of trips undertaken and total kilometres driven per month

numbers displayed below. The largest proportion of trips (between 40 and 50 %) has been performed with Opel Amperas. Nevertheless, it has to be taken into account that this model has been overrepresented in the RUHRAUTOe fleet. Furthermore, the location of the vehicles has had a significant effect on the individual vehicle utilization. As mentioned before, vehicles that have been placed downtown at major public transport hubs are generally more frequently utilized than vehicles in peripheral areas.

A comparison of Figs. 3 and 4 illustrates a consistent dominance of the medium-sized PHEV Opel Ampera in the total monthly distances driven. An explanation for this result could be the hybrid electric drivetrain technology, which allows covering longer distances without the need for recharging through running in a conventional fuel mode. Apparently, the applied Opel Amperas are consciously chosen by RUHRAUTOe clients to cover longer travel distances. Interestingly enough, the illustration of the average booking frequency per vehicle per month in Table 4 demonstrates that smaller all-electric vehicles (e.g. Nissan Leaf and smart fortwo electric drive) are more frequently booked than PHEVs. Thus, the integration of different vehicle models that vary in size and electric drivetrain technologies appears to be a sensible strategy, in order to comprehensively cover different travel distance demands within an all-electric carsharing fleet (Fig. 5).

6.5 Customer Structure

By early April 2014, RUHRAUTOe had a total number of 1,070 users. Whereas 560 customers directly subscribed to RUHRAUTOe, the additional 510 users came

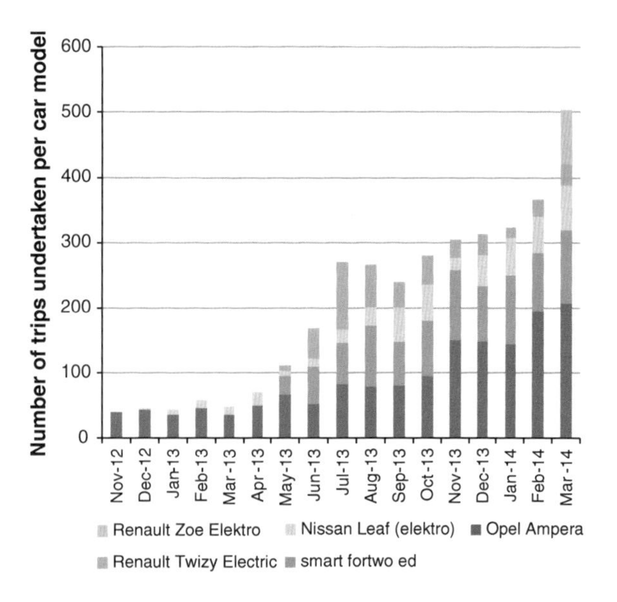

Fig. 3 The number of undertaken trips per vehicle model

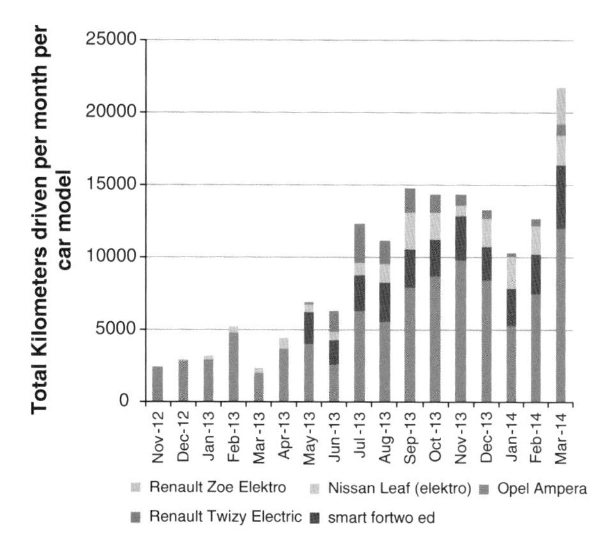

Fig. 4 Total monthly kilometres driven per vehicle model

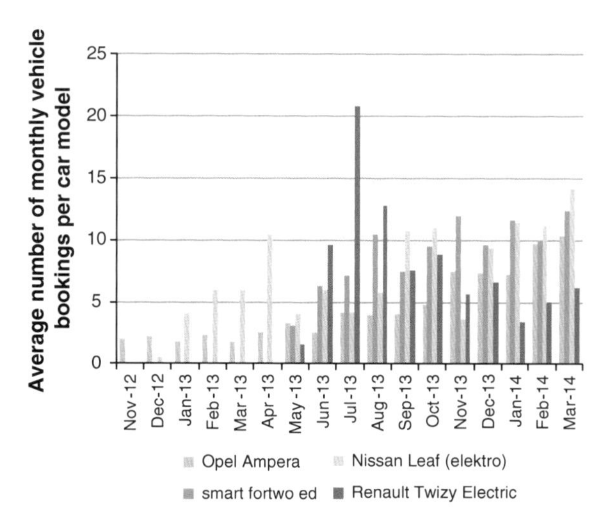

Fig. 5 Average number of monthly vehicle bookings per model

from other carsharing organizations RUHRAUTOe is affiliated to. Due to the open platform approach, such users enjoy unrestricted access to RUHRAUTOe-vehicles and are therefore termed "cross users". Table 4 displays the customer structure divided into both user groups in mid-April 2014.

Since "cross users" have not passed through the RUHRAUTOe-registration process, knowledge of the socio-demographic characteristics of this user group is strictly limited. Thus, the socio-demographic characteristics illustrated in the Fig. 6 apply to the RUHRAUTOe user group only. The data sets additional socio-demographic characteristics out, partially obtained from the research study which was conducted by CAR in collaboration with Fraunhofer ISI in the period from September 2013 to March 2014 (Figs. 6 and 7).

Table 4 Customer structure (effective: April 2014)

Group of users	Number of users	Male/commercial/ female in %	Proportion of total trips (%)	Proportion of total km (%)	Average age
RUHRAUTOe	560	69/8/23	56	48	36.7
"Cross-user"	510	76/13/10	44	52	No data

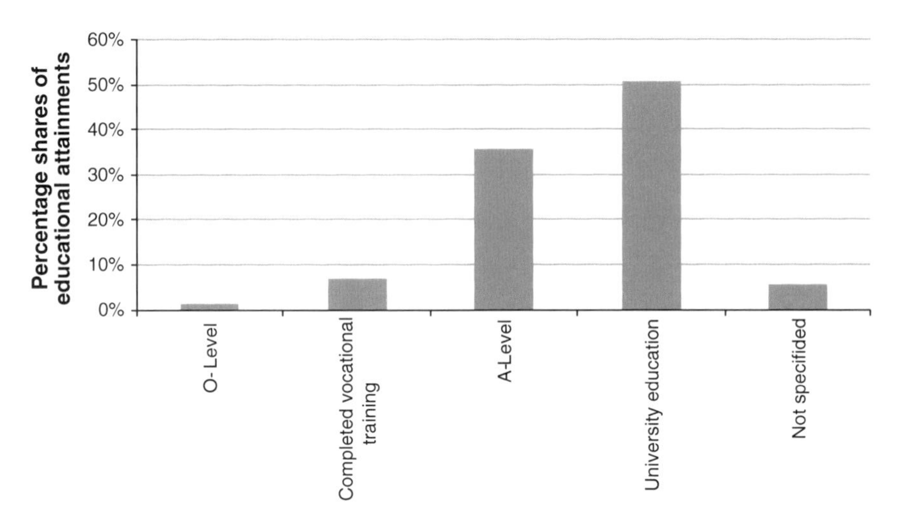

Fig. 6 Educational attainment of RUHRAUTOe users

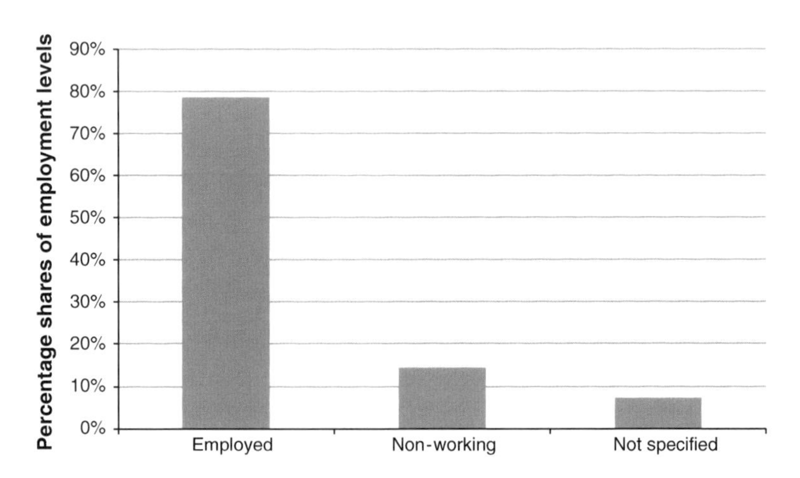

Fig. 7 Employment level of RUHRAUTOe users

The displayed data illustrates that RUHRAUTOe users are essentially well educated, since the level of employment is clearly above national average, and the average user age is significantly below the average and medium age of the German population. A rather surprising result is the gender distribution: 3 of 4 non-commercial, i.e. private users are male. The gender distribution has been persistent over the whole life span of the project. A cross-comparison with other carsharing research studies confirmed that these figures closely resemble general carsharing user characteristics (Shaheen et al. 2000; Cervero and Tsai 2004).

6.6 Interconnectedness with Public Transport

As mentioned before, the RUHRAUTOe concept is designed to function as a thoroughly interconnected, flexible mode of transport, which enhances the overall attractiveness of public transport by providing an additional mobility service for applications and areas that current public transport modes cannot cover. By pursuing the station-based carsharing approach it was ensured that the standard public transport remains the key means of transportation within the intermodal passenger transport.

Within the process of vehicle booking, users were encouraged to state the purposes of their trips. Effectively, a close inspection of the user feedback statements revealed that trip purposes like "bulk purchases", "need for immediate flexibility" and "last mile coverage" have been frequently mentioned. Such statements indicate that the concept has predominantly been used for the purposes it was meant to be, and not operated at the expense of traditional public transport. The mean booking period of RUHRAUTOe vehicles accounts for 6.5 h. The average distance travelled per booking is 44 km. In case of free-floating carsharing models, the average undertaken trip distances and the booking periods are generally much shorter (Weikl 2013). Thus, free floating-approaches probably pose a more serious threat to traditional public transport than RUHRAUTOe. The finding that vehicles which were placed at locations nearby public transport hubs showed generally higher levels of utilization can be seen as another affirmation of the intermodal RUHRAUTOe concept. One-third of the users who have directly signed up for RUHRAUTOe own an electronic VRR-subscription ticket or a subscription to other public transport providers. Hence, one can presume that the RUHRAUTOe system has been considerably adopted by local public transport customers. Nevertheless, the full integration of RUHRAUTOe into the electronic timetable of the VRR, which is expected to boost the usage among VRR-customers, is yet to be implemented.

6.7 Interconnectedness with Other Carsharing Organizations

As mentioned before, the open platform approach of RUHRAUTOe includes the collaboration and networking with other carsharing organizations as well. As illustrated in Table 4, about 50 % of the RUHRAUTOe users rank among the "cross user"-group. This group accounts for half of the total driven distance and booked trips as well. The carsharing operations of Deutsche Bahn (German Railways), Flinkster, turned out to be a dominant source of "cross users". Flinkster operates a similar station-based carsharing approach mainly in the surroundings of rail stations, whereas the vehicle fleet predominantly comprises conventional vehicles with internal combustion engines.

6.8 Monetary Consideration

The objective to transform the concept of RUHRAUTOe into an economically viable, self-sustained business model over the course of project duration has not been achieved as yet. Comparatively high vehicle purchase costs, a variety of e-mobility- and carsharing-related expenditures, and huge personnel costs encounter insufficient operational revenues. As of April 2014, vehicle rental revenues and public sponsorship allow it to operate RUHRAUTOe economically, whereas the doubling of the currently experienced vehicle utilization levels is necessary for a self-sustained operation.

The main vehicle costs to drivers have turned out to be the monthly leasing rates, vehicle insurance premiums and the carsharing-related technology with which the vehicles were equipped. The comparatively high purchase prices have in turn led to correspondingly high leasing rates and vehicle insurance premiums, which are generally high for vehicles applied in carsharing operations due to over-average wear and tear. However, recently falling prices for electric vehicles could ease financial burdens in this field. The costs related to the technical equipment which is necessary to integrate the vehicles in carsharing operations have not changed thus far. The installation costs of a special on-board unit which allows vehicle access by means of an RFID card as well as its monthly rent amount to 1.000 Euros per vehicle over 18 months.

The 18-month launch and operation of RUHRAUTOe was performed by 7.5 full-time employees. The specific fields of work, among others, cover vehicle and customer management, subordinate project coordination, public relations activities and the overall marketing of the project, as well as the accompanying business and technical research efforts. In addition, seven student assistants were in employment, mainly entrusted with promotional activities. After the completion of the start-up phase of the project and the finalization of accompanying research, personnel expenses can be reduced significantly.

Further incurred costs have arisen from the deployment of the online-based booking platform and utilization of the public charging infrastructure. The software that allows customers to register for RUHRAUTOe and reserve vehicles online has been provided by DBrent for a moderate fixed monthly charge. For an e-carsharing system that relies on the public recharging infrastructure the vehicle placement at highly frequented and visible places is vital. Such attractive parking spaces must be acquired. In general, the associated local authorities have generously not claimed any financial compensation for the provision of parking spaces, since RUHRAUTOe is a public sponsored research undertaking. If a commercial business model is implemented, additional costs for the exploitation of public parking space have to be taken into account.

The price structure has to be another import element of the monetary consideration of RUHRAUTOe. The structure has been changed twice over the project duration. Within the period from November 2012 to May 2013, the Opel Ampera was priced at 9.25 Euros per hour. Disappointing levels of vehicle utilization trigged a first price cut to 7.25 Euros in May 2013. Still, the smart fortwo electric drive and Nissan Leaf models showed much better vehicle utilization. Subsequently, the project consortium jointly decided to reduce further by the vehicle rental prices, and implemented another cut to 4.90 Euros in July 2013. Since then, the vehicle revenues have substantially risen due to higher vehicle utilizations for all deployed models.

Apart from this, users' reaction to the pricing structure was basically positive. At the regular local public relations events, it was a common reaction of interested parties to note a reasonable pricing. For instance, one pedestrian stated that the hourly rent of the Renault Twizy is "much cheaper than a bus ride". In a survey conducted in October 2013 with 120 city residents in Essen, merely 15 % of the people who had used RUHRAUTOe at least once stated that a lower pricing of the vehicles would raise their personal intention to use RUHRAUTOe more frequently (Table 5).

6.9 General Acceptance

A series of surveys on the general acceptance and evaluation of RUHRAUTOe was conducted by CAR, both in-house and in collaboration with Fraunhofer ISI.

The deployed electric vehicles and the general concept of RUHRAUTOe were predominantly seen as "positive" by more than 80 % of the RUHRAUTOe users. High levels of satisfaction (above 70 %) were obtained in terms of the assessment of the handling of the deployed electric vehicles, vehicle range, vehicle reservation/ booking process, availability of services, selection and variety of vehicles, accessibility of the charging stations, and accounting and costs. Beyond that, 4 of 5

Table 5 RUHRAUTOe price structure since July 2013

	Opel Ampera	Nissan Leaf, smart ed, Renault ZOE, Peugeot iOn	Renault Twizy
Rate per hour	4.90 Euro	4.90 Euro	1.95 Euro
Rate for the first 30 km	–	–	0.05 Euro/km
Rate until the 31 km	0.20 Euro/km	0.10 Euro/km	0.05 Euro/km

RUHRAUTOe customers who participated in the survey intended to further integrate vehicles with an electric drivetrain in their personal mobility behaviour, even after the termination of RUHRAUTOe. These results were obtained from the analysis of an online questionnaire, which was sent to all RUHRAUTOe users who had directly subscribed to RUHRAUTOe, with 104 returns.

Another study conducted by CAR focused on users and non-users in the locality of selected RUHRAUTOe stations in the city of Essen. In particular, the study participants were questioned on suggested improvements to RUHRAUTOe which would lead to a higher personal use intensity or use intention. The most striking result of this inquiry was that almost 100 % stated they would favour a more flexible e-carsharing system that allowed one-way trips.

7 Conclusions

After an 18-month period of field and desk work including conducting extensive analyses of driving data and various user surveys, it can be assumed that there is a potential market for commercial operation of an e-carsharing system in the Ruhr agglomeration. The reasons that support this assumption are a steadily growing customer base, consistently increasing vehicle utilization, and high levels of user satisfaction and public as well as private support from various sources.

In the early stages, RUHRAUTOe had to break down major structural barriers since it was an unprecedented undertaking. Nevertheless, the concept has paved a way for itself and successive e-carsharing projects. The required start-up phase and the necessary public relations efforts were initially underestimated by the project consortium. Thus, it has become clear that even a widely visible e-mobility concept requires extensive public relations measures in order to raise a considerable client base and anchor the project in people's minds. Contrary to this, the users' willingness to pay was overestimated. Price skimming is the wrong price strategy in the field e-carsharing, since price sensitivity appears to be high even among potential innovative users.

Furthermore, it can be concluded that e-carsharing vehicles need to be placed in inner city areas and nearby public transport hubs, if a high level of vehicle utilization is intended. If a vehicle is placed in such an area, it is likely to be utilized more frequently as at stations in residential or peripheral areas, where local residents still prefer to rely on their own car. Thus, the integration of an e-carsharing operation into an intermodal passenger transport concept appears to be a sensible undertaking. The accessibility of vehicles increases with the quality of the locally available public transport interfaces. The consequence is that vehicles at major public transport hubs are frequently used by commuters, city visitors and foreigners as well.

The identified relevant target group enfolds public, commercial and private users. Especially men between the age of 20 and 45 years with high levels of education, businesses in various branches and municipalities should be targeted for a quick build-up of a workable user base. Moreover, the pursuit of an open platform approach that aims to collaborate with various parties of its social and economic surroundings turned out to be a key success factor. More specifically, the involvement of open-minded public administrations is indispensable.

Based on the RUHRAUTOe-undertaking, valuable recommendations for the actual composition of all-electric vehicle fleets with different electric drivetrain technologies can be deduced. With regard to costs, the dominating usage environment and the vehicle applications, RUHRAUTOe has demonstrated that smaller all-electric vehicles are the key rentals. Nevertheless, one out of nine trips undertaken (11, 3 %) required vehicle ranges above 79 km which cannot be safely obtained with all-electric vehicles all year thus far. Hence, the inclusion of larger plug-In hybrid vehicles should be considered, whenever an all-electrical carsharing fleet is composed.

References

Barth, M., & Shaheen, S. A. (2002). Shared-use vehicle systems: A framework for classifying carsharing, station cars, and combined approaches. *Transportation Research Record, 1791*, 105–112.

Bundesregierung. (2014). *Energiewende - Elektroautos attraktiver machen. Press release*. Retrieved April 23, 2014, from http://www.bundesregierung.de/Content/DE/Artikel/2014/03/2014-03-27elektromobilitaet-hauptstadtkonferenz.html.

Bundesverband CarSharing e.V. (2014). *Was ist CarSharing*? Retrieved April 29, 2014, from http://www.carsharing.de/alles-ueber-carsharing/faq.

Canzler, W. (2010). Mobilitätskonzepte der Zukunft und Elektromobilität. In B. F. Hüttl, B. Pischetsrieder & D. Spath (Eds.), *Elektromobilität—Potenziale und wissenschaftlich-technische Herausforderungen, acatech—Deutsche Akademie der Technikwissenschaften* (pp. 39–62, 46). Berlin, Heidelberg: Springer.

Cervero, R., & Tsai, Y. (2004). *San Francisco City CarShare: Second-Year travel demand and car ownership impacts*. Retrieved April 24, 2014, from http://www.communauto.net/images/TRB2004002025.pdf.

Deffner, J. (2011). Schneller, öfter, weiter: Herausforderungen für eine mobile Gesellschaft von morgen. In H.-P. Hege, Y. Knapstein, R. Meng, K. Ruppenthal, A. Schmitz-Veltin & P. Zakrzewski (Eds.), *Schneller, öfter, weiter? Perspektiven der Raumentwicklung in der Mobilitätsgesellschaft. 13. Junges Forum der ARL 13.bis 15. Oktober 2010 in Mannheim, Arbeitsberichte der ARL 1, Hannover* (pp. 15–27). http://shop.arl-net.de/media/direct/pdf/ab/ab_001/ab_001_03.pdf.

Edwards, J. B., McKinnon, A. C., & Cullinane, S. L. (2010). Comparative analysis of the carbon footprints of conventional and online retailing: A "last mile" perspective. *International Journal of Physical Distribution & Logistics Management, 40*(1/2), 103.

European Federation for Transport and Environment (2010). *How clean are Europe's cars?—An analysis of carmaker progress towards EU CO_2 targets in 2010*. Retrieved April 29, 2014, from http://www.viviconstile.org/upload/vivi-con-stile/materiali/2011-09-car-company-co2report-final.pdf.

European Federation for Transport and Environment (2013). *How clean are Europe's cars?—An analysis of carmaker progress towards EU CO_2 targets in 2012*. Retrieved April 29, 2014, from http://www.transportenvironment.org/sites/te/files/publications/2013_09_TE_cars_C2_report_web_final.pdf.

Huwer, U. (2002). Pilotstudie zur Modellierung einer Schnittstelle zwischen ÖPNV und CarSharing. In: Grüne Reihe, Fachgebiet Verkehrswesen der Universität Kaiserslautern, Nr. 55. Kaiserslautern.

Jahns, F. (2009). CO_2-Bilanz von Elektroautos—Strom ist nicht gleich Strom: Eine Präsentation zur Vortragsreihe „Neue Entwicklungen auf den Energiemärkten" am Institut für Energietechnik, p. 4.

Katzev, R. (2003). Car Sharing: A new approach to urban transportation problems. *Analyses of Social Issues and Public Policy, 3*(1), 65–86.

Kley, F., Lerch. C., & Dallinger, D. (2011). New business models for electric cars—A holistic approach. *Energy Policy, 39*(6), 3392–3403.

Knapp, W., Schmitt, P., & Danielzyk, R. (2006). RhineRuhr: Towards compatibility? Strategic spatial policies for a specific configuration of polycentricity. *Built Environment, 32*(2), 137–139. (Special Issue: Reflections on the Polycentric Metropolis (2006)).

Loose, W., Mohr, M., & Nobis, C. (2006). Assessment of the future development of car sharing in Germany and related opportunities. *Transport Reviews, 26*(3), 365–382.

Müller, C., Benad, H., & Rennhak, C. (2011). *E-mobility: Treiber, Implikationen für die beteiligten Branchen und mögliche Geschäftsmodelle, Reutlinger Diskussionsbeiträge zu Marketing & Management*, No. 2011–09, p. 15. Retrieved April 23, 2014, from http://nbn-resolving.de/urn:nbn:de:bsz:21-opus-58269.

Nobis, C. (2006). Carsharing as key contribution to multimodal and sustainable mobility behavior. *Transportation Research Record: Journal of the Transportation Research Board, 1986*, 89–97. (Transportation Research Board of the National Academies, Washington, D.C.).

Rogers, E. M. (2003). *Diffusion of innovations* (5th ed., pp. 169–183). New York: Free Press.

Shaheen, S., Wright, J., Dick, D., & Novick, L. (2000). *Carlink—A smart carsharing system field test report—Prepared for partners for advanced transit and highways memorandum of understanding 380. Institute of Transportation Studies—University of California at Davis, Davis*. Retrieved April 29, 2014, from http://www.carsharing.net/library/PRR-2000-10.pdf.

Tukker, A. (2004). Eight types of product-service system: Eight ways to sustainability? Experiences from SusProNet. *Business Strategy and the Environment, 12*(4), 246–260.

Weikl, S. (2003). *Free-Floating Carsharing Systeme—Wirkung und Optimierungsstrategien*. Retrieved April 23, 2014, from http://www.ivt.ethz.ch/news/archive/20130929_ivt_tagung/20130930_weikl_free_floating_carsharing.pdf.

Cohousing and EV Sharing: Field Tests in Flanders

Sidharta Gautama, Dominique Gillis, Giuseppe Pace and Ivana Semanjski

Abstract This paper investigates the potential of electric vehicles (EVs) in a context of a pilot test in Belgium, consisting of car sharing services managed and exploited in small communities. Part of a broader testing activity in the framework of the e-Mobility NSR project, the test had the objective of metering EVs' charging and consumption in real daily transport operations, but soon it acquired new meanings. It shaped EV sharing services at a small scale, directly managed by the users, which also share energy and maintenance costs. The chosen context was cohousing, a special type of collaborative housing, and four ones were selected in Flanders: two urban units and two larger semi-urban ones. The two urban communities received a prepaid card for reserving and using EVs provided by Cambio, a Belgian car sharing company. The other two cohousings received two EVs and a charging box, organising and running an internal car sharing system for the duration of 1 year. During the tests, quantitative and qualitative data were collected. The paper reports the intermediate results, identifying potential EV sharing consumers, based on their behaviour and attitudes in relation to the condition of the local context in which they live.

Keywords Cohousing · Car sharing · Electric vehicle monitoring · Consumer insights · Field tests

---

S. Gautama (✉) · D. Gillis · G. Pace · I. Semanjski
Department of Telecommunications and Information Processing (TELIN), Ghent University, St-Pietersnieuwstraat 41, 9000 Ghent, Belgium
e-mail: sidharta.gautama@ugent.be

D. Gillis
e-mail: dominique.gillis@ugent.be

G. Pace
e-mail: giuseppe.pace@ugent.be

I. Semanjski
e-mail: isemanjski@fpz.hr

W. Leal Filho and R. Kotter (eds.), *E-Mobility in Europe*,
Green Energy and Technology, DOI 10.1007/978-3-319-13194-8_12

1 Introduction

The technological evolution of the electric vehicles (EVs) and their recent launch on the car market has opened new frontiers for sustainable mobility. However, it is a common view that a simple change from conventional to EVs cannot bring about a condition of sustainability (Bannister 2005; Dennis and Urry 2009; Gilbert and Perl 2008; Newman and Kenworthy 1999; Schiller et al. 2010; Cox 2010; Hickman and Banister 2014). The introduction of EVs will certainly contribute to moderate some environmental problems (i.e. oil dependence, pollution and global warming), but alone will not reduce the main economic and social problems (congestion, loss of urban land and accessibility) related to the 'hypermobility' phenomenon (Adams 2000). To move towards a sustainability condition, a renewing of culture and consciousness must support the emerging of new technologies, such as EVs, and new systems of relations, including policy, user practices, infrastructure, industry structures and symbolic meanings too.

Although the cultural problem is observable in all technological innovation, for the so-called "car system" (Dennis and Urry 2009)[1] it is even more evident. Having that system become a way of life and not just a means of transport from one place to another. Today, car culture is a dominant culture, a literary and visual icon explored through modernist literature and car movies. Since the last century, the car has turned into an object of desire, whose ownership and possess provides status to its owner through speed, security, safety, sexual success, career achievement and freedom (Dennis and Urry 2009, p. 36). Many interdependent effects have promoted the 'car culture', such as its being an exemplary manufactured object produced by leading industries and linked with other institutions, industries and related occupations,[2] whose profits are associated with those producing and selling cars and related infrastructure, products and services. Nonetheless, the car system has been associated to the growth of a socialisation based on the 'freedom of the road', enabling people travelling at any time in any direction. On the other hand, the car culture's success generated an overdependency on automobility in daily life, which bred an array of environmental, economic and social problems (Schiller et al. 2010, pp. 7–20). In particular, it causes serious problems of social injustice and inequity, not guaranteeing the principle of the 'access for all', typical or at least the stated normative ambition of the public transportation system, though even there pricing issues can lead to issues (Hutton 2013).

Today, no studies, researches and projects aiming at investigating the potentialities of EVs in terms of sustainable mobility can underestimate the cultural

---

[1]"Such a system consists of cars made of steel and weighing about 1 ton, powered by petrol, each seating at least four people, personally owned, and each driven independently of others", Dennis and Urry (2009, p. 28).

[2]"Licensing authorities, traffic police, petrol refining and distribution, road building and maintenance, hotels, roadside service areas and motels, car sales and repair workshops, suburban and greenfield house building sites, retailing and leisure complexes, advertising and marketing, and urban design and planning", Dennis and Urry, pp. 36–37.

problem at the basis of collective and individual choices in transport. As Sheller (2004, p. 222) wrote, "car consumption is never simply about rational economic choices, but is as much about aesthetic, emotional and sensory responses to driving, as well as patterns of kinship, sociability, habitation and work". However, the main cultural target for studies and surveys is considering the car as a market product and defining potential clients for EVs (e.g. Bunce et al. 2014), which should be "at least as effective as the current car at meeting people's economic, aesthetic, emotional, sensory and sociability requirements" (Dennis and Urry 2009, p. 64). However, remaining at the car level, those surveys typically miss out on exploring the more comprehensive 'system of connections', at the basis of new mobility patterns for EVs, which can provide value added in terms of significance and utility. Thanks to the shift to EVs, the new system should perform better and/or be more meaningful. Moving the focus from the product to the system requires for more integrated actions, based on interrelated elements, such as technology, policy, economy, society and culture change on both the demand and supply sides, stakeholders' involvement and long-term processes (Geels 2012). Though many surveys have analysed the policy side (e.g. changing prices, tax rates or technology), very few of them have investigated individual and collective behaviours and defined what many mobility gurus call simply 'the people' (i.e. Bunce et al. 2014). It is important to have clearly in mind that not all societies have the same cultural values, and even in the same societies, not all users have the same culture and/or socio-economic potentialities. Some population clusters, whose lifestyles are characterised by specific geographical, economic, social, and cultural factors, could have a pioneering role for testing sustainable mobility patterns, where EVs could replace conventionally fuelled vehicles (c.f. Hoogma 2002).

2 The Background

When the Interreg IVB project "E-Mobility North Sea Region" activities started in October 2011, the task of the Flemish team was to collect data and analyse smart grid models for supporting a sustainable use of the electric mobility in the North Sea Region (NSR). Planned activities included laboratory and field tests, aiming at measuring EVs' performances in terms of charging and consumption, possibly in real daily transport operations, and then defining some user cases relevant for the NSR. Nevertheless, the need of metreing EVs' charging and consumption in real daily transport operations invested the field tests with the possibility of acquiring new meanings. It was a remarkable opportunity for investigating new potential cultural changes related to the imminent EVs market penetration, and assisting the project's aim of "fostering the diffusion of the electric mobility and stimulating the use of public and private electric car transport as well as freight across the NSR".[3]

[3]http://e-mobility-nsr.eu/scope.

The choice of the tests' typology was very delicate, because the team's objective to test not only EVs performances, but also new mobility systems, giving centrality to transport connections[4] and promoting a shift from 'economies of ownership' to 'economies of access'.[5] Therefore, instead of accepting the most common option, that is, supplying different EVs to a selected number of test participant, the team decided to use the field tests as an opportunity for developing a new service, able to compete with the private owned car in terms of convenience and cost structure and, at the same time, contributing to reduce negative externalities in the cities.

The mobility pattern selected for the test was the car sharing, generally identified as "a short-period automobile rental services intended to substitute private vehicle ownership" (UTIP Secretary General 2002). For its set up, the test could not follow existing large urban or regional car sharing experiences, too big for the size of the project and the maximum available number of EVs. It was hence decided to design a service, which could fully meet the needs of small communities. The challenge was relevant and meant moving from an unpretentious but easily manageable car free-wheel test to the development of a new system, searching for the sampling population, training them and them monitoring their activities. On the other hand, the opportunity to obtain future collective behaviours answering to the hypermobility topic and to define a wider sampling of driving behaviours balanced the risks this entailed.

In order to perform as well as or better than the private car system, car sharing has to offer access to a vehicle whenever test participants require it. Its efficiency depends on the vehicle accessibility (within easy walking distance of people's homes), affordability (reasonable rates, suitable for short trips), convenience (vehicles that are easy to check in and out at any time), and reliability (available vehicles and a reliable booking and access system). The first option was to provide EVs, charging boxes, and monitoring applications to large condominiums, which internally had to organise their booking and car access system. However, practical experience says that condominiums have very limited capacity in sharing common goods, and the test could fail because internal conflicts on the EV use. In order to avoid that risk, the team decided to experiment with a car sharing approach in "cohousing"[6] communities.

---

[4]One of the main goals of the sustainable transportation is represented by the shift from the system based on the private car (apparently stable and unchanging), people (the cornerstone of all mobility systems, as drivers, passengers, and pedestrians), machines, materials, fuel, roads, buildings and cultures to a system where transport connections assume the primary importance based on needs/demands and a range of criteria around accessibility and use.

[5]The 'economy of access' develops the principle of "paying for access to travel/mobility services rather than the outright ownership of vehicles" (Dennis and Urry 2009, p. 97).

[6]The cohousing idea originated in Denmark, close to Copenhagen, where a Danish architect and a psychologist built the first cohousing community in 1972 for 27 families, influenced by Bodil Graae's 1967 article, "Every child should have 100 parents". Since then the cohousing movement has spread rapidly. Worldwide, there are now hundreds of cohousing communities, having expanded from Denmark into the U.S., Canada, Australia, Sweden, New Zealand, the Netherlands, Germany, France, Belgium, Austria, Japan and elsewhere. For European examples see, for instance: Institute for Creative Sustainability (2012).

Cohousing is a special type of collaborative housing in which residents actively participate in the management of their own neighbourhood. Cohousers are consciously committed to live in a community and to take care of common property. That builds a sense of working together, trust and support (c.f. Ruio 2014), which was a guarantee for the success of the car sharing tests. In addition, the new generations of cohousers are at least assumed to be getting much "greener" and, in general, are committed to develop photovoltaic or wind energy production and to start investigating EV's use too. They generally aspire to 'improve the world, one neighbourhood at a time' (c.f. www.cohousing.org). This desire to make a difference often becomes a stated mission, as the websites of many cohousing communities demonstrate.[7] While a certain flexibility characterises the cohousing design and organisation, easily adaptable to people's needs in different cultural contexts, two main typologies are predominant, the urban community and semi-urban/rural village. The first type, located right in the city centre, is organised in vertical buildings with common rooms (dining room, sport rooms and other facilities) but is lacking parking facilities and green areas. The size is very variable, from one building with about 10–12 people up to 184 apartments in 13 buildings accommodating more than 400 people (e.g. Stoplyckan in Linköping, Sweden) (see Krause 2012). The second type, on the contrary, is a village-like community, usually organised in attached or single-family homes along one or more pedestrian streets or clustered around a courtyard. In this type too, the size range is very variable, from seven to 67 people, but the majority of them houses 20–40 households (see, for instance, Institute for Creative Sustainability 2012).

In Flanders, cohousing is a growing way of inhabiting, but few ones are effective and many are still under construction.[8] The team selected four different cohousings, two small urban communities in Ghent—Papegaaistraat and Sint-Pietersaalststraat —without parking places, and two larger semi-urban communities, one located near the city of Brussels (La Placette in Wezenbeek-Oppem) and the other located near the city of Ghent (Vinderhoute), both characterised by open common spaces and parking facilities.

[7]See http://www.cohousing.org (USA), http://www.spatialagency.net/database/how/empowerment/co-housing (international), http://www.living-organically.com/cohousing.html (UK) and http://www.cohousing.org.uk/ (UK).

[8]Reasons can be found in the Belgian legal and administrative framework of housing ownership, structured on the single-family model. But as reported in a recent journalistic overview (http://www.flanderstoday.eu/living/co-housing-arrangements-win-popularity-flanders; as well as plans for the future: Blyth (2014), http://www.flanderstoday.eu/politics/flemish-cities-allot-land-co-housing-projects), some important successes have been achieved. The Flemish Parliament approved a resolution to encourage, stimulate and provide support for group housing projects (2009), the publication of a preliminary study on group housing titled Samenhuizen in België (Jonckheere et al. 2010) and the launch of the Samenhuizen Charter, stating that signatory cities and communities will support local group housing initiatives. Ghent was the first to sign in 2012, followed a year later by several Flemish cities, including Bruges, Kortrijk and Oostende, and then the province itself (Goodwin 2014). See web page of some cohousing projects in Flanders and mobility aspects: http://www.cohousingprojects.be/index.php/diensten-cohousing-projects/157-nieuw-ontdek-onze-deelauto-s.

Vinderhoute,[9] a nearly new cohousing not far from Ghent (about 10 km), was the first to receive the invitation to participate to the project tests. It is the association of seventeen families, mainly composed of young couples with children. The community shares a large parking area and a community building with a spacious kitchen, a large dining space, offices, workshops, a music room, a children's playroom and some guest rooms. Cohousers can easily book online the facilities and their joint management strengthens social contact and encourage spontaneous encounters between members. In addition, the community shares also a photovoltaic system for the production of energy (10 kW), used for supplying the common facilities, and all buildings are passive and low energy constructions. They were already thinking about a charging box in the parking, and accepted enthusiastically to test EVs with a car sharing approach. Furthermore, they informed the cohousing of 'La Placette' in Wezenbeek-Oppem about the possibility to take part to the test.

Built in 1986, La Placette is the result of the association of 11 families wishing to live together in a cohousing, based on the principles of "non-violence, self-management and social cohesion".[10] Each family owns a house, a private garden and a common garden. In addition to the garden, they have multiple common premises: a mini Amphitheatre, located on the side of the housing, a common house, a place of temporary home and a parking. In the interviews for the project, they acknowledged their interest in EVs and in participating in the test.

The two small urban communities, both situated in the centre of Ghent (Papegaaistraat and Sint-Pietersaalststraat), have private apartments and common rooms, but no parking facilities. Living in a mostly pedestrian and biking area, where parking is expensive and limited, only few of them owned a car and others were already customers of 'Cambio', the main commercial Belgian car sharing company.[11] After some internal meetings, both communities accepted to take part to the tests.

---

[9]see http://www.ic.org/directory/cohousing-vinderhoute/, http://gentintransitie.com/2013/08/07/cohousing-vinderhoute/, https://www.facebook.com/cohousingvinderhoute.

[10]see Eeman (2009), p. 32, and the following links: http://www.brusselnieuws.be/nl/nieuws/cohousing-de-lift; http://www.duwobo.be/media/Leer2%20COWO7%20Samenhuizen%20%5BCompatibiliteitsmodus%5D.pdf.

[11]Cambio is a car sharing organisation and operates in several Belgian cities. Wallonia was the first Belgian Region to start in 2002. Then in May 2003, Brussels followed and in September 2004 the Cambio car sharing system also started in Flanders. According to their website (http://www.cambio.be/), Cambio Belgium has more than 15,000 users, a car fleet with more than 500 cars spread across 220 stations in 27 Belgian cities. The company cooperates closely with VAB (the largest Flemish automobile association), De Lijn (public transport operator in Flanders), MIVB/STIB (the public transport operator in the Brussels metropolitan area) and TEC (the public transport operator in Wallonia). In 2009, the NMBS-holding (Belgian railways) also decided to participate in the project. This completes the cooperation between Cambio and public transport. Furthermore, local, regional and federal authorities participate: they help with the financing, give policy support and provide the necessary car sharing stations (parking places).

During a test period of about 1 year,[12] all the cohousings had to guarantee to organise a system for sharing the EVs among their members, to maintain and charge the EVs, to answer to the project questionnaires and to participate to events or demonstrations organised by the project. On the other hand, the project team had to lease EVs, stipulate a contract with each cohousing community, training the participants on EV driving, charging and identification systems, installing GPS loggers and other ancillary monitoring systems in each vehicle, organising an emergency number for any car default, designing and elaborating on questionnaires to deliver to the participants, advancing monitoring tests, and providing feedback to the test participants, about any technical and organisational topic.

3 Test Methodology

The subsequent step was to define the methodology for running the tests and collecting data, combining quantitative data about EV charging and consumption dynamics and qualitative data about perceptions of EVs and experiences with online questionnaires filled out by test participants.

The initial idea was to provide the communities with leased EVs, one or two according to the population size, and to install charging boxes inside their parking facilities. However, the participation of the two urban cohousing without parking facilities needed and demanded a different approach. That opened the way to another very fascinating investigation, made possible thanks to the support of Cambio, which was also starting to operate a car sharing service with EVs. The two urban communities received prepaid cards (with a distinct ID per each participant) for using EVs supplied by Cambio as part of their fleet.[13] Cohousers had to follow all Cambio rules for the booking, parking and charging,[14] and EVs were available connected to a charging box in parking facilities not far from the cohousings. In that way, thanks to the quantitative data collected by Cambio and delivered to the team, it was possible to compare the performance of a small community car sharing with a traditional one, and evaluating their results in terms of energy consumption, car performance and battery ageing.

Four EVs, three Nissan Leaf and one Peugeot Ion, were leased and delivered to the two semi-urban cohousings, together with a charging box installed in each

[12]Urban cohousings started in July 2013, earlier than semi-urban ones (October 2013), because for these last tests it was necessary to launch an open tender for the EVs leasing, and to install charging boxes in the cohousings parking premises and that took more time than getting prepaid cards from Cambio.

[13]The Cambio fleet includes the following EVs: Nissan Leaf, Renault Kangoo ZE, Mitsubishi i-MiEV, and Opel Ampera (only the last two available in Ghent).

[14]In particular for EV drivers, Cambio introduced specific rules, such as the obligation to mention the number of km planned to drive (with a max for 60–75 km per trip), and the obligation once finished of putting the EV in charge (see www.cambio.be).

cohousing parking facility. Quantitative data was collected by means of five different tools: a data logger installed in the car, a GPS, the CAN bus, the RFID scans of participants' badges, and finally the charging pole meter. Thanks to the real-time data collection, it was possible to monitor cohousers' different travel behaviours through the MOVE-platform.[15]

In order to evaluate in a dynamic way cohousers' perception of the EVs', and then to investigate of their potential change in culture and consciousness, the team prepared three questionnaires to be filled out online before, during and after the test period by each participant. At the time of writing this paper, the questionnaire survey cycle was not completed yet, but there is enough data to provide a comparison between the results of the first questionnaires, where respondents generally had a blurred idea of EVs, and the intermediate ones, where test participants started to know more about EVs than the majority of the population.

4 Qualitative Dynamic Analysis

The questionnaires intended to provide a survey of the 78 cohousers (47.5 % men, 52.5 % women) living in the selected four Flemish cohousing units, which had accepted to participate in the tests.

In terms of the test population composition (Table 1), the urban and semi-urban cohousing communities displayed relevant differences in size, marital status and number of children, which suggested clustering test populations in two groups and comparing behaviours and needs. In fact, the two urban communities have a small number of cohousers (5–7), predominantly with an age range from 18 to 35 (only one is older), singles (though two are cohabiting), and all with no children. The two semi-urban ones have a bigger population (33–35), composed mainly by families with children (under 18 not counted in the test population). At Vinderhoute, participants' age ranges from 26 to 50 years, while La Placette has a very mixed composition of elder and younger people (as it already consists of two generations of familiar groups).

In terms of education of the stakeholders, i.e. school diplomas and professional status (Table 2), the differences between the two clusters are not very relevant, with 95 % of the participants having a high school or university educational attainment level and 77 % of the participants a part- or full-time job. That confirms the picture of the cohouser as a highly educated and professionally integrated person, searching for a more social and liveable way of inhabiting. In addition, most of the working participants have a daytime job and a regular working address out of the home (77 %).

Subsequently, the survey investigated participants' mobility behaviours for different purposes, in terms of frequency and travel mode. In terms of trip frequency

[15]See http://move2.ugent.be/index.php/en/.

Table 1 Composition of the sample population per cohousing community

	La Placette	Papegaaistraat	Sint-Pietersaalststraat	Vinderhoute	Total
Number of inhabitants	35	7	5	33	80
Participants in the survey	34	7	4	33	78
Gender					
Male	16	3	2	16	37
Female	18	4	2	17	41
Age					
18–25	8	2	0	0	10
26–35	7	5	3	13	28
35–50	0	0	1	16	17
51–65	19	0	0	4	23
Marital status					
Single	9	6	3	3	21
Married	19	0	0	25	44
Cohabiting	6	0	2	5	13
Number of children					
0	7	6	5	7	25
1	7	0	0	4	11
2	4	0	0	13	17
3+	16	0	0	9	25

Table 2 Education attainment level and professional situation of the sample population (n.)

	La Placette	Papegaaistraat	Sint-Pietersaalststraat	Vinderhoute	Total
Highest educational attainment					
Secondary school or lower	4	0	0	2	6
High school	15	2	2	16	35
University	15	4	3	15	37
Professional status					
Student	6	0	0	0	6
Inactive	2	1	0	4	7
Part-time job	9	2	1	8	20
Full-time job	15	3	3	19	40
Blank	2	0	1	2	5

per purpose (Table 3), all working participants travelled more often to work (daily or at least several times a week), less frequently for shopping or for recreation (from once to several times a week). The only relevant difference between urban and semi-urban cohousings was the frequency of trips to take or collect other people, where urban cohousers travelled monthly or never and semi-urban ones daily or weekly, clearly because of their different familiar composition.

In terms of travel mode, participants had to specify the mode predominantly used for different purposes. In general, the car is the dominant travel mode (Table 4), but differences between urban and semi-urban clusters are very relevant, mainly

Table 3 Frequency of trips for different purposes (n.)

	To work	To a shop	To take or collect other persons		For recreation
			Urban	Semi-urban	
Daily	46	4	0	14	5
several times a week	26	28	0	13	37
Weekly	1	35	1	21	27
Monthly	1	11	9	12	9
Never	4	0	2	6	0

Table 4 Most frequently used transport mode for different purposes

	To work		To a shop		To bring or get other persons		For recreation	
	Urban (%)	Semi-urban (%)	Urban (%)	Semi-urban (%)	Urban (%)	Semi-urban (%)	Urban (%)	Semi-urban (%)
Car	17	47	8	68	41	64	8	59
Private car	17	32	8	56	41	53	8	44
Company car	0	15	0	12	0	11	0	15
Shared car	0	1.5	0	9	25	9	8	11
Bicycle	33	26	83	14	17	9	75	24
Train	50	6	0	0	8	1	0	0
Tram or bus	0	7.5	8	4.5	0	3	0	4.5
On foot	0	9	0	1.5	8	11	8	0
Other	0	3	0	3	0	3	0	1.5

because of their geographical location and their familiar status. For commuting to work, urban cohousers mainly use the train (50 %), with the bicycle as the main alternative (33 %). For semi-urban cohousers, the car (private and company car) is the main means (47 %), although with a lower rate than for other purposes, and bicycle (26 %) is the main alternative. For shopping, almost all urban participants use the bike, being already in the shopping area (83 %), whereas semi-urban ones tend to take the car (68 %), with the bicycle as a remoter alternative (14 %). Only for taking or collecting other persons (i.e. children, parents, friends, colleagues), the two sample populations have similar behaviours (although with completely different frequencies) with the car being the most common travel mode (41 % for urban people, 68 % for semi-urban). It is interesting to remark that 25 % of urban cohousers use shared cars for such a more sporadic activity. The main alternative for semi-urban cohousers is walking (11 %), using a shared car (9 %) or cycling (9 %). Finally, for recreational trips, urban cohousers use the bike (75 %) in contrast with the semi-urban ones, which most often use the car (59 %), with biking (24 %) and shared cars (11 %) as the most common alternatives.

A second part of the questionnaire asked for project-related questions, starting by their motivations to participate to the tests. As reported in Fig. 1, for more than 70 % of the participants the main motivations are to help the environment, to contribute to the development of electric cars, and the belief in electric cars as the

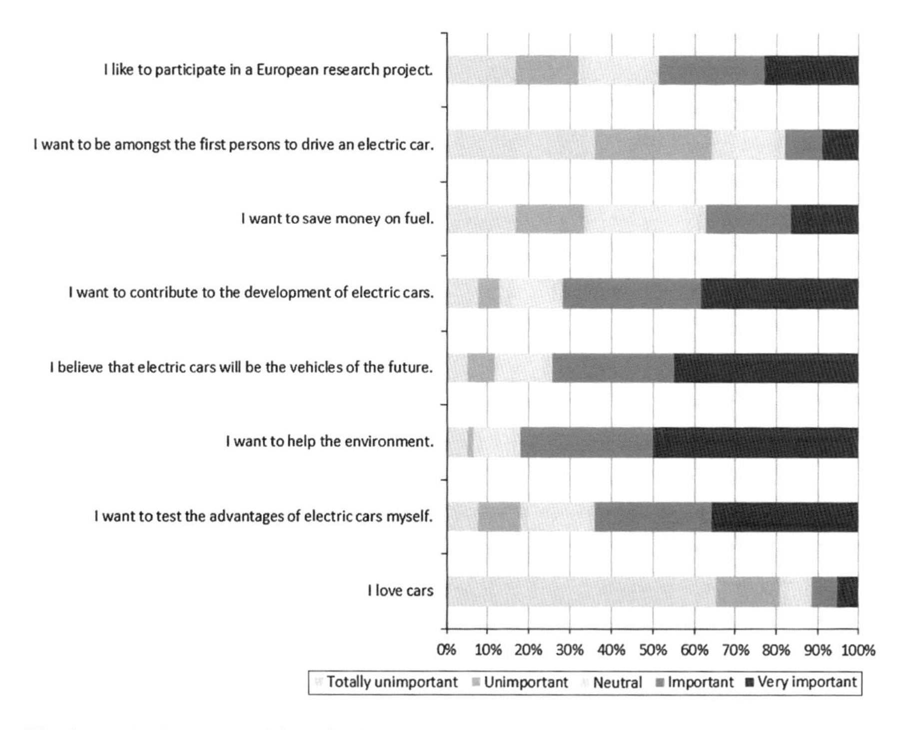

Fig. 1 Motivations to participate in the e-Mobility project

vehicles of the future. That emphasises participants' strong environmental and social commitment.

A confirmation of the common cultural approach comes out from the participants' answers on what aspects are determining their choice of a car (Fig. 2). Both groups answered similarly, assigning priority to environmental impact criteria (emissions, safety) and tangible car values (reliability, price and fuel consumption). More than 70 % of the participants find these attributes (very) important.

Over 60 % of the participants judge luxury criteria, such as car appearances, brand and technology gadgets as unimportant. The two cohousing types also provided homogenous answers about aspects that (may) keep them from purchasing an electric vehicle (Fig. 3). The most important barriers to buying an EV are their actual high purchasing price (more than 80 %), the limited driving range (more than 60 %) and the problems related to battery charging, such as time needed, and charging point availability (ranging from 30 to 50 % of the participants). Other possible topics, such as the EVs' limited performances or the limited number of brands and types, are not a problem for over 50 % of the participants. In addition, unfamiliarity with electric cars or safety doubts are not considered an issue at all.

When asked about which measures could stimulate the purchase of an electric car, most participants answered that they believed in the effectiveness of some type of public financial support (Fig. 4). Over 60 % think that exemption from taxes, free

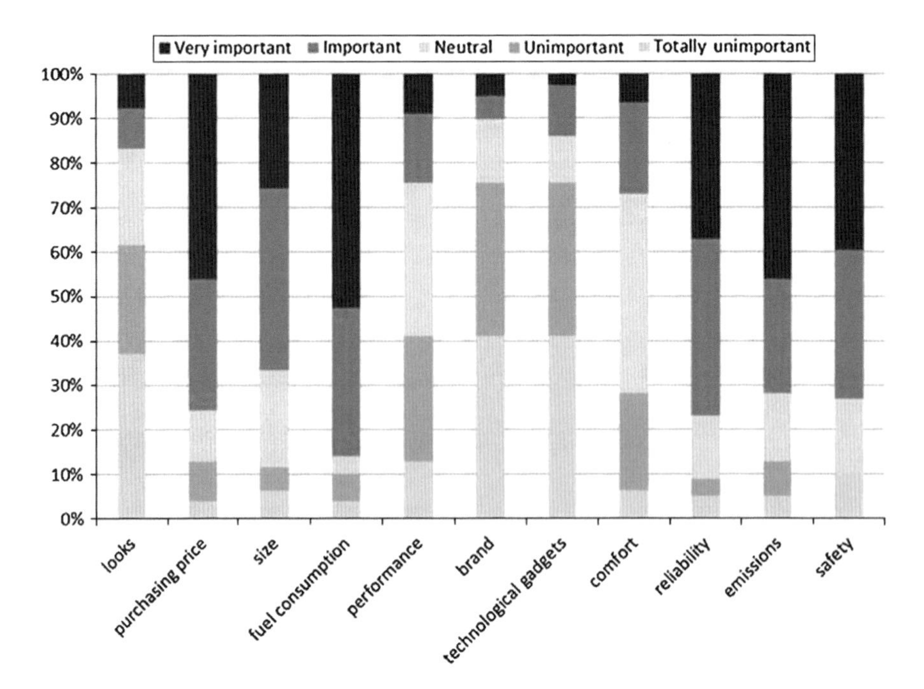

Fig. 2 Attributes' relevance in the choice of a car

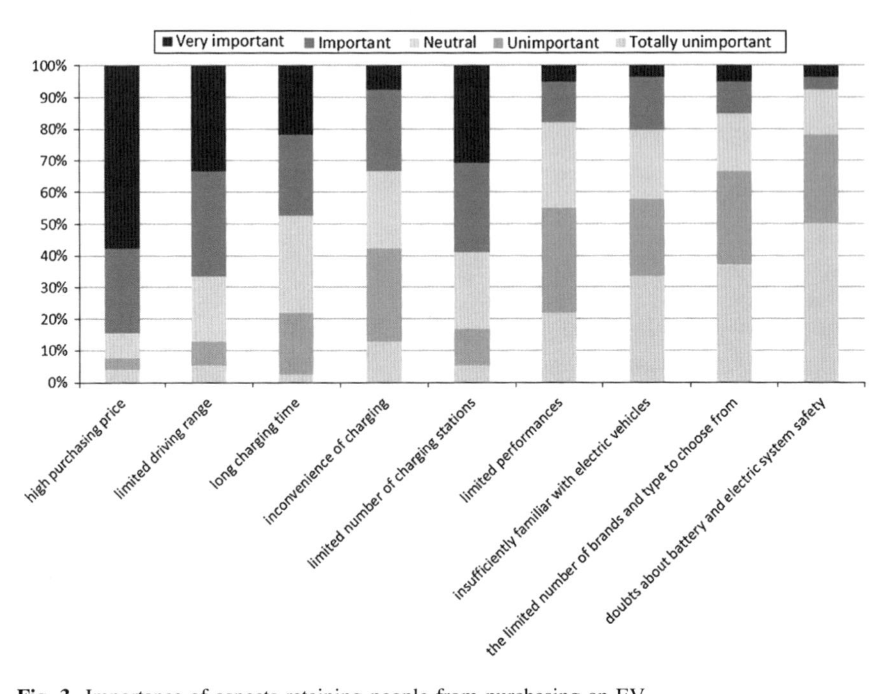

Fig. 3 Importance of aspects retaining people from purchasing an EV

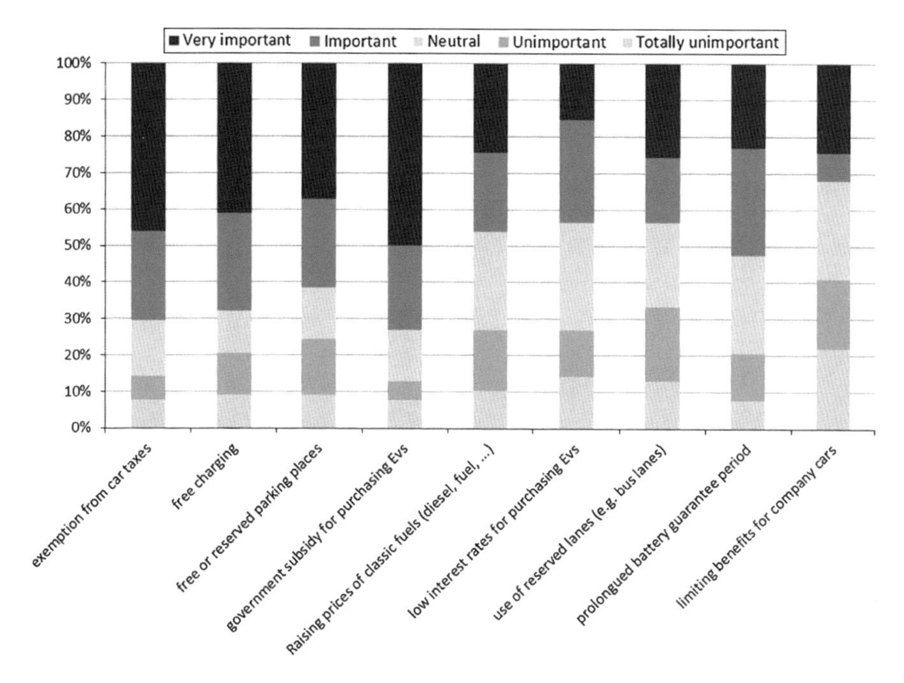

Fig. 4 Belief in government actions to stimulate the purchase of an electric vehicle

charging or an EV purchasing subsidy could be (very) important governmental actions.

As a proxy for their initial expectations about the EV, participants answered as to how the EV would score for a number of criteria, in comparison to a conventional car (Fig. 5). They expected that EVs could outperform conventional cars in terms of facility to use and acceleration. On the other hand, people assumed that conventional cars would score better in terms of top speed, the ease of charging (refuelling) and the availability of charging stations (fuel stations). Concerning safety, design and car interiors, there is no clear preference and a large share of the participants express no opinion. It is interesting to see that for the overall impression, people tend towards the EV, although 50 % participants state no opinion.

In terms of expectations about the EVs' performances, in absolute terms, they are higher for energy consumption, environmental score and vehicle noise: over 80 % have high to very high expectations about all these items. Also about the ease of driving, the reliability and the safety of the EV over 50 % have high to very high expectations. On the other hand, over (40 is correct) 50 % of the participants have (very) low expectations about the cars' top speed and acceleration and about the options and gadgets in the car (Fig. 6).

In addition, participants were asked to answer about their expected use of the EV sharing in the cohousing. Again, their answers expressed relevant differences between urban and semi-urban cluster. Urban cohousers expect a limited use of the shared car: only 25 % at least twice per week, and more than 33 % less than once

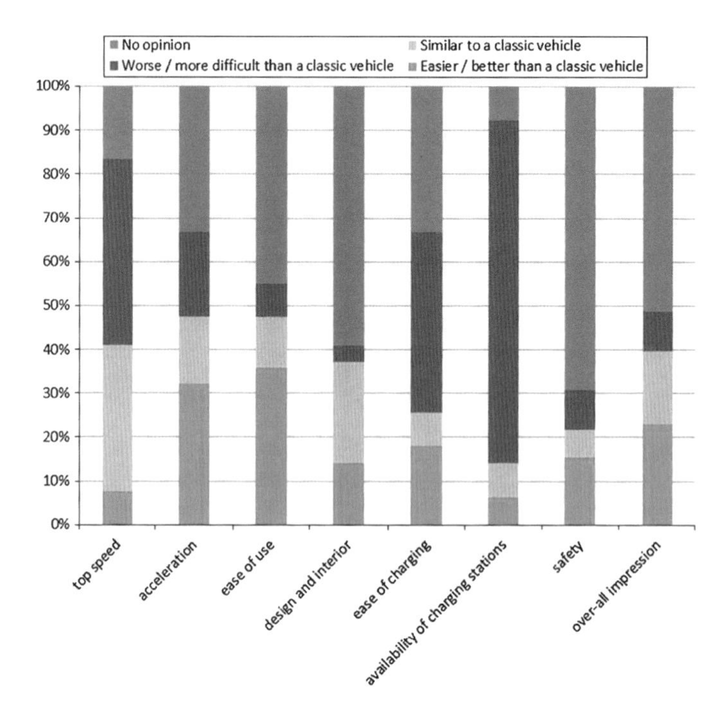

Fig. 5 Expectations about the performances of the electric car, compared to a classic car

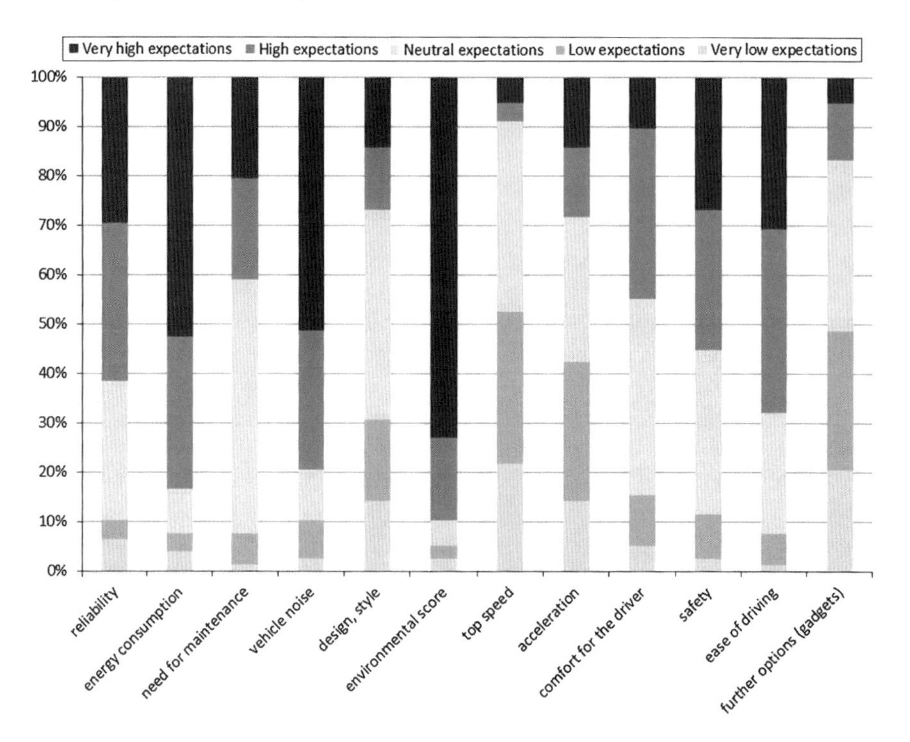

Fig. 6 Expectations about the performances of the electric car

per week. The semi-urban ones, on the contrary, expect a higher use of the EVs, at least twice per week (71 %) or once (27 %). To the question as to when they planned to use the shared EV, on weekdays or weekend, and during daytime or in the evening, both clusters anticipated weekend days during daytime as the busiest period (Table 5). Only 17 % of the urban participants planned to use the car during the week during daytime, revealing a certain tendency to consider the car sharing as a part of the free time lifestyle. The semi-urban cluster, on the contrary, gave the idea of a certain willingness to integrate this new car sharing into their daily activities.

The reason for that come out from the participants' answers to the question "how do you plan to use the shared vehicle" (Table 6). In the urban cluster, only 33 % of the participants own a car and consider the shared car as an extension of their mobility, and the others did not intend to change their pre-existing behaviours. In the semi-urban cluster, although 59 % of the participants only considered the shared car as a second car, very promisingly 20 % considered disposing of their current car and replacing it with the shared vehicle.

After 6 months of EV sharing, cohousers responded to a second round of survey. Results show how the participants' perception of EVs have changed due to their tangible experience. Participants answered questions about the main aspects relevant for the choice of an EV, the actual barriers to purchasing one, EVs' performances in absolute terms and compared to conventional cars, and reasons for not using them during the tests (but one should note peer communication amongst cohousers may well still have influenced their perceptions of EVs after 6 months). The two clusters, this time, did not show relevant differences, featuring a common cultural background.

The main relevant aspects in choosing an EV (Fig. 7) are comfort, reliability and emissions (84 % each), followed by safety (77 %), performance (75 %), fuel

Table 5 Expected use of the shared vehicle: when do you plan to use the shared vehicle?

When do you plan to use the car?	Urban (%)	Semi-urban (%)	Total (%)
On week days, during the daytime	17	35	32
On week days, during the evening	25	30	29
On weekend days, during the daytime	58	48	50
On weekend days, during the evening	42	27	29

Table 6 Expected use of the shared vehicle: how do you plan to use the shared vehicle?

How do you plan to use the car?	Urban (%)	Semi-urban (%)	Total (%)
As an addition to my current car (as a second car)	33	59	55
I don't own a car	67	14	22
To replace my current car	0	20	17
(blank)	0	8	6

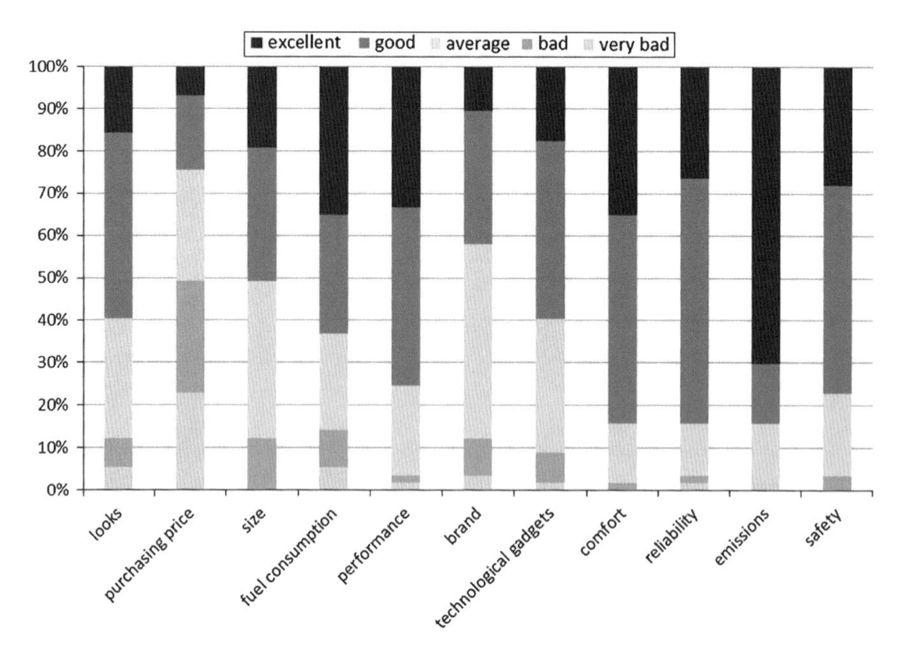

Fig. 7 Attributes' relevance in the choice of an EV

consumption (63 %), design and technology (60 % each). The only aspect relevant for not preferring an EV is its high purchasing price (only 25 % rated this as excellent or good). Comparing these participant responses with those in Fig. 2 referring to a generic car, the attributes matching their car choice priorities are reliability, emissions, safety and fuel consumption.

Participants' experience confirmed their opinion about barriers to EV purchasing. Furthermore, the EV score compared to a conventional car has not changed in a relevant way. However, they developed an awareness and 'no opinion' responses disappeared completely. Major changes can be seen in the EV performance evaluation (Fig. 8). Their EV appreciation has increased in terms of reliability (60–80 %), maintenance (40–60 %), vehicle noise (80–95 %), design and style (27–60 %), top speed (less than 10–65 %), acceleration (30 to almost 80 %), driver comfort (45–85 %), safety (55–70 %), ease of driving (65–90 %) and accessories and gadgets (15–65 %). Reversely, it has diminished in terms of energy consumption (85–70 %) and it is stable on the environmental score.

The intermediate survey was also a good opportunity to investigate the cohousers' car sharing experience, as well as to compare differences between small community car sharing and the traditional one, as perceived by the cohousers. Consequently, the two clusters have been analysed separately and then compared. The first question is about barriers to the use of the EV in a shared mode (Fig. 9a, b). Both urban and semi-urban cohousers, in a mainly positive evaluation, consider the insufficient driving range of the EVs as the main barrier (important to very important

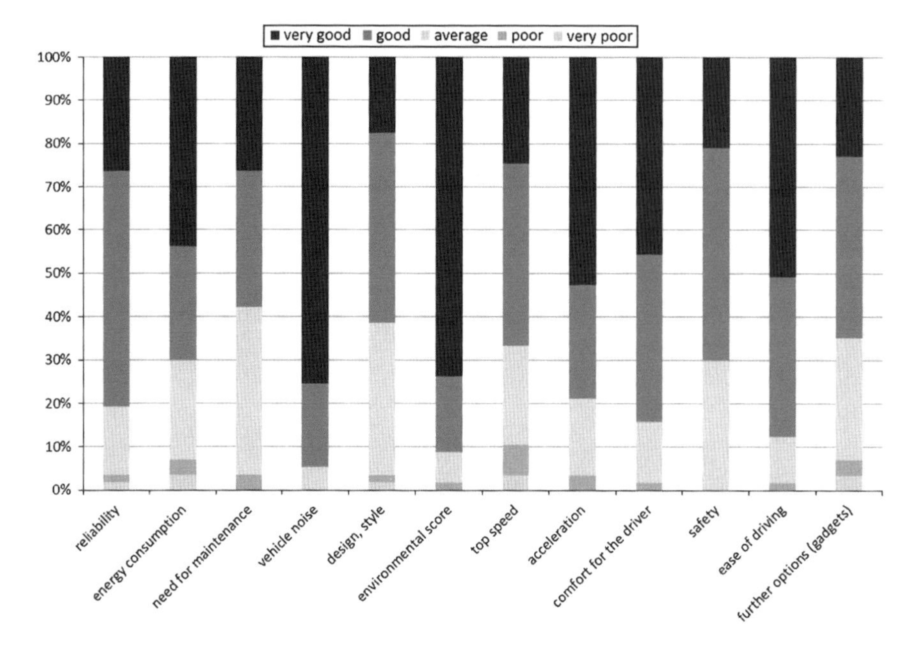

Fig. 8 Evaluation of the performances of the electric car after 6 months use

for 50 % of urban cluster and for 70 % of semi-urban cluster). Ranked in the second place of importance, both groups mention the problem of the non-availability of an EV (40 % urban and 35 % semi-urban). For the large majority of urban participants, the other potential barriers, such as car reservation, delays in obtaining the car from the previous user or insufficient charge, were not important.

The semi-urban participants reveal wide-ranging opinions, with some of them considering as important the need for car reservation (for 30 % important to very important), the preference for the independence of owing a car (30 %), the non-practicality of the car (23 %) and the EV often not sufficiently charged (21 %).

The second question was about how they used the shared car, in order to have a comparison with their initial expectations. The small number of answers received (return of 57 out of 78) do not permit a comparison yet; nonetheless, seven of them confirmed the willingness to replace their current car with an EV. Finally, participants were asked to answer as to how they would make their trips, if the shared car had not been there (Table 7). The car is the first option for the 70 % of urban cohousers and 87 % of semi-urban ones. However, in connection with the second option the two clusters provided very relevant differences. Without car sharing, 40 % of urban cohousers would have used public transport or bikes to make their trips. On the contrary, 19 % of the semi-urban would not have been able to do the trip, and only 13 % would have used a bike and even less public transport (6 %).

If confirmed by the quantitative data, that information could be the most relevant one for developing a sustainable transportation approach. Car sharing, on the one

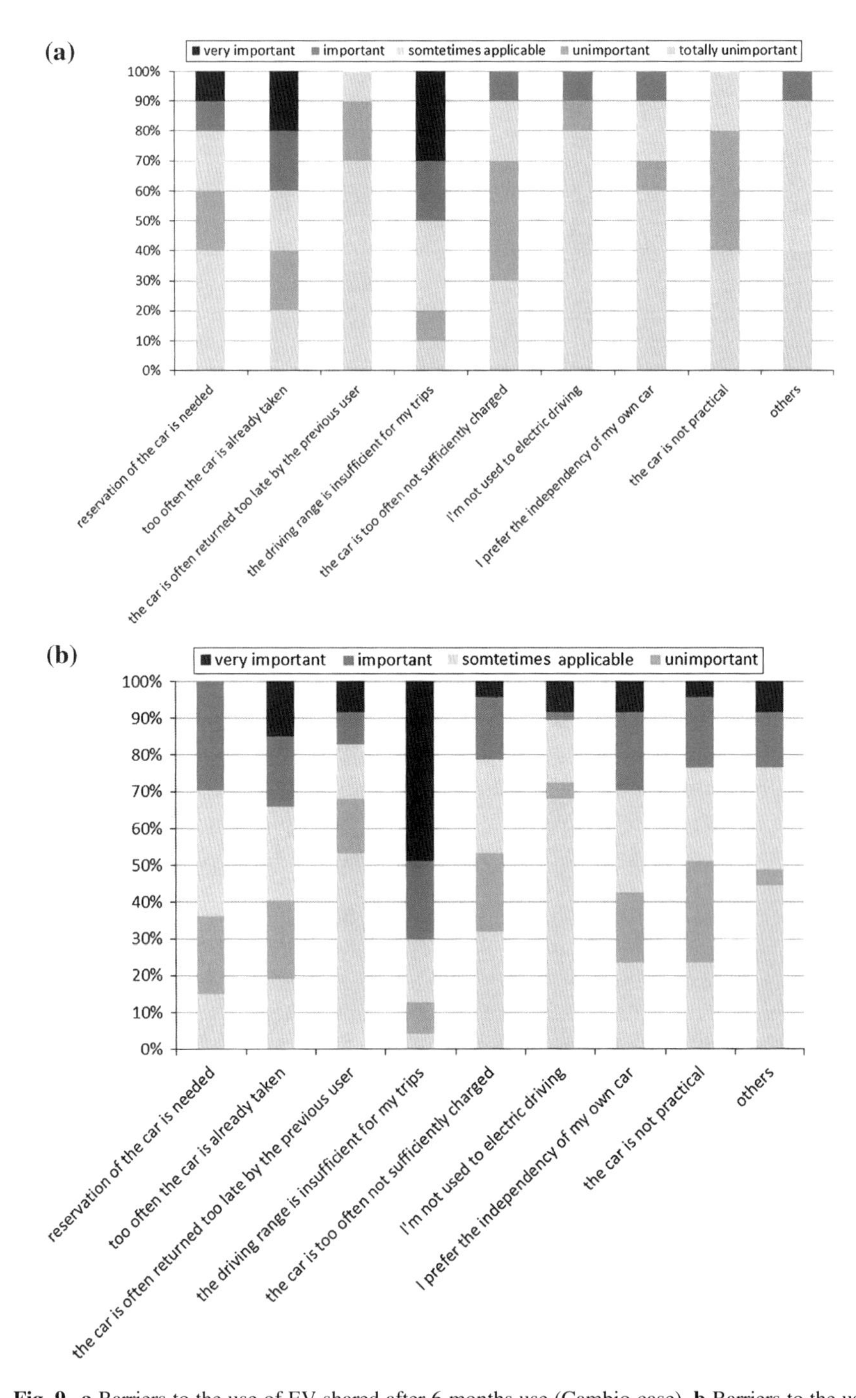

Fig. 9 **a** Barriers to the use of EV shared after 6 months use (Cambio case). **b** Barriers to the use of EV shared after 6 months use (small community EV sharing)

Table 7 If the EV sharing would not be available, how would you make those trips?

	Urban (%)	Semi-urban (%)	Total (%)
Not do those trips	10	19	17
Use another car	70	87	84
Use public transport	40	6	12
Use the bike	40	13	18
Choose a closer destination	10	4	5

hand, for the urban cohousing rivals with public transportation and soft mobility (active travel or reduced mobility). On the other hand, for the semi-urban communities it definitely rivals with the private car and represents a relevant extension of the users' mobility.

5 Analysis of Travel and Charging Behaviours

Although tests are still running and data collection is not yet completed, it is already possible to have a preliminary insight into how cohousing communities use and share EVs, by analysing trip distances and charging behaviours. For the charging data, participants are clustered into urban and semi-urban cohousers, similarly to the qualitative data presented above, in order to filter results according their different geographical positioning and car sharing organisation. In particular, although this is not a critical characteristic for understanding the way they use EVs, the different organisation of the EV sharing service provided to cohousing community surely influences the interpretation of some of the findings. In particular, having to pay the electric bill for the EV charging, semi-urban cohousers pay more attention to the energy consumption, charging EVs possibly night time and only when necessary. On the contrary, Cambio users are obliged to charge the car once they finished using it.

5.1 Trip Distance Analysis

Taking into account official statistics for the whole Flanders region, the average trip distance is 28,795 km. However, the numbers are quite different for the Ghent urban area, where trip distance average is only 11.964 km (Reports OVG Flanders 2013). When measuring cohousing trips, for one trip we mean a round journey to destination and back to the starting point (including all intermediate stops made during this trip). Therefore, if a user goes to pick up kids from school and on the way to school stops at supermarket to buy groceries for dinner, that all would be considered one trip (not three separate trips). Table 8 provides a summary of

Table 8 Trip distances descriptive statistics (km)

Mean	Median	Minimum	Maximum
27.044	20.00	0.00	172.00
Lower quartile	Upper quartile	Quartile range	Standard deviation
11.00	37.00	26.00	25.055

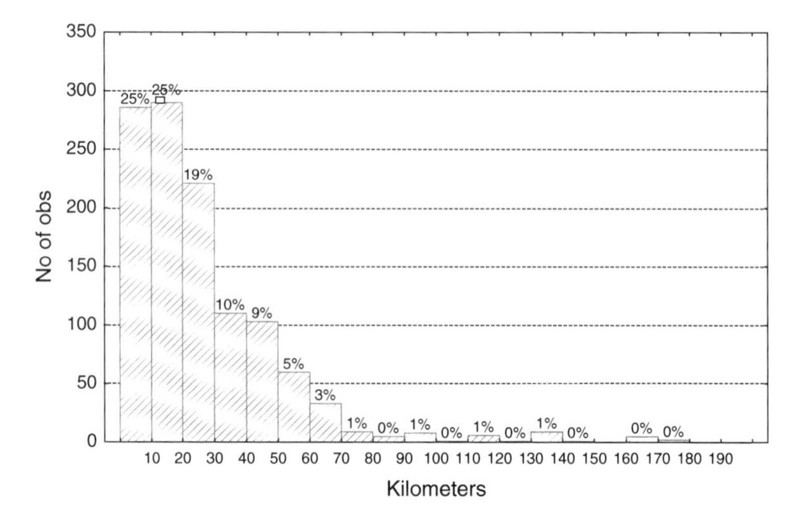

Fig. 10 Distribution of trip distances (n. of obs)

descriptive statistics for trip distances travelled by the cohousers during the period of 9 months and Fig. 10 shows the distribution of trip distances over the same period, with trip distances measured in km, and number of observations (=obs) to mean number of trips as defined as above.

The average trip length is around 27 km, with a median of 20 km and a standard deviation of around 25 km. The shortest trip recorded is less than 1 km long, while the longest trip is 172 km long. 25 % of all trips are less than 11 km long, and 25 % of them are longer than 37 km, which means that half of all trips made are over a distance of between of 11 and 37 km. When taking a closer look at the graphical representation of trip distances, 50 % of all trips are shorter than 20 km and just 4 % of trips are longer than 70 km. If we consider that the autonomy of a fully charged EV ranges between 80 and 140 km, it is possible to have at least three trips with one charge, which means that EVs do not need to be fully charged before each trip.

The trip distance analysis is only at the beginning and, once completed the data collection, it will compare differences between urban and semi-urban clusters' travel behaviours, also segmenting participants in terms of age and familiar status. In addition, there will be a check of the daily number of trips per EV and per community, frequency of use per participant and in which part of the day/week. All those data will be compared with the qualitative surveys and provide information on the level of the service acceptance for both different car sharing approaches.

5.2 Charging Behaviour Analysis

Charging behaviour is analysed based on two observed values, battery State of Charge (SoC) and the time of the day when the vehicle is plugged in for recharging (and unplugged). Figure 11 shows the distribution of battery SoC value at the start of recharging event for the Cambio service users, while Fig. 12 shows the same distribution for cohousing community where vehicles were provided for research purposes.

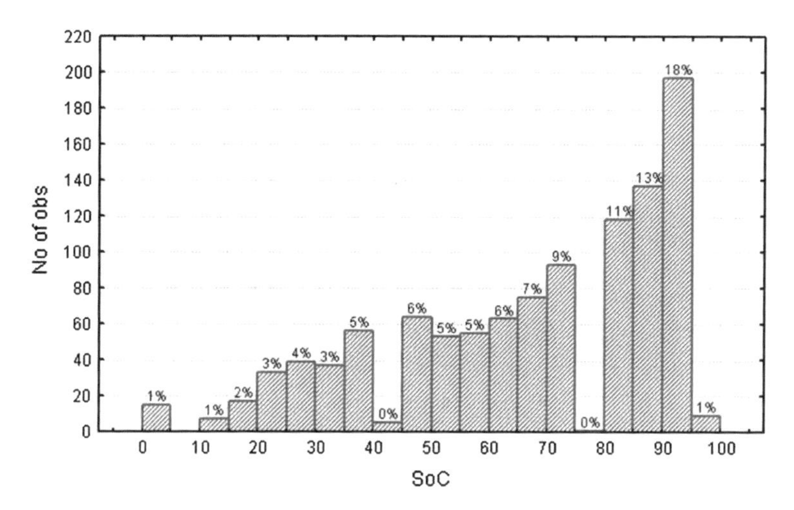

Fig. 11 Distribution of battery SoC at the start of recharging event for Cambio users

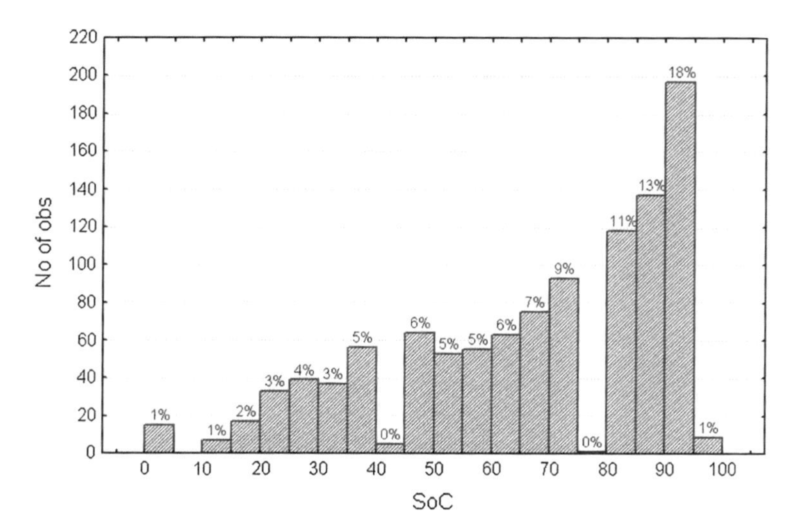

Fig. 12 Distribution of battery SoC at the start of recharging event for Wezenbeek-Oppem cohousing

In Fig. 11 it can be seen that values are quite oriented towards a higher battery SoC, and that the highest value for SoC, at the beginning of the recharging event, is between 90 and 95, meaning that 18 % of times recharging stated when SoC was in this interval. This is due to the Cambio car sharing rules/terms of conditions for users that after every trip users should start recharging of the battery for the next user so as to have as high as possible battery's SoC at the start of the trip. Therefore, Fig. 11 gives a good overview of how much battery is degraded during the individual trips. Figure 12 gives a better description of cohousing members' behaviour regarding when they consider themselves that they should recharge their car's battery. The distribution of values of the SoC at the start of recharging event in Fig. 12 is also oriented towards a higher battery SoC but with highest frequencies of SoC values in interval between 80 and 90. Also, 55 % of users will consider that they should recharge even if the SoC is higher than 70, while only 5 % of users will wait till SoC gets below 15 and none of the users waited till SoC is lower than 10 to start recharging electric car's battery.

Figures 13 and 14 provide an overview of the SoC value at the end of recharging process. For Cambio users it can be seen that batteries are charged to its maximum capacity and that users unplug vehicle in 89 % of cases when it is full. This also means that in between vehicles users prefer to share the one with highest SoC, while leaving others to recharge, and it can only happen if there is always sufficient number of EVs available for the users to share.

As the number of available EVs for non-Cambio users is five times less than for Cambio users, the distribution of SoC values are different in the case of the Wezenbeek-Oppem cohousing community. In Fig. 14 it can be seen that just in 43 % of cases battery's SoC was recharged to its highest value and this is less than half of the percentage for the same values at Papegaaistraat and Sint-Pietersaalststraat.

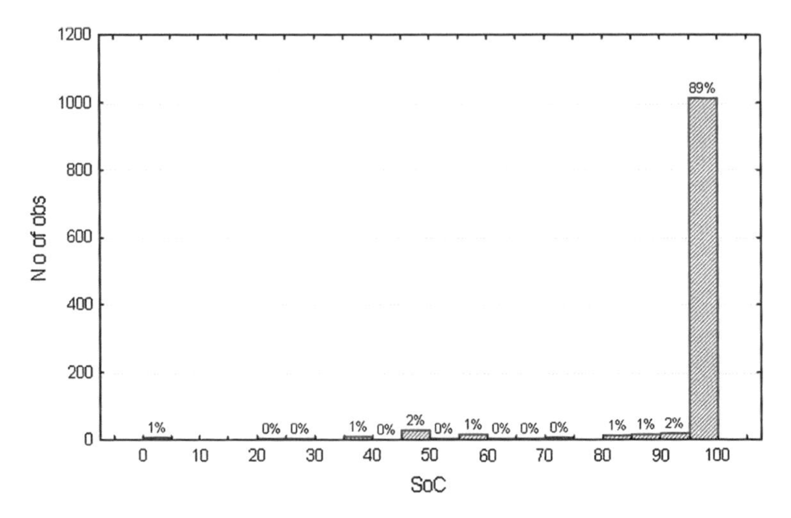

Fig. 13 Distribution of SoC values at the end of recharging event for Cambio users

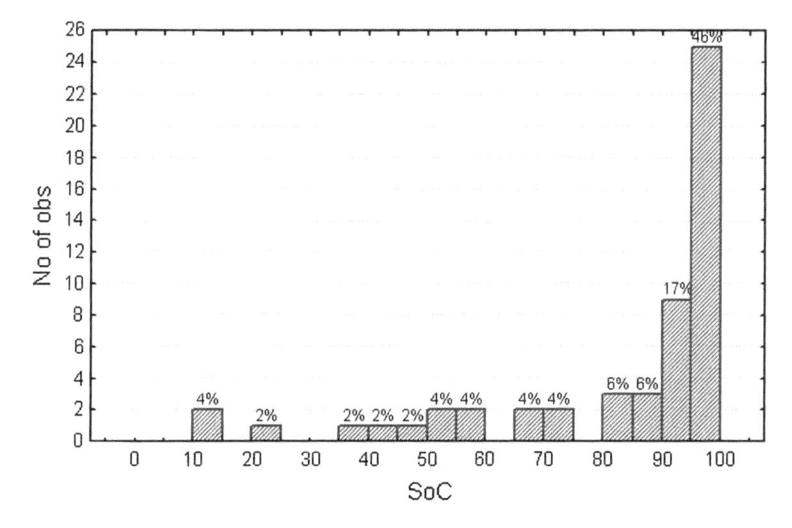

Fig. 14 Distribution of SoC values at the end of recharging event for Wezenbeek-Oppem cohousing

In Wezenbeek-Oppem cohousing, 12 % of times recharging was stopped although SoC value was below 50. Differences in the times of the day when users start (Fig. 15) and stop (Fig. 16) the recharging process suggest that different SoC distributions can not only be explained by the availability of shared vehicles, but also by the differences in cohousing's population characteristics.

For Cambio users, the average time of the day when they return and plug in the car is 13:17 h, though the time window is wide enough, and the distribution of starting time of day for recharging time's distribution has quite normal shape. For Wezenbeek-Oppem cohousing, the time window in which recharging starts is more narrow (from 10 until 23 h) and has an average value of 15:45 h, with peaks at 11,

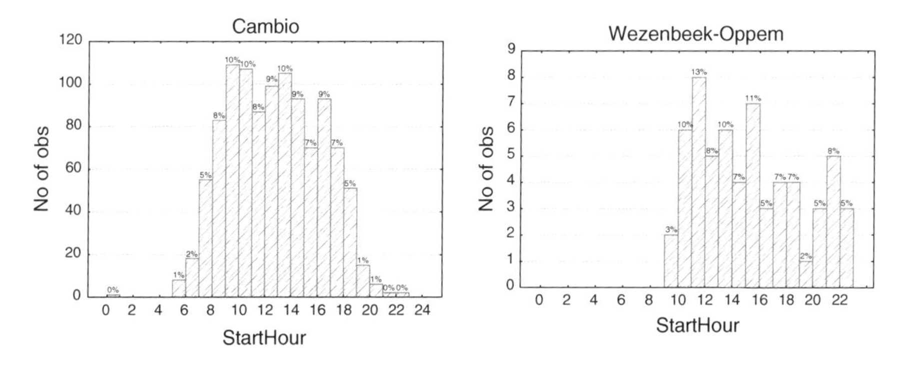

Fig. 15 Time of day when recharging is started (*left* Cambio users, *right* Wezenbeek-Oppem cohousing)

15 and 21 h. On the other hand, the time of day when recharging is finished for Cambio users has a rather similar shape as the distribution of start of the recharging times though with a bit more of a shape towards the end of the day. The average time of day when Cambio users end battery recharging is at 15:53 h, and this has two modal values (15 and 19 h); for the Wezenbeek-Oppem users the average time of day when recharging is finished is quite similar (15:17) but with two peaks, a late morning peak at 11 h and an afternoon peak at 17 h.

When considering how long the battery is charged (Fig. 17), Wezenbeek-Oppem users most of the times recharge for an hour, with 67 % of times less than 2 h and never longer than 13 h. Cambio users on average recharge the vehicle for a duration of 4:18 h, with 36 % of times less than two hours and 2 % of times the vehicle stays in charge more than a day. In line with that, Cambio users recharge the vehicle on average 0.44 times per day, while users in Wezenbeek-Oppem averagely recharge three times more (1.3 times per day), as shown in Fig. 18.

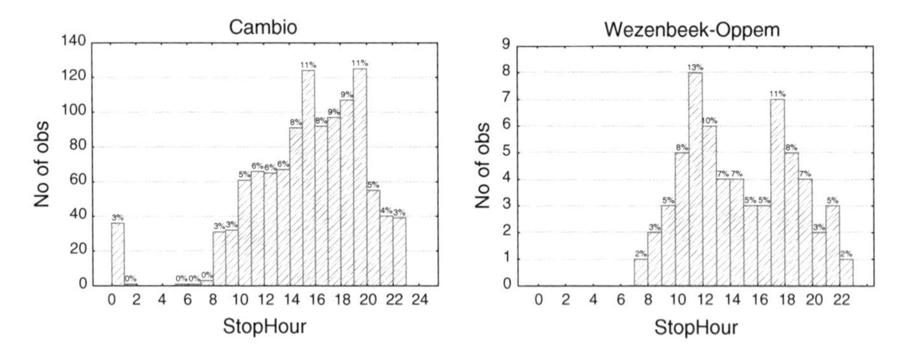

Fig. 16 Time of day when recharging stops (*left* Cambio users, *right* Wezenbeek-Oppem cohousing)

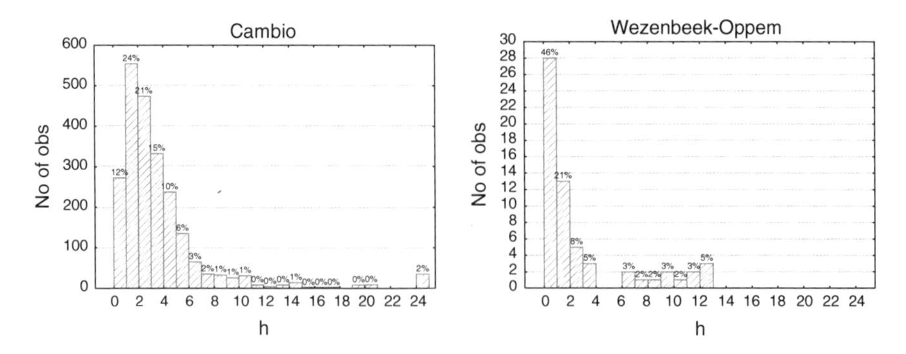

Fig. 17 Duration of recharging (*left* Cambio users, *right* Wezenbeek-Oppem cohousing)

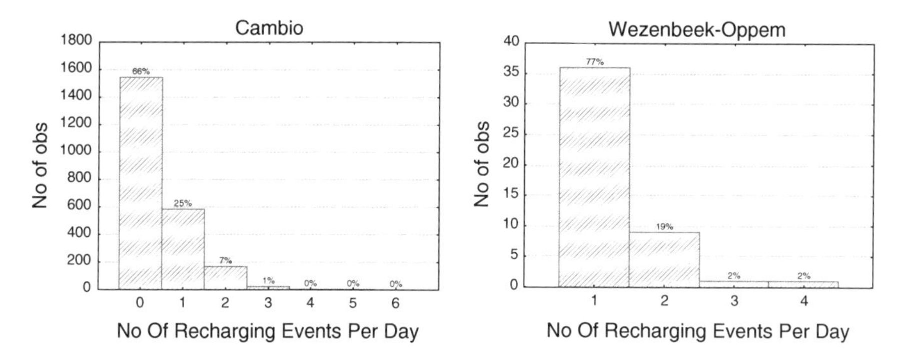

Fig. 18 Number of recharging events per car per day (*left* Cambio users, *right* Wezenbeek-Oppem cohousing)

6 Conclusions

The car sharing test conducted with the cohousing communities, although not complete yet, has provided some interesting findings. First, it defines a potentially new car sharing model, which can provide value added to EVs in terms of significance, utility and performance. The test population has been segmented into at least two different geographical clusters, urban and semi-urban. That selection points out the importance of variables, such as culture, socio-economic status, familial composition and geographical location. Noticeably, geographical location and familiar status seem influencing mobility choices more than educational level, common cultural background on EVs and shared "green" attitude.

Secondly, from the qualitative data, it emerges that the urban cohouser is highly educated, green oriented, predominantly single, does not own a car, and uses the train for commuting to work and the bike for shopping and leisure. For rare events, s/he uses the car (mainly a shared one). The semi-urban cohouser, also highly educated and green oriented, on the contrary is married with children, and owns a car, which is his or her main means of transport. In the Wezenbeek-Oppem cohousing, there is a relevant percentage of youngsters, whose behaviours could provide an interesting insight and validate their potential as target group.

When analysing cohousers behaviours in the tests (though not concluded yet, as noted above), some relevant topics emerged. The urban cohousers use car sharing as a secondary mode of transport to increase their mobility and, therefore, accessibility. Nonetheless, for them the EV sharing, being 'green', risks being an alternative to soft mobility, biking and walking, and public transport, and not to the privately owned car. On the contrary, semi-urban cohousers replaced their private car with the shared EVs, developing daily and weekly repetitive car sharing behaviours (semi-organised), in contrast to the completely non-organised behaviours of the urban cohousers. In addition, some of them demonstrated an interest to continue the car sharing also after the end of the test, and others to buy the leased EV.

All of them felt having a pioneering role in testing sustainable mobility patterns, where EVs could replace conventionally fuelled vehicles. By developing new mobility patterns, the test benefited from cohousers towards the exploration (new technology, developing the future, charging at home with renewable energy, etc.…) and provided them in exchange with the possibility to verify by themselves the EV convenience (low refuelling costs, less maintenance possibility of using self-produced energy, etc.…).

Looking at the charging experience, urban cohousers with Cambio were obliged to behave not very efficiently, charging when not necessary and using a more expensive energy (related to energy prices varying between peak and off-peak supply). On the contrary, semi-urban cohousers, which have to pay for energy consumption, mainly charged only when necessary and if possible at night time.

Finally, although based only on pre-experience and 6 months intermediate surveys and driving/charging data, it is nevertheless possible to remark that sharing EVs amongst small communities represents a powerful tool for promoting their zero-emission approach, and their potential of lower charging and maintenance costs.

References

Adams, J. (2000). Hypermobility. *Prospect* (March), pp. 27–31.

Bannister, D. (2005). *Unsustainable transport: City transport in the 21st century*. London: Routledge.

Blyth, D. (2014). Flemish cities allot land to co-housing projects. *Flanders Today*. Retrieved March 20, 2014 from http://www.flanderstoday.eu/politics/flemish-cities-allot-land-co-housing-projects.

Bunce, L., Harris, M., & Burgess, M. (2014). Charge up then charge out? Drivers' perceptions and experiences of electric vehicles in the UK. *Transportation Research Part A, 59*, 278–287.

Cox, P. (2010). *Moving people*. London: Zed Books.

Dennis, K., & Urry, J. (2009). *After the car*. Cambridge: Polity Press.

Department of Mobility and Public Works. (2013). *Reports OVG Flanders 4.4* (September 2011–September 2012). Retrieved April 28, 2014, from http://www.mobielvlaanderen.be/.

Eeman, C. (2009). *l'Habitat Groupé par ses limites, Mémoire MASTER architecture et cultures constructives*. Grenoble: Ecole Nationale Supérieure d'Architecture de Grenoble. Retrieved from http://www.habitat-participatif-lr.org/app/download/8729845993/Memoire+habitat++groupe+ecole+archi+Grenoble.pdf?t=1396379605.

Geels, F. W. (2012). *Automobility in transition? A socio-technical analysis of sustainable transport*. New York: Routledge.

Gilbert, R., & Perl, A. (2008). *Transport revolutions: Moving people and freight without oil*. London: Earthscan.

Goodwin, D. (2014) Cohousing arrangement win popularity in Flanders. *Flanders Today*. Retrieved March 13, 2014, from http://www.flanderstoday.eu/living/co-housing-arrangements-win-popularity-flanders.

Hickman, R., & Banister, D. (2014). *Transport, climate change and the city*. London: Routledge.

Hoogma, R. (2002). *Experimenting for sustainable transport: The approach of strategic niche management*. London: Spon.

Hutton, B. (2013). *Planning sustainable transport*. London: Routledge.

Institute for Creative Sustainability. (2012). *Cohousing cultures: Handbook for self-organized, community-oriented and sustainable housing*. Berlin: Jovis Verlag.

Jonckheere, L., Kums, R., Maelstaf, H., & Maes, T. (2010). *Samenhuizen in België: waar staan we, waar gaan*. Beersel (BE): Samenhuizen vzw.

Krause, L. (Ed.). (2012). *Sustaining cities: Urban policies, practices, and perceptions*. New Brunswick, NJ: Rutgers University Press.

Newman, P. W. G., & Kenworthy, J. R. (1999). *Sustainability and cities: Overcoming automobile dependence*. Washington, D.C.: Island Press.

Ruio, M. L. (2014). Differences between cohousing and gated communities: A literature review. *Sociological Review, 22*(2), 1–20.

Schiller, P. L., Bruun, E. C., & Kenworthy, J. R. (2010). *An introduction to sustainable transportation: Policy, planning and implementation*. London: Earthscan.

Sheller, M. (2004). Automotive emotions. *Feeling the car, Theory, Culture and Society, 21*, 221–242.

UITP Secretary General. (2002). *Bremen paper. Public transport and car-sharing: Together for the better*. Brussels: UITP.

New Electric Mobility in Fleets in the Rural Area of Bremen/Oldenburg

Dirk Fornahl and Noreen Wernern

Abstract In the last decades several case studies took place to discover the potential of e-mobility in car fleets in Germany. However, the results vary according to the employed data and the specific context (e.g. the sector). The project NeMoLand in the model region Bremen/Oldenburg focuses on the rural area to gain significant experiences and develop recommendations concerning the handling of e-mobility in commercial and public fleets. The hypothesis is that fleets in rural areas have a high potential for the use of e-mobility because of advantages related to a higher average of driving distance and frequency of car use and available charging infrastructure in combination with renewable energies. To identify mobility patterns of different enterprises a survey combined with the application of GPS data loggers is conducted. The results indicate that e-mobility has a high potential in the near future in the analysed fleets. The study points out that due to a high amount of planned trips and fitting mobility patterns, nearly 80 % of the conventional vehicles could be substituted by battery and hybrid electric vehicles for economic reasons until 2020. However, there are still some problems which have to be solved (e.g. the psychological effect of public charging infrastructure) until e-mobility diffuses in rural areas. Considering the modal split of most manufacturers it seems important to stress the positive effects and advantages of e-mobility to achieve a higher impact of low-emission technologies.

Keywords E-mobility · Transport sector · Mobility pattern · Potential

An erratum of this chapter can be found under DOI 10.1007/978-3-319-13194-8_21

D. Fornahl (✉) · N. Wernern
Centre for Regional and Innovation Economics (CRIE), Bremen University,
Wilhelm-Herbst-Str. 12, 28203 Bremen, Germany
e-mail: dirk.fornahl@uni-bremen.de

N. Wernern
e-mail: n.werner@uni-bremen.de

W. Leal Filho and R. Kotter (eds.), *E-Mobility in Europe*,
Green Energy and Technology, DOI 10.1007/978-3-319-13194-8_13

1 Introduction

Spatial configurations and characteristics have an impact on the genesis of region-specific mobility patterns. Population density, spatial settlement structures and functional characteristics of the community cause different traffic volumes. The political aim is to establish an efficient and sustainable future mobility system in Germany. Therefore, the interest in e-mobility solutions increased due to the aim of reducing greenhouse gas as well as noise emissions, the dependence on oil and the rising oil price as well as the expansion of renewable energies. Especially in cities the air quality can be improved by employing electric vehicles for individual and public transport as well as in the commercial transport and logistics field. Furthermore, many cities suffer from noise pollution at day and nighttime. Again electric vehicles provide an option to reduce this pollution and the potential negative effects for the inhabitants. Since CO_2 emissions have a global effect, the use of electric vehicles does not directly and locally improve the environmental conditions, but if electric vehicles are employed at a larger scale and renewable energies are used for charging, all regions can gain from the reduction of CO_2 emissions.

However, the development of new mobility concepts and especially the electrification of vehicles is still a niche technology which competes with other technologies. Currently e-mobility solutions come along with disadvantages in price, driving distance and practicability. Furthermore, the acceptance is affected by design, image and safety aspects (Sammer et al. 2008). Under these circumstances current research deals mostly with private users and focuses on urban regions as the most promising areas for the diffusion of e-mobility innovations. The authors here aim to show that this perspective has to be extended in order to promote the introduction and diffusion of e-mobility on a larger scale.

This paper thus intends to shed some light on the question of how electric mobility can be introduced in commercial fleets in rural areas. There are several reasons why we selected rural areas and commercial fleets, as will be shown below. But the core argument is that such fleets, in the view of the authors here, have a very high likelihood to adopt e-mobility because they have the best ability to amortize the high purchasing price of battery electric vehicles (BEV). Thus, these fleets can serve as catalysts for the overall diffusion of electric mobility.

The periphery regions offer several advantages for e-mobility (Fornahl et al. 2011). For example, the necessity for individual motorized mobility is higher. First, several households tend to have two or more vehicles and the option to choose the most suitable vehicle according to their particular needs in a specific situation (Follmer et al. 2010). Secondly, in rural areas the access to public transport is restricted. Furthermore, it is easier to install and use private charging infrastructure since most vehicle owners have their own parking space or a garage and maybe they even can employ self-produced renewable energies to charge the vehicles. By comparing inhabitants' mobility patterns in urban and rural areas, researchers conclude that there are significant differences in average driving distances, which gives reason to also focus on non-urban areas for the diffusion of e-mobility because in urban areas the higher purchasing price can be quicker amortized due to the longer distances driven and the lower variable costs of battery electric vehicles (Fornahl et al. 2011).

Since vehicles in commercial fleets on average have longer driving distances than private vehicles, the cost reduction potential in these fleets is relatively high. Thus, the adoption rate of e-mobility in commercial fleets should be higher than for private users.

In the authors' research the focus was on the passenger transport sector, because in Germany commercial (trading) transport has a high economic impact. Nearly 33 % of all driving kilometers on German roads are caused by commercial and trading transportation (WVI 2012). The mobility patterns of commercial users particularly are different to private ones. Almost 90 % of all trips by commercial passenger cars are shorter than 100 km with an average daily driving time of 1 h and 17 min (ibid.). More than 70 % of trips ended on private premises, e.g. the own workplace, costumer households or on building sites (ibid.). Also, 35 % of all BEVs in Germany are owned and operated by commercial organizations (ibid.). Under these conditions enquiring into the mobility requirements of commercial and public fleets is relevant for a successful market penetration of alternative vehicle technologies.

The analysis is part of the project "Neue Mobilität im ländlichen Raum: Angewandte Elektromobilität—Technologiekonzepte—Mobilitätseffekte"—or new mobility in rural areas, applied electric mobility, technology concepts and mobility effects—which was conducted in the model region of e-mobility Bremen/ Oldenburg between October 2011 and March 2014. The paper is structured as follows: First, the authors give a short overview of the literature dealing with the potential of e-mobility in commercial transportation in periphery regions. Second, they briefly outline the methodology of the study and consider the main results. Finally, they provide a critical view on the outcomes.

2 Literature Review

In Germany, the commercial transport sector accounts for a share of about 60 % of all newly registered vehicles per year (Gnann et al. 2012). The average length of daily routes is shorter than 100 km, which is well below the achievable driving distance of battery electric vehicles (BEV) (WVI 2012; Dijk et al. 2013). Currently, the operating costs of BEV, such as cost per kilometers, are already lower than for internal combustion engine vehicles (ICEV). The difference between the price of electricity and the price of diesel is obvious. It is assumed that commercially used vehicles run mostly on fixed routes with a predictable mileage, which could result in a faster amortization of vehicle costs. Especially, fleet operators are more sensitive to this advantage of low operating costs as well as to the "reputational benefits from the decarbonisation strategies" (Dijk et al. 2013, p. 139) than private consumers.

Despite of the relevance of the transportation sector in the diffusion of electric mobility in Germany, most of the existing studies refer to individual passenger transport when it comes to identifying potential customers (e.g. Peters and Dütschke 2010). As stated by Gnann et al. (2012), the commercial transportation sector is under-represented in the scientific literature because of data availability and the heterogeneity of the commercial transportation sector itself. Furthermore, general statements regarding commercial transport are rather difficult to make because most studies only focus on data of specific industries and single case studies, with the exception of "Kraftverkehr in Deutschland" (e.g. Gnann et al. 2012).

Existing studies estimate a rather small potential of e-mobility in commercial transportation applications (Gnann et al. 2012). A significant increase in the number of electric vehicles in German commercial and municipal fleets can be expected in about 12 years (Gnann et al. 2012). For the year 2013, an e-mobility potential of 12 % can be determined for commercial vehicle fleets (FfE 2011). Another study also identifies a higher potential of electric mobility in commercial fleets compared to private owned vehicles due to the high share of compact cars in commercial fleets. This can lead to a faster diffusion in particular industries (Öko-Institut 2011; FfE 2011). Especially small enterprises, which typically own compact vehicles, make short trips suitable for e-mobility. As main target groups able to buy BEV until 2020, FfE (2011) identified food distributors, couriers, postal services and taxi enterprises. The diffusion of electric vehicles as commercially used passenger cars, craftsmen vehicles and car-sharing applications are most likely to occur considerably after 2020 (FfE 2011). Again these results are based on single case studies and should not be considered as generally valid statements.

From the authors' point of view, there is a research deficit in the field of electric mobility in fleets in rural areas. Existing studies focus mainly on larger companies' fleets (e.g. DHL, Daimler AG, Siemens AG, Volkswagen AG) in urban areas (e.g. Daimler AG and Vattenfall Innovation GmbH 2011; Volkswagen AG et al. 2011). Although rural areas are considered to induce a higher average travel distance and hence users of BEV achieve a faster amortization of investments along with larger space capacities for infrastructural concerns, no studies regarding the application of electric mobility in rural areas are currently available.

3 Methodology

The current study is empirically grounded through a survey of mobility patterns, e.g. the usage of cars for official purposes by employees in rural areas. The questionnaire was conducted from May 2012 to July 2013. It consisted of data acquisition using standardized questionnaires and the recording of mobility patterns via drivers' logbooks and Global Positioning System (GPS). The focal point of the study was the examination of relevant industries with a fleet mainly consisting of passenger vehicles. As a starting point, the assumption was made that future electric mobility concepts will develop their highest potential in commercial traffic applications. Due to the settlement structure of the model region Bremen/Oldenburg, the study focuses on rural areas of the model region, as peculiar economic and demand-driven configurations can be expected in comparison to urban areas.

For the selection of participants, the authors considered the following criteria. First, the participants should be located in the rural region of Bremen/Oldenburg. Second, the composition regarding the commercial sectors should be as heterogeneous as possible. Moreover, the survey participants should own commercial passenger cars and most of the trips should be below 100 km per vehicle and day.

The questionnaire asked for general information on the company and fleet, the utilization profiles of company vehicles, the trip disposition, the vehicle procurement as well as personal opinions on economic aspects and user acceptance of electric mobility. The GPS data loggers were provided by the Fraunhofer Institute for Systems and Innovation Research from May to June 2012. For a time period of at least 3 weeks the data loggers were installed in the participants' company vehicles. The data loggers automatically recorded information such as date, time, trip length and duration as well as the vehicle's location. Additionally, a driver's logbook was used to gather information that could not be recorded by the data loggers, such as the trip's purpose, the number of passengers, the vehicle payload and if the trip could be scheduled or not. The logbooks that were kept without the use of data loggers contained additional questions such as the date and time of routes, the mileage in kilometers as well as the vehicle's location at the start and end of a trip and the frequency of refueling. The participants had the option to receive a BEV for a time period of at least 2 months in exchange for a fixed charge. The available vehicles were provided by the Fraunhofer Institute for Manufacturing Technology and Advanced Materials IFAM and the German Research Center for Artificial Intelligence DFKI.

A total of 22 companies participated in the questionnaire. Eight of these companies were willing to utilize 31 GPS data loggers along with keeping drivers' logbooks. Another six companies agreed to keep a driver's logbook, so that the mobility patterns of 49 vehicles were acquired and analysed. In the study's fleet test, a total number of seven companies and local authorities were willing to participate.

4 Empirical Results

The answers obtained by the questionnaire illustrated several interesting aspects regarding the mobility patterns of the participants.

Company and fleet information The organizations were asked to estimate the average individual travelling distance of their vehicles per year and day in order to compare this information with the average range of current BEVs of approximately 100 km. The answers were analysed with regard to the number of employees of each participating enterprise. The results showed that medium-sized enterprises

(50–250 employees) have the lowest yearly travelling distance of about 11,500 km per vehicle. Small enterprises (less than 10 employees), small to medium enterprises (10–50 employees) and large enterprises (more than 250 employees) have an average yearly travelling distance of around 25,000 km per vehicle. The analysis of the daily travelling distance showed that all participants, with the exception of small enterprises, have an average travelling distance below 100 km per vehicle and day. Small to medium sized enterprises achieved 79 km, large enterprises realized 90 km and small enterprises achieved 120 km per vehicle and day. An explanation for the longer travelling distances of these small enterprises might be the limited number of available vehicles in the fleets as well as combined journey purposes in these fleets: typically, vehicles are utilized by only one person for commercial as well as private purposes.

Trips, locations and daily parking It can be assumed that schedulable trips have a positive influence on the charging management of BEVs because the charging duration currently adds up to several hours. Only 15 % of the participants have mainly spontaneous trips, with the majority of the participants (73 %) stating that their daily trips could be scheduled. Furthermore, the participants were asked to estimate at which locations their vehicles are parked during different times of the day (Fig. 1). The results show that the majority of vehicles are parked on company owned parking spaces, garages or premises. During the morning about 42 % of the vehicles are located at the companies' premises while during noon and afternoon this applies to about 32 % of all vehicles. In the evening hours about 58 % of the vehicles return back to the company owned parking space. About 69 % of all vehicles are located at the companies' premises during nighttime. The available

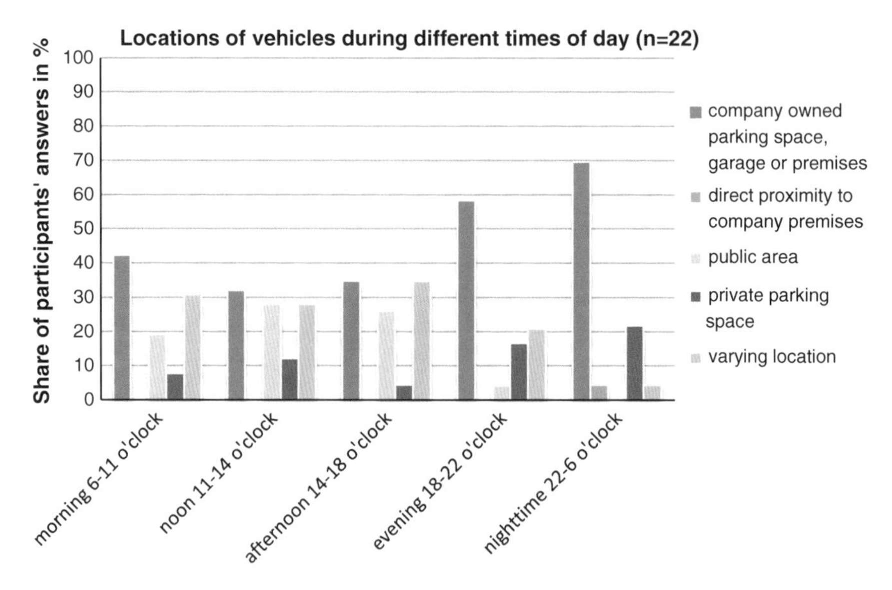

Fig. 1 Locations of vehicles during different times of day

time period for charging purposes adds up to approximately 8 h. This result is similar to the conclusion of the Öko-Institut (2011).

About one-third of the examined vehicles change their location frequently in the morning. At noon, a share of 40 % is parked in public areas or a private parking lot. In the afternoon, the share of vehicles with varying locations increases to 34 %. In the evening and at night the vehicles are parked either on or in direct proximity to company premises or in private parking lots. Nearly all enterprises (21 of 22) have access to permanent parking space for their own fleet. Eight participants would be able to charge electric vehicles using a solar power system. This would generally support the possibility of installing charging stations on company premises. Therefore, the technically and regulatory complex public charging infrastructure might only be necessary at strategically important spots for intermediate charging purposes.

This leads to several conclusions concerning the required charging infrastructure. Nearly 70 % of the participants' vehicles are parked at least once a day for a time period of 8 h on companies' premises, 22 % of all vehicles are parked on private parking lots. This implies that at least one BEV could be completely charged at a company owned charging station per day.

Mobility patterns Nearly 47 % of the commercial vehicles in our study are in the segment of small cars. Almost 28 % of the vehicles are middle-sized cars while 25 % of the tested vehicles are transporters, compact cars and cars of the premium segment. The main purposes of utilization are service transports with a maximum additional load of 100 kg (85 %). Further, the study examined the share of trips within communities, on overland routes and motorways for the reason that the speed has an impact on the battery charge. Almost 27 % of the trips are within cities or municipalities. Trips on overland routes have a share of 65 %. According to BBR (2007) the speed ranges from 20 to 40 km per hour on routes in municipalities, while on overland routes the travelling speed is higher. For example, the average speed increases from 45 in municipalities to 70 km per hour on overland routes. The results show that the amount of trips which are less compatible with BEVs, especially trips on motorways (8 %), are negligible.

Mobility and access to charging infrastructure Figure 2 shows a map of north-western Germany in which the GPS coordinates of frequent destinations of all participants' vehicles are visible. Because of the amount of trips and point locations, the data was classified and concentrates on highly frequented locations. Furthermore, a cruising radius of 50 km is shown around each of the companies' premises. The radius of 50 km was chosen to account for a safe travelling distance of 100 km of a fully charged electric vehicle. Therefore, it was assumed that the vehicle will always return to a company's premises for charging purposes. It can be seen that most destinations of all participants are located in the given 50 km radius along main transportation corridors of the model region. Additionally, the figure shows that there are only a few charging stations (marked green, desired charging stations marked blue) located in the periphery of Bremen, Bremerhaven, Oldenburg and Wilhelmshaven.

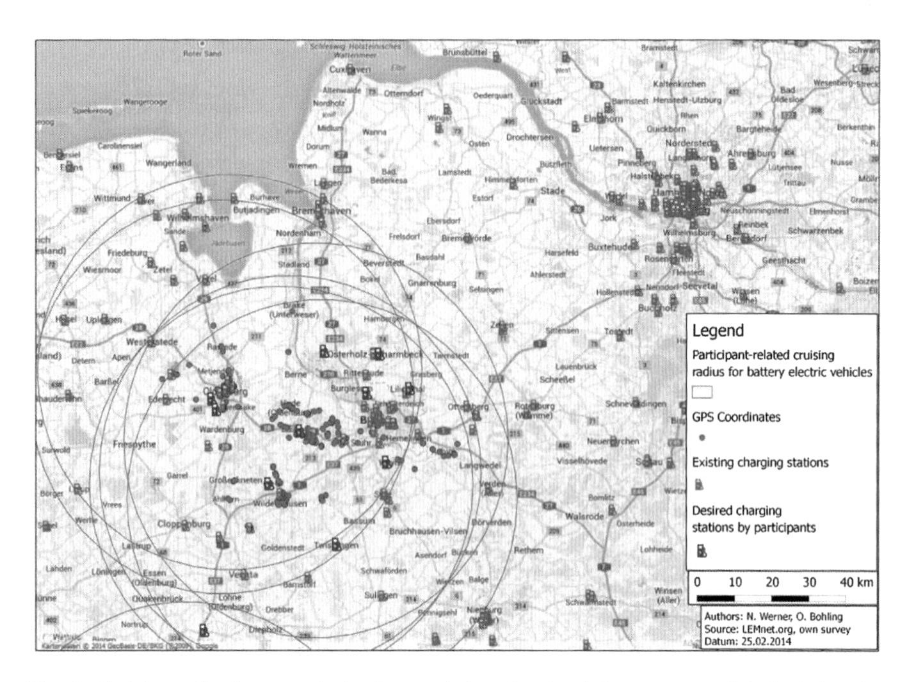

Fig. 2 Cruising radia, frequent destinations, existing charging infrastructure and desired charging stations by participants

Access to information The lack of information concerning e-mobility in general and charging infrastructure in particular as well as the low willingness of consumers to collect the available information may lead to low adoption rates of e-mobility solutions because the consumer does not know the characteristics of the new technology. Aspects affecting the adoption of new technologies are: Relative advantage, compatibility, complexity, trialability and observability (Rogers 2003). Network externalities are important aspects of the relative advantage. For example, an adopter's relative advantage by using a BEV is influenced by the degree of complementary components available (e.g. charging infrastructure garages). This degree is relatively low in most rural areas. In addition, there are organizational aspects (and of course technological aspects) that affect the adoption of BEVs in fleets. For this, the implementation of e-mobility should be embedded in a firm-specific mobility and fleet management. The complexity of an e-mobility system has to be compensated by a variety of information provision activities (e.g. leaflets, newspaper reports, presentation of fairs or websites) or the supply of possibilities to test drive a BEV. Both is rare in most rural areas. It must be pointed out that some solutions like the introduction of public charging infrastructure mainly has a psychological effect on adopting e-mobility and is economically not sensible in rural areas. In these areas the charging infrastructure should be based on private solutions or publicly available private or semi-private charging points. Such measures are necessary to overcome the main innovation barriers and to increase the adoption rate of e-mobility.

Vehicle purchasing criteria Besides the individual mobility patterns, the participants were asked to evaluate different vehicle purchasing criteria to identify relevant vehicle characteristic. The criteria were divided into four categories, namely “image”, “vehicle emissions”, “vehicle features” and “costs”. The participants could choose whether a criterion was regarded as important, neutral or not important in their individual vehicle purchasing preferences.

As expected, the category “costs” is most important for most of the participants (76 %). The category “costs” consists of the criteria “leasing rate, rent or acquisition costs”, “fuel costs”, “total cost of ownership” and “maintenance requirement”. For the majority of the participants, the total cost of ownership is the most important criterion in their vehicle purchasing preferences. While the criteria “leasing rate, rent or acquisition costs” and “fuel costs” are also regarded as important by nearly 80 % of the participants, the “maintenance requirement” is the least important criterion with an importance rating of only 64 %. However, participants criticized the accessibility to information about driving, charging and repairing costs of the electric vehicles.

The second important category is the “image” of vehicles, especially in relation to manufacturers and brands, which is rated as important by 68 % of the participants and consists of the criteria “representativeness”, “environmental safety” and “reliability”. The reliability of a vehicle is the most important criterion in this category (96 %), followed by environmental safety (72 %). More than half of the participants use German manufactured cars, which are considered to stand for reliability and efficiency. The representativeness is the least important criterion to the participants with an importance rating of only 36 %. The fact that most of the examined vehicles were small passenger cars or vans may explain the low importance rating of representativeness. The purchased vehicles were chosen to fit the needs of the application area rather than being used for representative purposes.

The category “vehicle emissions” is regarded as important by over half of the participants. The category consists of the criteria “CO_2-emissions” due to the introduction of CO_2-dependent automobile taxes, “noise emissions” and “pollution”, such as fine particulates, nitrous gases or odour emissions. 60 % of the participants regarded both of the criteria “CO_2-emissions” and “pollution” as important; the other 40 % considered them to be neutral. The results for “noise emissions” reached an importance rating of 50 %. Especially the tax ratings play a key role for the purchasing process of new vehicles.

“Vehicle features” was the least important category; only about 24 % of the participants considered it to be important. It consisted of the criteria “performance”, “luggage space”, “maximum vehicle load capacity” and “comfort features”. While the luggage space and maximum load capacity of a vehicle were important to some participants (46 and 40 %), the vehicles performance and comfort features were widely regarded as neutral or not important.

Substitution potential In addition to the collected GPS data, the study analysed the economic and technical potential to substitute conventional fleet vehicles by

BEV and PHEV (plug-in electric vehicle). With regard to the economic and technical aspects the authors based their analysis on the calculation model of the Fraunhofer ISI. The authors focused on the observed driving distance per day and introduced a threshold of 100 km. Only one real trip above this threshold leads to the exclusion of the vehicle since we assumed that most electric vehicles are limited to this distance without recharging. Although the maximum distance is higher for many vehicles under ideal conditions, the study's threshold takes into account that these conditions are not always ideal because for example heating is necessary in winter. To analyse the economic potential, the authors considered capital costs (battery, purchasing price) and operating costs (fuel and energy cost, vehicle tax rating, maintenance and repair) for different vehicle segments: Vehicles in the small and medium sized vehicle segment as well as the premium car segment and transporters. In addition to the purchasing price the authors considered fuel costs, energy costs, vehicle tax ratings in relation to other vehicles in the segment and costs for maintenance and repair for the years 2011 and 2020. The capital costs are discounted by 3 % in 12 years. The resale value amounts to 50 % for a vehicle holding period of 4 years. The rest of the purchasing price is discounted by 3 % and a holding period of 4 years (see Table 1). Firstly, the study has a sample of 49 logged vehicles with relevant information about daily driving distance, days of travelling and travelling time, so it is possible to calculate the average daily driving distance and travelling time. In the next step the driving distances of the given time period are extrapolated to the yearly driving distance for each car. The authors calculated the yearly operating costs for each tested car based on the distinctive assumptions for the year 2011 and 2020. From the technical point of view, which only considered the trips per day, the authors derived a substitution potential of 37 %. Due to the range restriction of 100 km for one trip, the authors expect that the potential could be higher in reality. Planning trips and the use of fleet vehicles, for example time for charging stops, could increase the technical potential to substitute conventional fleet vehicles by BEVs.

Apart from the technical potential, the authors calculated the economic potential with regard to the underlying assumptions, distinguishing between four kinds of alternatives: BEV, PHEV (economical), PHEV (technical) and ICEV (internal combustion engine). In the category BEV all vehicles from the study's sample are included for which the replacement by battery electric cars is economically as well as technically more feasible than the use of a conventional vehicle. For the category PHEV (economical) the authors calculated a higher economically feasible potential for plug-in electric vehicles than for conventional vehicles or battery electric cars. In some cases it would make sense from an economic point of view to use BEVs, but the individual driving behaviour does not fit the technological characteristics of a BEV (range of 100 km). In these cases the authors assign a potential of PHEV (technical). At last, there are vehicles with no economical or technical substitution potential, which is the ICEV category. While Fig. 3 shows the substitution potential for 2011, Fig. 4 gives an overview of changes for the year 2020.

For 2011 we identified no potential for the use of BEVs in the examined fleets (Fig. 3). The main reason is certainly a high purchasing price through high battery

Table 1 Assumptions for the substitution potential of the tested fleet vehicles

Vehicle segment	Small			Middle-class			Luxury-class			Transporters		
2011	BEV	PHEV	ICE	BEV	PHEV	ICE	BEV	PHEV	ICE	BEV	PHEV	ICE
Battery (kWh)	15	6	0	20	10	0	40	14	0	40	14	0
Purchase price (€)	23.915	18.470	10.804	37.847	32.266	20.655	71.400	52.452	39.017	65.244	46.997	29.339
Consumption (kWh/100 km)	17.4	17.4	0	22.4	22.4	0	27.4	27.4	0	34.8	34.8	0
Consumption (l/100 km)	0	4	6	0	5	8	0	8	14	0	10	15
Vehicle tax rating (€/a)	23	114	114	30	224	224	36	428	428	36	132	132
Maintenance (€/km)	0.008	0.013	0.017	0.01	0.017	0.02	0.017	0.027	0.026	0.1	0.122	0.121
Total cots (€/km)	0.046	0.069	0.01	0.059	0.087	0.132	0.077	0.139	0.222	0.177	0.262	0.331
2020												
Battery (kWh)	15	6	0	20	10	0	40	14	0	40	14	0
Purchase price (€)	16.775	15.227	10.804	28.327	26.911	20.655	52.360	44.955	39.017	46.204	39.500	29.339
Consumption (kWh/100 km)	16.53	16.53	0	21.28	21.28	0	26.03	26.03	0	33.06	33.06	0
Consumption (l/100 km)	0	3.8	5.7	0	4.75	7.6	0	7.6	13.3	0	9.5	14.25
Vehicle tax rating (€/a)	6.6	114	114	8.4	241.5	241.5	10.3	428	428	10.3	132.22	132.22
Maintenance (€/km)	0.008	0.013	0.017	0.01	0.017	0.02	0.017	0.027	0.026	0.1	0.122	0.121
Total costs (€/km)	0.044	0.081	0.012	0.057	0.103	0.157	0.074	0.163	0.265	0.173	0.293	0.378
	2011	2020										
Fuel cost (€/l)	1.5	1.8										
Energy (€/kWh)	0.22	0.22										
Battery BEV (€/kWh)	700	300										
Battery PHEV (€/kWh)	800	350										

BEV battery electric vehicle, *PHEV* plug-in electric vehicle, *ICE* internal combustion engine, *kWh* kilowatt-hour, *l* liter, *km* kilometer, *€* euros
Source Data provided by Fraunhofer ISI

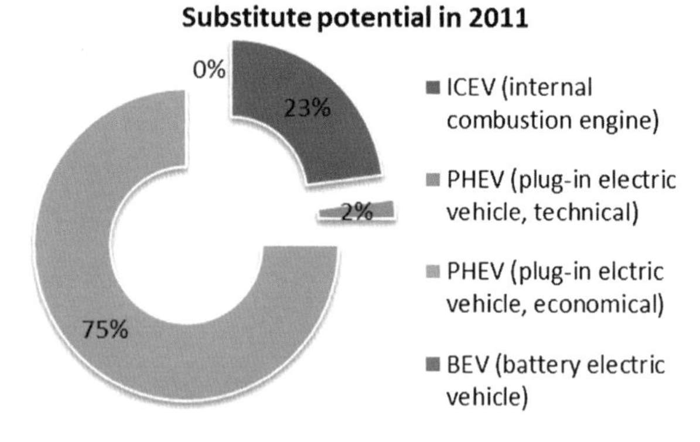

Fig. 3 Substitution potential for the tested conventional fleet vehicles in the model region of Bremen/Oldenburg in 2011

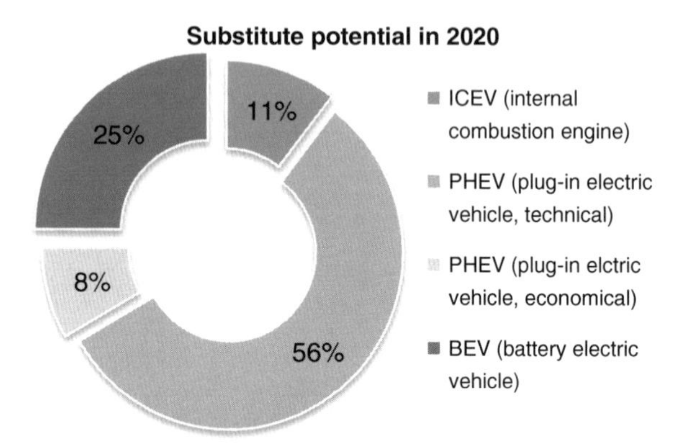

Fig. 4 Substitution potential for the tested conventional fleet vehicles in the model region of Bremen/Oldenburg 2020

costs. The use of hybrid cars has more economic advantages than the use of conventional cars. Especially the lower purchasing price of about 18.000 Euro in the small segment against 24.000 Euro for BEV and the lower battery costs have positive influences on the substitution potential towards PHEV. The advantages compared to conventional cars are mainly derived from the lower operating costs. Since the share of small vehicles was high in this study, this probably positively affects the substitution potential. Therefore only 23 % of the tested cars are not replaceable by alternative drive trains.

In 2020 (Fig. 4) the potential of BEVs increases with the decrease of purchasing price, battery costs and vehicle tax ratings for alternative engines to 25 %. The technical potential for hybrid cars is still high with 56 %. This category represents the potential, when the mobility pattern indicates long trips above 100 km. The authors expect that the potential of BEVs is higher if firms introduce a system of mobility management and planned trips. Only 11 % of the vehicles could not be replaced by alternative drive trains. These vehicles are either in the premium segment with a high travelling distance or are transporters or busses for more than four car passengers or with a high vehicle load capacity, which are not available on the e-mobility market yet.

5 Discussion and Conclusion

This study was conducted in the model region of Bremen/Oldenburg between 2011 and 2014. Under the assumption that several advantages, e.g. driving distance, practicability and amortization of costs, facilitating the use of e-mobility in commercial transport fleets in rural areas, the authors analysed the mobility patterns and substitution potential for BEV of municipalities and firms in the rural area. In contrast to existing studies, the authors identified a considerably high potential for alternative engines in commercial fleets until 2020. The adoption of hybrid vehicles is already appropriate in 2011, but with sinking costs and fleet mobility management systems the use of alternative engines could increase even further. In fact, hybrid and battery electric vehicles are even superior to conventional vehicles in many cases. The answers obtained through the study's questionnaire indicated a lack of information regarding the participants' own mobility patterns, which results in an underestimation of the e-mobility potential in fleets. Additionally, there are a range of advantages and disadvantages, which play a key role for the adoption of electric vehicles, such as the mobility behaviour or technical aspects. These kinds of prospects could not be taken into account. Most of the participants have a positive attitude towards e-mobility, which might affect the answers, too.

Despite these critical factors, the study shows that nearly 85 % of all trips are predictable. The authors found that the majority of fleet vehicles park on private property and return daily to their home or firm base. In contrast to public charging infrastructure there are advantages for private charging infrastructures through lower installation costs, an easy maintenance and less administrative barriers during the build-up (Trümper 2014). Nevertheless, it is necessary to apply publicly or semi-publicly available charging points for a successful market launch of e-mobility in the rural area—mainly for psychological reasons. The authors suspect that a centralized corporate charging infrastructure, e.g. in industrial parks, can be installed as long as the density of public charging points is underdeveloped. Nearly 90 % of all evaluated trips are inside communities or on overland routes, which has strong effects on the battery status and driving distance of BEV.

Problems of e-mobility in fleet vehicles mainly originated from a lack of fleet management and insensitive handling with vehicles. As an example, the participants of the study pointed out that the management of battery status, charging time as well as the continuous availability of vehicles during daytime is very complex. The breakdown of a fleet vehicle is difficult to compensate for spontaneously, especially in small fleets like in this study's sample. The authors conclude that the diffusion of e-mobility in fleets depends on the practicability of the vehicles, a firm-specific fleet and professional mobility management. The fleet managements differ in their ability and resources to evaluate the potential of e-mobility for their fleets and to introduce and operate e-mobility solutions. Thus, the organizations have to decide whether they want to build up these capabilities in-house or whether they employ external professional fleet management services. Small and medium sized enterprises, just as many or even most municipalities, are mostly unable to pay the high purchasing price for an electric vehicle at present. Hence, it is more important for most fleets to reduce the price difference to conventional cars than employ soft instruments such as free parking lots or driving on bus lanes. Additionally, the information and consulting offers should be extended, which might be the most important instrument. Nevertheless, the regional access to e-mobility solutions, such as vehicles, charging points, best practices or an integrated mobility and transport system, could be a step to establish synergies for a sustainable German e-mobility system.

Although this research generated some insights that can be employed by policy-makers and fleet managers, there are still several research questions that have to be answered in order to support the diffusion of electric vehicles on a larger scale. These are, for example: Producers of renewable energies have to consume or market an increasing share of their produced energy on their own. Although this offers a potential for the use of electric vehicles, it still remains unclear to which extent such a combination affects the amortization of both investments. Despite the positive conditions offered by fleets, one could still ask whether such conditions cannot even be further improved by cooperation activities between fleets or by the opening of fleets for private use. Such approaches might offer possibilities for increased efficiency but how such a cooperation might look like and how the efficiency is really affected has to be examined by future research.

References

BBR. (2007). Erreichbarkeitsmodell des BBSR. Bonn. Retrieved from http://www.bbsr.bund.de/BBSR/DE/Raumbeobachtung/UeberRaumbeobachtung/Komponenten/Erreichbarkeitsmodell/erreichbarkeitsmodell_node.html.

Daimler AG and Vattenfall Innovation GmbH. (2011). Abschlussbericht zum Verbundvorhaben Elektrifizierung von Mercedes-Benz Kleintransportern in Entwicklung und Produktion im Rahmen des FuE-Programms. “Förderung von Forschung und Entwicklung im Bereich der Elektromobilität”. Retrieved from http://www.pt-elektromobilitaet.de/projekte/foerderprojekte-aus-dem-konjunkturpaket-ii-2009-2011/wirtschaftsverkehr-feldversuche/abschlusberichte-wirtschaftsverkehr/abschlussbericht-emkep.pdf.

Dijk, M., Orsato, R. J., & Kemp, R. (2013). The emergence of an electric mobility trajectory. *Energy Policy, 52*, 135–145.

FfE (Forschungsstelle für Energiewirtschaft e.V.). (2011). (Hrsg.): eFlott. Wissenschaftliche Analysen zu Elektromobilität. Retrieved from http://www.ffe.de/download/article/333/eFlott_Abschlussbericht_FfE.pdf.

Follmer, R., Gruschwitz, D., Jesske, B., Quandt, S., Lenz, B., Nobis, C., et al. (2010). Mobilität in Deutschland 2008: Ergebnisbericht Struktur—Aufkommen—Emissionen—Trends. Report No. FE-Nr. 70.801/2006. Bundesministerium für.Verkehr, Bau und Stadtentwicklung. Bonn & Berlin: infas & DLR. Retrieved from www.mobilitaet-in-deutschland.de.

Fornahl, D., Meier-Dörzenbach, C., Santner, D., Werner, N., Kahle, S., & Bensler, A. (2011). Mobilitätsstrukturen in der Modellregion Elektromobilität Bremen/Oldenburg. Unpublished manuscript. Bremen.

Gnann, T., Plötz, P., Zischler, F., & Wietschel, M. (2012). Elektromobilität im Personenwirtschaftsverkehr—Eine Potenzialanalyse. Working Paper “Sustainability and Innovation”, Karlsruhe. Retrieved from http://hdl.handle.net/10419/62593.

Öko-Institut e.V. (2011) (Hrsg.). Betrachtung der Umweltentlastungspotenziale durch den verstärkten Einsatz von kleinen, batterieelektrischen Fahrzeugen im Rahmen des Projekts „E-Mobility“. Schlussbericht im Rahmen der Förderung der Modellregionen Elektromobilität des Bundesministeriums für Verkehr, Bau- und Wohnungswesen. Retrieved from http://www.oeko.de/oekodoc/1344/2011-007-de.pdf.

Peters, A., & Dütschke, E. (2010). Zur Nutzerakzeptanz von Elektromobilität. Analyse aus Expertensicht. Karlsruhe: Fraunhofer ISI. Retrieved from http://edok01.tib.uni-hannover.de/edoks/e01fb11/672256282.pdf.

Rogers, E. M. (2003). *Diffusion of innovations*. New York: Free Press.

Sammer, G., Meth, D., & Gruber, C. J. (2008). Elektromobilität—Die Sicht der Nutzer. e&i. *Elektrotechnik und Informationstechnik, 125*(11), 393–400.

Trümper, S. C. (2014). Coping with a growing mobility demand in a growing city: Hamburg as pilot region for electric mobility. In D. Fornahl & M. Hülsmann (Eds), *Evolutionary paths towards the mobility patterns of the future* (pp. 283–306). Heidelberg: Springer.

VW AG (Volkswagen AG), Deutsche Post DHL and Hochschule für Bildende Künste Braunschweig. (2011). Abschlussbericht zum Verbundvorhaben Erprobung nutzfahrzeugspezifischer E-Mobilität—EmiL im Rahmen des FuE-Programms “Förderung von Forschung und Entwicklung im Bereich der Elektromobilität”. Retrieved from http://www.pt-elektromobilitaet.de/projekte/foerderprojekte-aus-dem-konjunkturpaket-ii-2009-2011/wirtschaftsverkehr-feldversuche/abschlusberichte-wirtschaftsverkehr/abschlussbericht-emil.pdf.

WVI Prof. Dr. Wermuth Verkehrsforschung und Infrastrukturplanung GmbH. (2012). Mobilitätsstudie „Kraftfahrzeugverkehr in Deutschland 2010“ (KiD 2010)—Ergebnisse im Überblick. Braunschweig. Retrieved from http://daten.clearingstelle-verkehr.de/240/9/KiD2010-Schlussbericht.pdf.

Part III
Technological Advancements and User-Friendly Strategies

To Cluster the E-Mobility Recharging Facilities (RFs)

Eiman Y. ElBanhawy

Abstract The world is witnessing an accelerating expansion of urban areas and intensive urbanisation. The robust relation between transport infrastructure and urban planning is reflected in how integrated and reliable a system is within the urban fabric. Designing an integrated infrastructure to support full electric vehicle (EV) use is a crucial matter, which worries planning authorities, policy makers, as well as current and potential users. Reducing range anxiety by facilitating access to public recharging facilities is designed to overcome the main barrier that stops potential users to utilise EVs. The uncertainty of having a reliable and integrated charging infrastructure also presents hurdles, and slows down the growing trend of smart ecosystems and sustainable urban communities as a whole. Automotive, battery and utility technologies have formed the cornerstone of the EV industry to compete with currently mainstream means of transport, and to gain more prominence within many regions. Strategically locating public EV charging points will help to pave the way for better market penetration of EVs. This paper analyses real information about EV users in one of these metropolitan areas. A case study of 13 charging points with 48 EV users located in the inner urban core (NE1 postcode district) of a metropolitan area in North East England, the city of Newcastle upon Tyne, incorporating space-time analysis of the EV population, is presented here. Information about usage and charging patterns is collected from the main local service provider in North East England, Charge Your Car (CYC) Ltd. The methodology employed is a clustering analysis. It is conducted as a dimensional analysis technique for data mining and for significant analysis of quantitative data sets. A spatial and temporal analysis of charging patterns is conducted using SPSS and predictive analytics software. The study outcomes provide recommendations, exploring design theory and the implementation of public EV recharging

E.Y. ElBanhawy (✉)
Department of Architecture and the Built Environment, Faculty of Engineering and Environment, Northumbria University, City Campus, Ellison Place, Newcastle upon Tyne, Tyne and Wear NE1 8ST, UK
e-mail: eiman.elbanhawy@northumbria.ac.uk; eiman.elbanhawy@open.ac.uk

W. Leal Filho and R. Kotter (eds.), *E-Mobility in Europe*,
Green Energy and Technology, DOI 10.1007/978-3-319-13194-8_14

infrastructure. The chapter presents a methodological approach useful for planning authorities, policy makers and commercial agents in evaluating and measuring the degree of usability of the public electric mobility system.

Keywords Clustering analysis · Design configuration · Electric vehicles · Spatial analysis · Recharging facilities

1 Introduction

EVs offer considerable potential to make progress with regard to a variety of wider environmental, societal and economic objectives, which accelerates the development of smarter cities (ElBanhawy et al. 2012; Lindblad 2012). Nevertheless, the transitional phase between using purely conventional means of transportations to state-of-the-art technological ones is a long way for development and awareness. Alternative means of transport target those who appreciate environmental benefits, values and aim at building smart integrated ecosystems. Those technology-focused and/or green consumers tend to form geographical clusters and urban communities. Within these ecosystems, in turn, the notion of owning, leasing or sharing smart cars is viable and the creation of smart transport recharging hotspots is potential. Car ownership, leasing or sharing is the reflection of a community's norms and values (Kahn 2007), which reflects the importance of raising awareness. Yet so far, EV is predominantly replacing the secondary car in multi-car owning households due to range limitation (Elbanhawy and Dalton 2013). The absence of a proactive and constructive approach and feasible scheduling for recharging infrastructure is a major impediment to mainstreaming full EVs due to the sole (full electric vehicles) dependency on batteries as a source of power, and hence range limits and longer recharging time (Namdeo et al. 2013). This is considered as a major issue, especially for full electric car passengers due to the sole dependency on batteries as a source of power, hence range limits and longer recharging time (Namdeo et al. 2013). The problem of sizing and placing RFs is a new topic that recently has been receiving some attention. A few important strides have been made to tackle this problem, particularly in late 2012 and 2013 (Chen et al. 2013).

Several documentations and reports have been published and released, containing phased plans, initiatives and long-term development of recharging infrastructure (OLEV 2011). However, these reports do not share how the presented size and location of RFs were determined (Wirges and Linder 2012). The very recent but slim literature covers previous studies which were conducted to solve the placement of charging stations. The location problem has been seen differently and each of the proposed solutions is concerned by an aspect; power, emissions, facility management or comfort zone. Lam et al. (2014) discussed the candidate locations and suggested sizes of RFs in different urban contexts and tackled the planning problem, following two previous studies. One study was based on the power system factors: power, voltage and current, capacity, e.g. where the authors adopted particle swarm optimization to compute the solution. A similar study was conducted by Liu et al. (2013) looking into the network losses and degradation in voltage profiles, which might happen as a result of poorly insufficient distribution of charging points in urban networks. Another one was based on circuit topologies and grids while discussing the power architectures and power electronics (Lam et al. 2014). Moreover, an enhanced study was conducted by Lam et al. (2014) using nodes and links method based on the charging stations coverage and the convenience of drivers. Chen et al. (2013) revealed a study of proposing solutions for the station location problem, which is a function of identified travel demand, possible parking lots and the time needed for charging. Facility location model aided by a GIS was developed by Xu et al. (2013) to identify locations for and size of the network of charging stations. The study proposed geometric reasoning method for identifying ideal charging location in urban areas. A macroscopic RFs planning model was proposed by He et al. (2013) to maximize social welfare associated with both transport and power networks. A mathematics-based model is developed there to locate the charging stations.

Another study (Lindblad 2012) was conducted to address the problem using grid partition method. It divides the urban layer into partitions, and calculates the electricity loads adding the charging demands of each partition. Genetic algorithm is deployed to optimally locate the RFs while considering the electricity and travel time costs, to optimise the travel cost finding an RF. A different study was conducted by Dong et al. (2014) where the genetic algorithm is applied to find optimal locations of siting public charging points. A similar study (Ge et al. 2011) was carried out where the traffic density, charging stations capacity were taken into account. Recently, another study was undertaken to find the optimal location and the number of refuelling stations for compressed natural gas-fuelled vehicles in toll roads in northeastern United States (Hwang et al. 2013). A relatively different research was undertaken in Germany regarding the CO_2 emissions associated with the transport sector and the use of hybrid model of simulation and optimization to find candidate-charging locations (Turan et al. 2012).

All the above studies were targeting the hard infrastructure; though yet another study highlighted the soft infrastructure of the EV system (Zabala et al. 2012).

It is important to address the EV system so as to clarify how the complex system is integrated and connected via communication and protocols among different players, systems, regulations and the vehicle itself. The complexity can and so far often does result in the lack of a strategic approach in the design of publicly available recharging network, which could otherwise increase the market penetration level. The proposition is that by improving the recharging experience, the probability of potential users to shift to EV market and households' loyalty to use the EV as their first choice would soar.

1.1 Comfort Zone

The limited range is projected into the maximum road trip driven by an EV. The older EV models have the capacity for driving at 60–80 km as a trip interval between two chargers. With advanced battery technology and use of Li-lon batteries, this range has increased to hit 120–180 km (Christensen et al. 2010). However, this is theoretical; in the real world, the practical range is different due to the physiological and technical factors involved. There is a discrepancy between the maximum available range and the maximum range the driver is comfortable reaching, which means the use of EV is not only governed by car specifications or experience, but the avoidance of the range stress is also a factor (Franke et al. 2014). The lesser and easier the charging event, the more it is used. It has been noticed that the use of fast chargers extends the road trip, hence widening the comfort zone of the EV driver for convenience and easiness. This indicates the potential use of fast charging, Type 3, though the utility cost will be the first hurdle. The supply of fast chargers costs almost three times more than Type 1, 7 KWh (Christensen et al. 2010), which does not make the investment in and reliance on fast charging feasible in the near future. However, installing fast chargers along the highway corridors connecting cities is essential to facilitate EV drivers to conduct intercity level (Nie and Ghamami 2013). Another valid type of charging is battery swapping. It is also seen as a lavish option for replenishing the batteries during the road trip due to the facilities requirements and high technology. This type is used now in Israel, though it is not commonly installed and used (ElBanhawy and Nassar 2013).

A major impediment to the EV mainstream market is the absence of a proactive and constructive planning approach for RFs. The limited range of the EV is reflected in the personal applicable comfort zone within which the drivers allow themselves to drive their EVs freely, without worrying about recharging. Figure 1 portrays this thoroughly. The centre of the figure is the starting point of the EV driver to begin his road trip, which is called the Origin. The Origin is the last place that has an RF that the driver has passed by. The first circle from inside is the

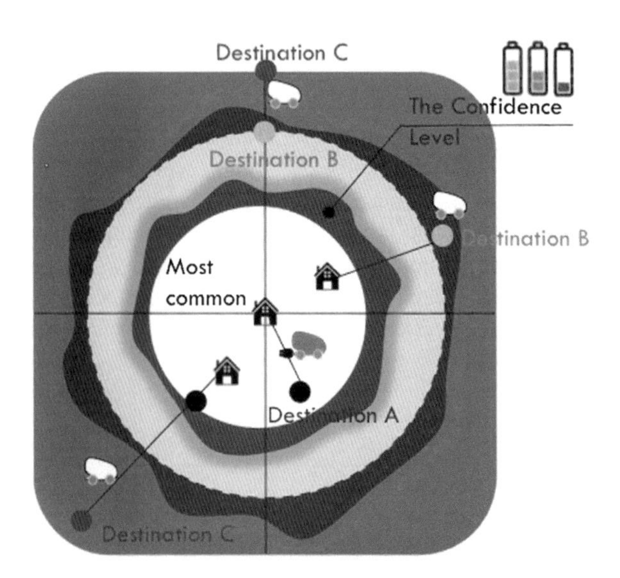

Fig. 1 The social aspect of the EV system (Color figure online)

comfort zone of the users. The road trip can be directed to any of the directions, as the destinations are denoted as black, green and red circles.

The comfort zone is relatively small compared to conventional means of transport, thus representing one of the main hurdles that stop potential users to join the EV market. The comfort zone is coupled with the confidence level of the users, which is the area between the first and the second circles. This area has an irregular curvature shape; it can be extended to cover beyond the boundaries of the second circle, in which case the confidence level scores the highest levels of certainty. The confidence level is a psychological state of the driver as it depends on how secure and certain he is to drive around the urban area without looking for an RF. It has an elastic nature; it gets bigger and extends with practice (Burgess et al. 2013; Franke et al. 2014). The black circle represents Destination A, which is a destination that falls within the comfort zone. The green circle is Destination B, which is relatively far compared to Destination A, and it might be reached if the comfort zone of the driver is wide enough to reach it. The red circles are the destinations where the EV driver needs to have access to public charging points throughout their road trip.

The wider the comfort zone, the less worried the driver would be. To get a wider comfort zone, the routes need to be supported by charging services so as to cover the routes to Destination C, for instance. In other words, covering the road network with necessary RFs will allow the EV drivers to commute longer road trips and go to more number of destinations during their daily routine without getting a flat battery. The sequential addressing of the implications of research outcomes presented and referenced here shall contribute to wider comfort zones which will allow the driver to reach farther destinations and hence promote e-mobility. Intelligent supply and positioning of RFs is the way out as the distribution of RFs is associated with the cost of supply. Unnecessary supply is to be avoided as otherwise it will be underused and will be counted by some at least as a wasted investment.

1.2 Importance of the Study

Smart key transport stakeholders such as planning authorities and policy makers need to understand the current system in order to evaluate the performance, avoid pitfalls, work on barriers and raise awareness of potential users. To do this, a reliable accessible monitoring system that reports all the charging events and transactions has to be available, which is challenging. Researchers, policy makers and utilities companies need to regularly monitor the market and work in-line with the up-to-date technologies, strategies, policy initiatives and supply to the markets from vehicle manufacturers. By improving the charging experience, the probability of potential users to shift to EVs and the loyalty of households and organizations to use the EV as their first choice should soar. The problem with planners and policy makers is that they deal with locating and sizing the recharging infrastructure network as a static location planning problem (Wirges and Linder 2012).

1.3 The Link

Previous research conducted by Gil et al. (2009) has shown the correlation between the design configuration of the built environment, and the users' behaviour (Elbanhawy 2014) reported on the correlation between the network design configuration and the charging behaviour. This study takes into account the spatial design features of the system as variables alongside other charging behavioural elements. The study should determine the best spatial set-up and design configuration for the EV system that generates profit and would work best for the service provider as well as the end user. Having access to the RF system database that records all the transactions made by drivers using RFs shall provide us with the raw data needed to run some statistical analyses. This data set is intrinsic as it reflects the usage records; however, it is not enough to indicate and quantify usability. The time of arrival, the time spent, the number of users, the postcode of the users' homes and the total energy used are different sides of the charging patterns, which are worth the study.

The goal of this study is to undertake exploratory analyses trying to interpret the current socio-behavioural configuration of EV systems by data summarization, inference and intuition about EV users charging path data. It draws on the drivers' charging behaviour and attempts to cluster their usage patterns via recording the transactions made to use the publicly available RFs. The transactions work as an indicator of usage, however, this is not enough.

1.4 Why Clustering Analysis

We can obtain meaningful observations by integrating the design configuration parameters with site location features and charging pattern. The use of clustering techniques to group different observations/cases while considering different continuous and categorical variables is a valid research tool. Clustering is based on mathematical tools, which are designed to cluster the data based on similarity and homogeneity. This should benefit the EV stakeholders as it identifies the main influential factors affecting the use of RFs, the main features and can potentially provide guidelines for smart RF network for a more optimized rollout of RFs.

The clustering outcomes give a description of what the prototypical shape of an EV recharging infrastructure should be. There are many factors considered in this study; behavioural, technical, spatial configuration and demographics.

Spatial configuration analysis is conducted to ultimately help developing design tools for planning authorities and policy makers.

A data spatial clustering technique is employed to design users' segments, and thus identify the best set-up and spatial configuration of desirable, reliable and usable RFs. Eventually, this is to increase the market penetration and to contribute to the mainstreaming of EVs. If the RFs can be grouped according to the available data, this task can be largely computed automatically, using data clustering analysis. This paper presents a methodological approach for planning authorities and policy makers to elucidate, evaluate and measure the degree of usability of the electric mobility system.

1.5 Structure of the Paper

The chapter starts with previous related work in analysing EV behaviour followed by data mining and data spatial clustering methodologies. Through using the selected technique, the EV system of the selected metropolitan area is analysed. Model outcomes are documented and followed by discussion and conclusion.

2 Literature on the Behavioural Side of the System

In 2010, a study was conducted in Denmark to investigate if EV is able to fulfil the travel behavioural needs of customers (Christensen et al. 2010). In addition, it investigates which type of charging is needed to meet the end-user demand. Due to data limitation, the study was carried out based on conventional-car passenger data. Behavioural and socio-technical aspects were taken into consideration to analyse the system. The problem with this study is the high random error we get if we rely on the study outcomes. The travel demand and driving pattern will change in case of EV; in addition, it will not be a random selection in the case of the current and likely near future EV population. EV owners and potential users form an elite sector of society who are environmentally friendly, arguably interested in image and can accommodate the charging pattern in their daily routes. The study was based on a time series database of driving pattern spanning 15 years and GPS data for 2 years, 2001–2003. The outcomes are based on a hypothetical model as it is based on the conventional means of transport travel demand. Nevertheless, it was good enough to reflect the importance of having a reliable charging infrastructure. The majority tends to rely or look for public charging access to replenish their batteries within the week. Some meaningful observations were included, such as where most likely the RFs are expected to be, charging location preference during daytime (workplace, malls, domestic), car ownership and how this affects the charging pattern urban economics analysis.

A few studies were conducted on the purchase motives of alternative means of transport, particularly concerning hybrid cars in developed regions (Heffner et al. 2007; Klein 2007; De Haan et al. 2006). Egbue and Long (2012) identified potential socio-technical barriers to consumer adoption of EVs through a web-based survey of and what affects their choice of purchase. The choice of purchase of a smart car is seen as a response to the increase in gas prices and governments' incentives, and a way to reduce carbon emissions and energy consumption (Ozaki and Sevastyanova 2013).

Another qualitative-based research was undertaken by Steinhilber et al. (2013) on how the public in UK and Germany perceives new technologies and innovations and sheds light on the adoption and diffusion of innovation. The study captures effective and behavioural responses observed by the key stakeholders when launching alternative means of transport within the automotive sector in the UK and Germany. It also identifies the strategies they employ and their opinion and perceptions of the current investments, regulations, standardization, and governmental incentives and schemes. Other studies were focusing on interviewing early adopters (Pierre et al. 2011; Franke and Krems 2013) to conclude the end-user feedback and document their experience driving an EV. Franke et al. (2014) through dairy methods and data loggers scanned 75 customers driving an EV over 3 months. The study sheds light on the adoption with limited resources and how the driver can still accept the technology and deal with the limited resources (availability of recharging facilities, range, capacity and speed). Another high quality research was conducted in Newcastle University launching SwitchEV trial 2011–2013. The final report documented the users' experience driving EV via their driving dairy using GPS, recording charging event through CYC database monitoring system. By 2013, rich analysis was conducted in users' charging profiles and energy consumptions (Robinson et al. 2013; SwitchEV 2013).

2.1 The Gap

As discussed, many studies addressed the socio-technical barriers, diffusion of adoption, purchase motives and introduction to new technologies and inventions. From the literature we can identify the gap in the research, which is the design features of RFs that would create an integrated network, which is utilized by the drivers and, in turn, generate profits to the government as tax revenue in the medium term, and which helps to build a successful business model to utility management companies. There is a link between EV infrastructure planners, users and also business needs, which would be reflected in and by an integrated and reliable system. This link is the design of an integrated spatiotemporal recharging network. The development of an integrated design via a spatial clustering approach and an end users' perspective paves the way for a mainstream market where the level of usability improves as the level of user satisfaction is improving.

EV system usability depends on several underlying principles, such as the expected daily mileage, the driving pattern, the number (and sequencing) of destinations and the charging pattern of the driver, which all differ (statistically aggregated) from one region to another. What will remain as common aspect is households' perception of having (routine use access to) an EV. This part of the paper investigates the relationships between the design characteristics of the recharging infrastructure and the usability state for its current drivers. The design of the recharging network is a function of its size and the distribution of the charging points within a given urban area, as well as the number of EVs in it.

Spatial analysis often employs methods adapted from conventional statistical analysis to address problems in which spatial location is the most important explanatory variable. The idea of configuration modelling revolves around the space theory that incorporates the space topological relationships and its relation with the movement. It has been also asserted that the principle of configuration models that are based on street segments with high accessibility indexes strongly present a high level of connectivity with other links and thus a high level of potential uses (Barros et al. 2007). It can be observed from the study presented by Barros et al. (2007) that configuration modelling in general and space syntax in particular can play a role in transport infrastructure design studies, especially in the early planning stages. This study applies a space syntax spatial analysis software, Depthmap. Depthmap basically transforms the street pattern into a network graph by disaggregating the network at the intersections (ElBanhawy et al. 2013). The travel cost between a pair of segments is measured by the shortest path approach. The distance is weighted by three key cost relations: connectivity, angular integration (topo-geometrical) and mean depth. The spatial analysis starts with generating a road network centreline mapping via using AutoCAD and converting it to a segment map using Depthmap. In Depthmap, different space syntax analysis is applied to the axial map generating more attributes to the district road network (Turner 2004) which will be used as dependent variables in the statistical analysis presented in the case study here.

3 Methodology

Data clustering is a continuous fine-tuned process of grouping sets of data. It is a convenient method for identifying homogeneous groups of objects, called clusters (Mooi and Sarstedt 2011). Clustering analysis is used for identifying groups within the data while also being able to analyse groups based on similar data observations instead of individual observations. It also works on simplifying the structure of the input data and showing a relationship not revealed here before (Caccam and Refran 2012). The crux is finding a group of similar objects sharing many characteristics and qualities, which are unrelated to other objects not belonging to that group and thus aiming at reducing the size of the large data sets. These objects (cases or observations) (Mooi and Sarstedt 2011) can be customers, products, employees, users, clients, etc. The goal is to analyse their behaviour, preference, pattern, usage or any other quantified parameter and classify it into groups (Larson et al. 2005). In cluster analysis one searches for patterns in a data set by grouping the (multivariate) observations into clusters. The aim is to find an optimal grouping for which the observations or objects within each cluster are similar, but the clusters are dissimilar to each other. The analyst hopes to find the natural groupings in the data, groupings that make sense to the researcher due to the contextual knowledge of the phenomenon. Cluster analysis differs fundamentally from classification analysis (Mooi and Sarstedt 2011). In classification analysis, one allocates the observations to a known number of predefined groups or populations. In cluster analysis, neither the number of groups nor the groups themselves are known in advance. To group the observations into clusters, many techniques begin with similarities between all pairs of observations. In many cases the similarities are based on some measure of distance (Mooi and Sarstedt 2011). Other cluster methods use a preliminary choice for cluster centres or a comparison of cluster variability within, and between, clusters. It is also possible to cluster the variables, in which case the similarity could be a correlation (Schaeffer 2007).

3.1 An Overview of Possible Algorithms and Techniques

To explain the approach chosen to conduct the data clustering mechanism, a quick overview of the possible available algorithms/techniques is presented now, highlighting the one being used. The classification of data clustering algorithms can be in different shapes. For the present study, the author presents the classification as per the platform being used. SPSS, a predictive analytics software, is the commercial platform being used. SPSS has three techniques with different algorithms: K-means (Partitioning or Flat-Hierarchical) clustering, Hierarchical clustering and TwoStep. The first algorithm works on dividing the data into non-overlapping subsets, see Fig. 2a. The second algorithm divides them into nested clusters organized as a hierarchical tree. The last method is a sort of combined technique that has two steps, partitioning and hierarchal.

3.2 K-Means Clustering

This is a flat hierarchical method which attempts to find a user-specified number of clusters (k) represented by their centroids. The centroid is the mean of the points in the cluster. The initial centroid is randomly assigned and keeps changing. The centroid's position is recalculated every time a component is added to the cluster. Iterations and computation take place until the centroids do not change, thus forming the final required number of clusters. Usually, convergence happens in the first few iterations and the Euclidean distance measures the closeness.

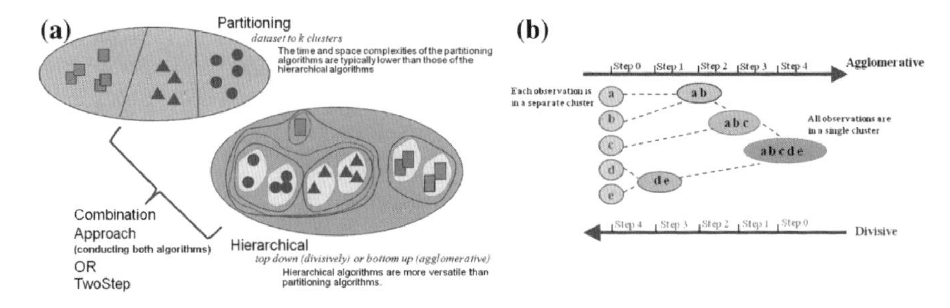

Fig. 2 **a**, **b** Significant spatial data clustering algorithms

3.3 Hierarchical Clustering

There are agglomerative and divisive techniques being used in this method, see Fig. 2b. The first one starts with creating clusters from individual objects. The longer the process continues, the bigger the size of the cluster as a merging process takes place. The convergence of similar objects continues in a bottom-up approach. The divisive approach is the other way round where the start is a single big-sized cluster, and the iterations gradually split up the cluster. This reflects the nature of hierarchical clustering: If an object is assigned to a cluster, it will not be reassigned to another cluster. The higher level of hierarchy always compasses the lower levels (Mooi and Sarstedt 2011).

3.4 Combining Both Methods

A combined approach can be implemented by employing a hierarchical approach, followed by flat hierarchal approach, see Fig. 2a. The first is used to determine the number of clusters and profile cluster centres (centroid) that would serve as initial cluster formation in the partitioning one. The second phase would take place to provide more accurate cluster membership (as the K-means clustering needs to identify the number of clusters as a first step). This enables the advantage of the hierarchal methods to complement the partitioning method in being able to refine the results by allowing the switching of the cluster membership.

3.5 TwoStep

The SPSS TwoStep clustering was developed for analysis of large data sets (Chiu et al. 2001). This is an exploratory tool designed to reveal natural groupings within the data set (see Fig. 2a). The algorithm employed has several features that give credit to this technique compared to traditional clustering techniques. This hybrid method creates clusters based on both continuous and categorical variables. It has the ability of automatically selecting the number of clusters as well as analysing large data files in an efficient manner (Caccam and Refran 2012). It requires only one pass of data and can produce solutions based on mixtures of continuous and categorical variables, and for varying numbers of clusters. The clustering algorithm is based on a distance measure that yields the best results if all variables are independent, and it deals with continuous and categorical data sets. Continuous variables have a normal distribution, while categorical variables have a multinomial distribution (Mooi and Sarstedt 2011).

The procedure consists of two steps: pre-clustering and clustering. The first is the formation of pre-clusters. It is a sequential approach, which is used to pre-cluster the cases. The goal of pre-clustering is to reduce the size of the matrix that contains distances between all possible pairs of cases. Pre-clusters are just clusters of the original cases that are used in place of the raw data in the hierarchical clustering. As a case is read, the algorithm decides—based on a distance measure—if the current case should be merged with a previously formed pre-cluster or whether to start a new pre-cluster. When pre-clustering is complete, all cases in the same pre-cluster are treated as a single entity. The size of the distance matrix is no longer dependent on the number of cases but instead on the number of pre-clusters (Mooi and Sarstedt 2011). The second step is where the hierarchical technique is applied. Similar to agglomerative hierarchical techniques, the pre-clusters are merged stepwise until all pre-clusters are in one cluster. In contrast to agglomerative hierarchical techniques, an underlying statistical model is used. Forming clusters hierarchically lets one explore a range of solutions with different numbers of clusters (Mooi and Sarstedt 2011).

3.6 The Case Study—NE1, Newcastle upon Tyne, England, UK

The case study is the inner urban core, NE1 postcode area of the city of Newcastle. Newcastle upon Tyne is one of the popular cities in North East England. This region is considered as one of the greenest cities in the UK, attempting to work towards sustainable development and implementing progressive plans towards resilience concepts. The NE1 area contains several express and arterial long roads, which vary in width, speed and capacity. The NE1 area is a suitable experimental area to be syntactically studied. It is an area rich in trip assignments and movements, which enables researchers to study the flow of the EV population and the behavioural characteristic of system, reflecting on the usage patterns. It accommodates around 3959 residents within a total area of 6 km^2. It contains the city centre of the metropolitan area, two universities, schools, shopping and recreational areas, commercial buildings, the central train station, squares, parks and partially, some of the busy residential wards of Newcastle area. The area has the main traffic arteries of the city and 25 sites of publicly available RF (NE1 2009).

3.7 Residential Versus Commercial

Nowadays, households are advised and encouraged to have a home installation of an EV charger. PodPoint EV chargers are supplied and installed free of charge to the household (FLO 2014) whether or not the householder has an EV or not, as they can take advantage of current government funding, and hence can have a charging point installed at their doors. Considering the fact that the majority of the

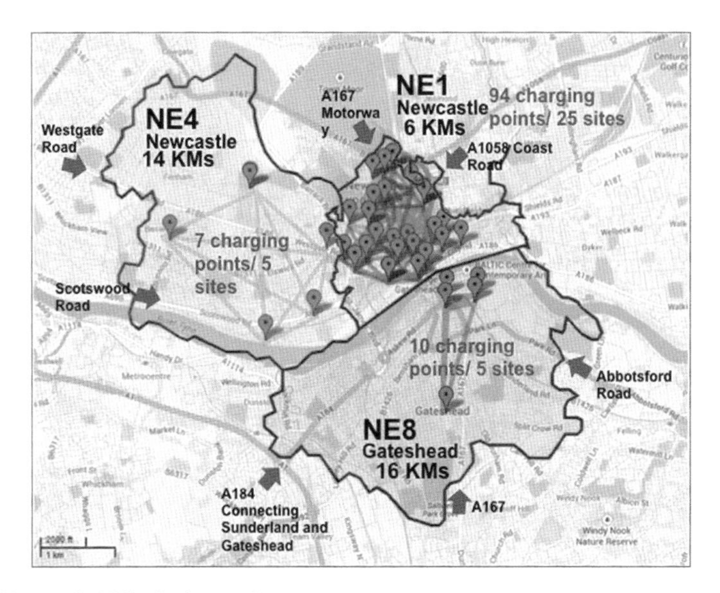

Fig. 3 Newcastle NE1, the inner urban core

users depend on domestic charging, this initiative is likely to boost up the number of points, and therefore, encourage potential users, and generate more trust for electrically fuelled means of transport. In particular, this scheme helps in residential neighbourhoods more than in commercial and central business districts. Therefore, the importance of the selected area here is clear. NE1, as shown in Fig. 3, is recognised as a vital commercial area. It contains a versatile land use with pretty busy avenues and streams of movement, whether pedestrian, car or bus passengers. The EV population includes inhabitants of NE1 and also commuters coming from neighbouring postal areas. Information about usage and charging patterns is collected from the main local service provider in North East England, Charge Your Car (CYC).

3.8 Perfect Timing

The North East is to become "UK's electric car capital" in terms of the density of the charging infrastructure, with plans to install up to 1,000 charging points around Newcastle and Gateshead metropolitan areas over the next 2 years with government funding. The planning with regard to the sitting of these needs to be well studied (Lumsden 2012). Newcastle is in need of proper planning, business model development (as these charge points will need to be maintained, and this is connected to usage) and the conducting of a feasibility study to study the market and plan economically for efficient sitting and operations of RFs. The distribution of these facilities should be well studied to avoid any waste of investment cost resulting from an underused charging point or unrecognisable facility.

3.9 Spatial Clustering of RFs

A spatiotemporal analysis of users' charging and driving patterns is now presented and discussed. There are 41 charging points in total distributed in the NE1, NE4 and NE8 postcode areas. The total number of registered drivers (CYC) in the year 2012 in the Newcastle and Gateshead area (the latter just over the river Tyne in very close proximity with a new urban centre also) is 420 users. The majority of the charging posts are located in NE1 (6 km^2), with 26 charging points. A total of 4 charging points are in NE4 (14 km^2), and 8 charging points are in NE8 (6 km^2), see Fig. 3. There are 48 users among the CYC users which charge their cars in NE1 (according to data from September 2012 to January 2013). This study focuses on these users and tries to investigate the charging pattern of some of these users and tries to correlate this with other dependent variables which signify the EV system. For the purpose of this study, 13 RF sites have been selected and 23 charging points in total are analysed, Fig. 4. The spatial clustering methodology is applied to the gathered data set. The data set contains different observations and cases that need to be analyzed and grouped in a way that has a well-defined meaning.

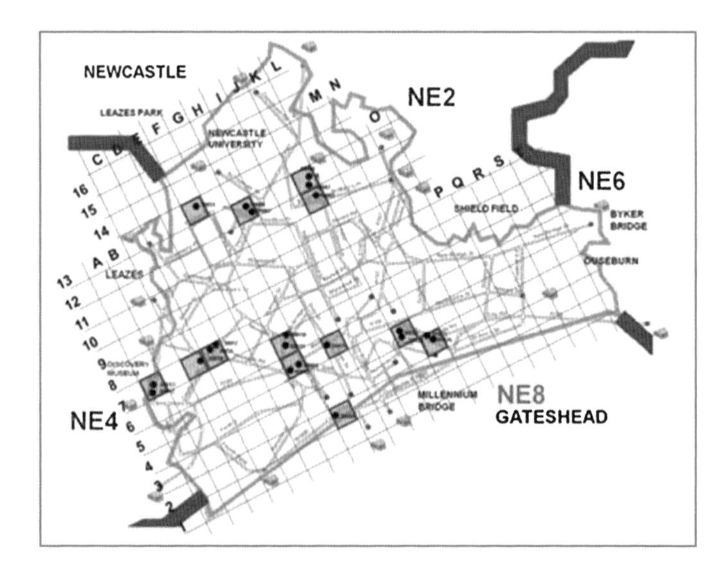

Fig. 4 Mapping the inner urban core with RFs

4 Analysis

4.1 Pre-clustering

The TwoStep clustering method is applied here. There are some key alterations that can be made to the selection of the variables, the display, and the maximum and minimum number of clusters, as well as the evaluation fields. Cluster analysis involves several procedures as summarized by Milligan (1996): selecting clustering objects and clustering variables (dependent and independent), deciding on the type of data, variable standardization, choosing the measure of association, selecting the clustering method, determining the number of clusters and interpretation, validation and replication. This includes a description of the administration procedure of data collection, data cleaning and a description of the data set. The data set is analysed and classified as per the predefined three pillars.

The profiles and usage patterns form mainly three groups: (a) the origins of commuters, (b) the popularity of the charging point within the network and (c) the charging behaviour. The origins are the residential postcodes of frequent EV commuters using the recharging infrastructure of NE1. This is identified by the first part of the postcode which can identify the postal zone, provided by the service operator in an anonymous fashion. NE1 is the centre of the city's territory; this group is classified as North zone, North East, South and all directions.

4.2 Clustering Analysis

The TwoStep clustering technique generates a report with some graphs and figures, see Fig. 5, showing the cluster quality, size, structure and influential variables. The clustering process took several iterations until the chosen one was reached. The decision is made based on the cluster quality, which is a function of the number of clusters, and the ratio of cluster sizes. The quality should not be poor, and the ratio should not exceed three. Figure 6 shows the main paradigms of the clustering process. The number of inputs (categorical and continues selected variables) is eight. The overall distribution of cluster is quite decent and balanced. There is no single dominant influential variable that affects the clustering process; connectivity, (off street/on street) and charging pattern distributed over the course of the week, and the number of transactions is the main predictor.

The clusters are organised based on the size of the clusters. The first group is the "The Connected" charging points. This group forms the biggest cluster, containing five sites with a total of 11 charging points. This cluster has very high connectivity values, which means that these charging points are spatially connected within the road network of the Newcastle metropolitan area. In the space syntax literature, connectivity measures the number of immediate neighbours that are directly connected to a space (space here is the line that represents the road with the charging

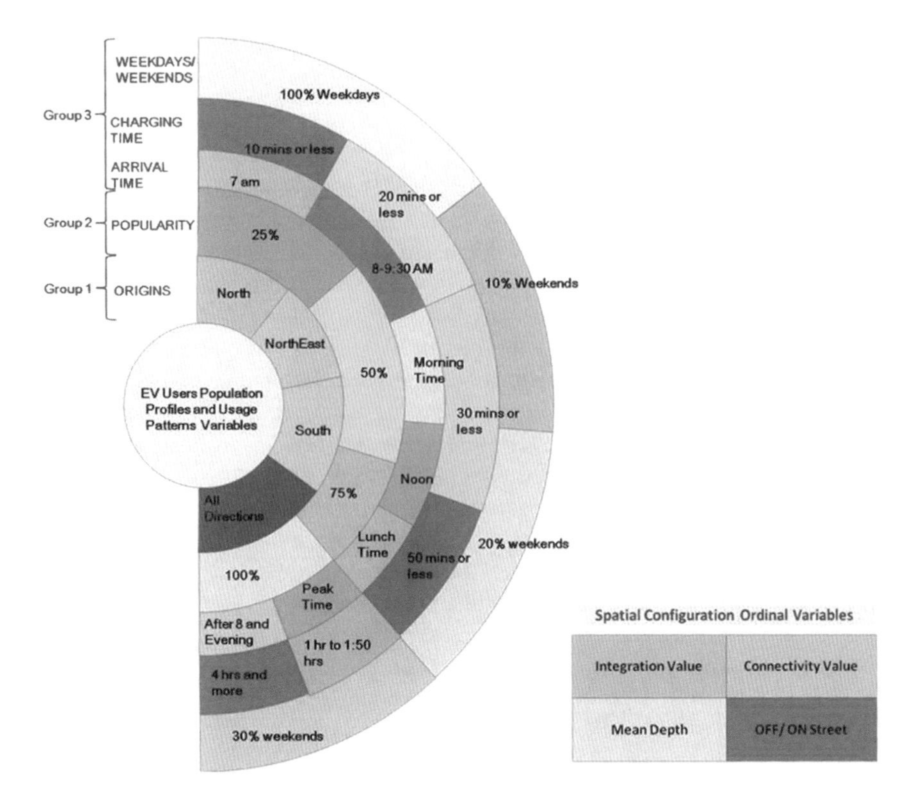

Fig. 5 Usage profile attributes

points) (Hillier and Hanson 1984). These five sites, based on CYC identification system, are 10002, 30058, 40012, 400025 and 20046. This cluster contains off street charging points, and they are used by the inhabitants and visitors coming from North East postal codes (NE7, NE23, NE26, NE30, NE31 and NE34). These charging outlets are doubled and sometimes tripled on the same site (such as 10002, 10005 and 10006). EV users come to replenish their batteries during the noon/lunchtime period, mainly on weekdays. 30 % of the transactions take place over the weekends. They commute around 8 miles as a minimum road trip to reach the charging station. Hence the charging time takes place during noon/lunchtime time; this happens during the day, which leads the researcher to suspect that the charging point is not the first destination of the day trip. It is worth mentioning that the charging points of this cluster are not popular within the recharging network as the frequency of usage of these points is not high among the network.

The second group is the "The Comfy" charging points. This group forms the second biggest cluster, containing three sites (20007, 30056 and 20059) with a total of five charging points. These charging outlets are doubled up (20007 and 20008, 30056 and 30057) and one single with 20059. This cluster's main feature is an outstanding number of transactions made by a relatively high number of EV drivers

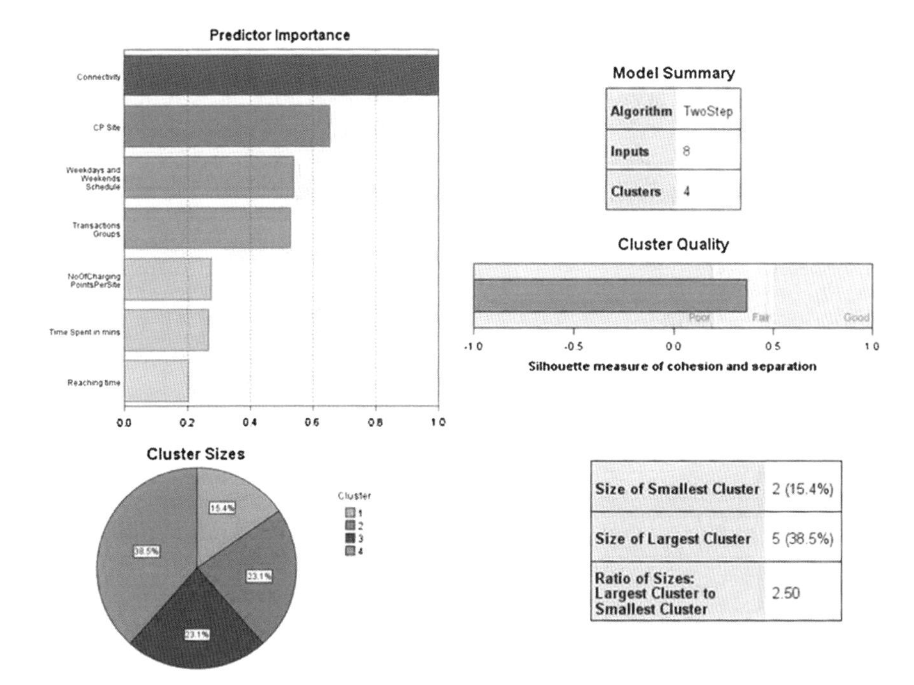

Fig. 6 Clustering analysis

compared to other cluster users. The second remarkable feature of this cluster is that of frequent EV drivers who use these charging points come from all neighbouring postal codes (NE1, NE3, NE5, NE7, NE8, NE10, NE13, NE31, NE34, NE40 and NE47). The charging points are off-street located. These points have high connectivity values, which reflect accessibility and ease of reach within the road network. EV users come to replenish their batteries during the noon/lunchtime period, mainly on weekdays as 10 % of the transactions take place over the weekend. This reflects a low demand during the weekend. EV users commute around 13 mile-road trips to reach the charging station, presumably having more than one destination during the daily route.

The third group is the "The Intermittent" charging points. This cluster has the same size of the second cluster in terms of sites though less number of outlets and with different paradigms. This cluster's main feature is that frequently EV drivers use its charging points only during the weekdays. This cluster has extremely low connectivity values and all sites are off-street. These three sites are 30050, 30051 and 20006, with a total of four charging outlets. This cluster contains off-street charging points and they are used by inhabitants and visitors coming from North including North East and North West postal codes (NE5, NE7, NE23, NE26, NE30, NE31 and NE34). The charging points of this cluster are single-located. EV users come here to replenish their batteries at noon. They frequently stop over at these recharging sites between 1:00 and 3:00 pm, and only during the weekdays, which is perceived as a different charging pattern considering the working hours. These commuters reach the sites after an average of 10 miles. It has already been stated that the time of charge is a bit later during the day, and some of the northern zones are remote. It is also worth mentioning that users do not spend ample time charging their vehicles at these points.

The fourth group is the “On the Go” charging points. This group forms the smallest cluster, as it has only two sites (40004 and 40018). However, it has four charging outlets (40004, 40005, 40018 and 400019). This cluster is ambiguous as it is the only cluster that has its charging sites on-street; nevertheless, they have an almost imperceptible connectivity value. EV commuters who use these sites charge for almost less than 10 min, which will electrify their batteries with enough power to take them home in case they arrived at the charging point with flat state, or they depend more on domestic charging and are topping up with a few kilowatts. The frequent use happened to be by the North East post code commuters between the times of 1:00 and 3:00 pm. It is worth mentioning that the total number of transactions made by frequent users is below the expectations compared to the number of users for charging stations.

5 Discussion

Four different clusters of RFs have been generated as the outcome of the TwoStep spatial built-in clustering algorithm. Each cluster has main features that identify and configure common RF usability attributes, recharging static design characteristics and spatial configuration values. This study is conducted to investigate the relation between spatial configuration of RFs and their usability. This relation can be observed from the formation membership of the clusters and the graphs show the quality, separation and distribution. The cluster quality bar displayed in Fig. 6 reflects a fair and close to good quality of cluster in terms of cohesion and separation. The quality could have been better with a higher number of cases and a variety of variables.

Accordingly, we can accept the hypothesis that spatial configuration and design characteristic of EV recharging facilities affect the system usability significantly.

As shown in Fig. 6, connectivity (a spatial configuration value) is the most influential predictor forming the different EV clusters. The location of the site (on-street or off-street) comes as the second most important predictor, while the frequency of use over the weekdays and the weekend ranked third with regard to influence as a predictor. Equivalent to the frequency of use, in terms of importance and influence, is the number of transactions made for each site.

6 Conclusion

In line with the worldwide goals and global technology attention, the recharging infrastructure should be well designed and sited so as to be reliable, stable, efficient, meeting the demands of users and also economically viable. The recharging experience should not be a worrying matter for EV drivers. The use of e-mobility is associated with a range-anxiety-syndrome, presenting hurdles for many potential users to electrify their vehicle use. Even for current users, so far, the EV is still replacing the secondary car in multi-car owning households due to range limitation (Elbanhawy and Dalton 2013). The present study aims at interpreting the users' data in a meaningful way regarding the data observations, and providing guidelines

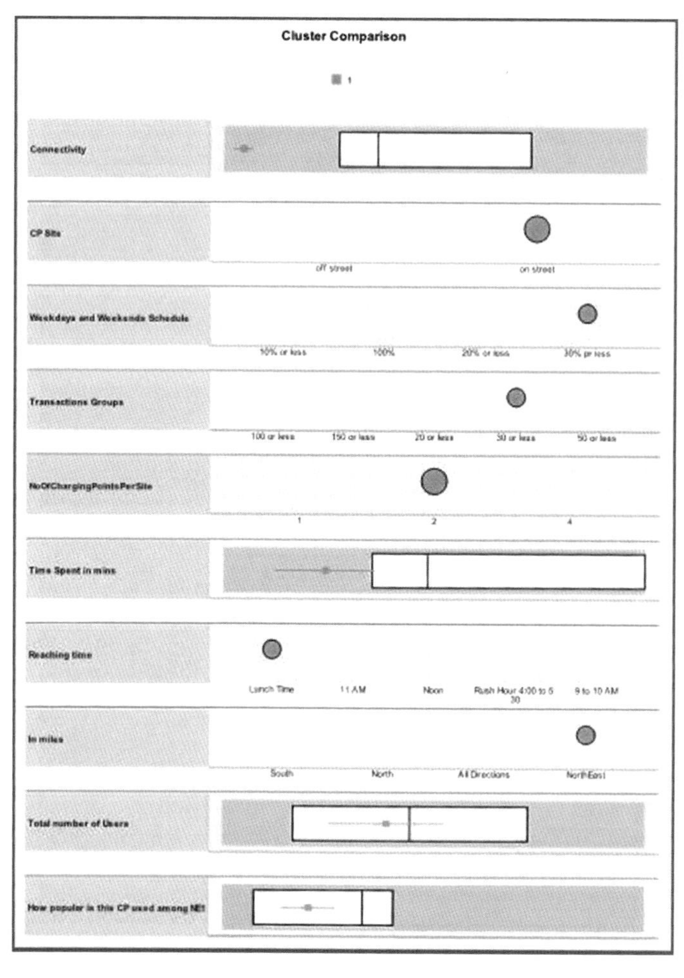

Fig. 7 Cluster 1_"the connected"-SPSS

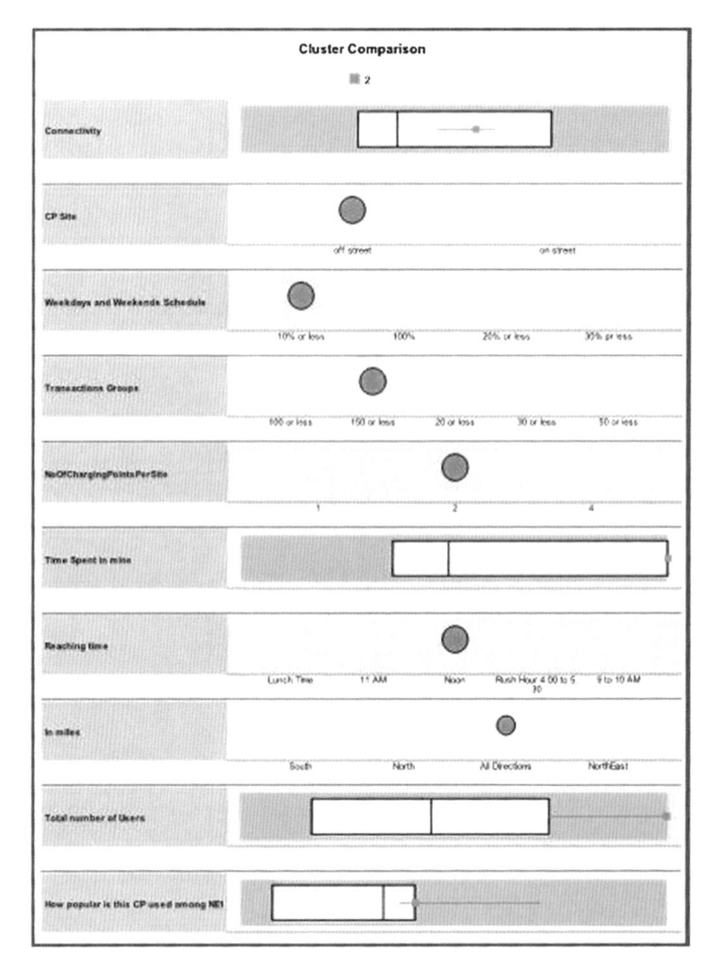

Fig. 8 Cluster 2_"the comfy"-SPSS

and recommendations with regard to the design and sitting of recharging facilities based on this. This should be of interest to researchers, planning authorities, policy makers and commercial service providers.

The study complements an agent-based model that is developed to denote and replicate EV real users in an interactive virtual environment. The present study looks at the features of RFs that can be observed by studying and analysing the charging patterns of the users spanning a period of time. The analysis incorporates many aspects of the system and the users themselves. It counts for technical assumptions and considerations of an EV and also other socio-economic, behavioural and demographic aspects of the users.

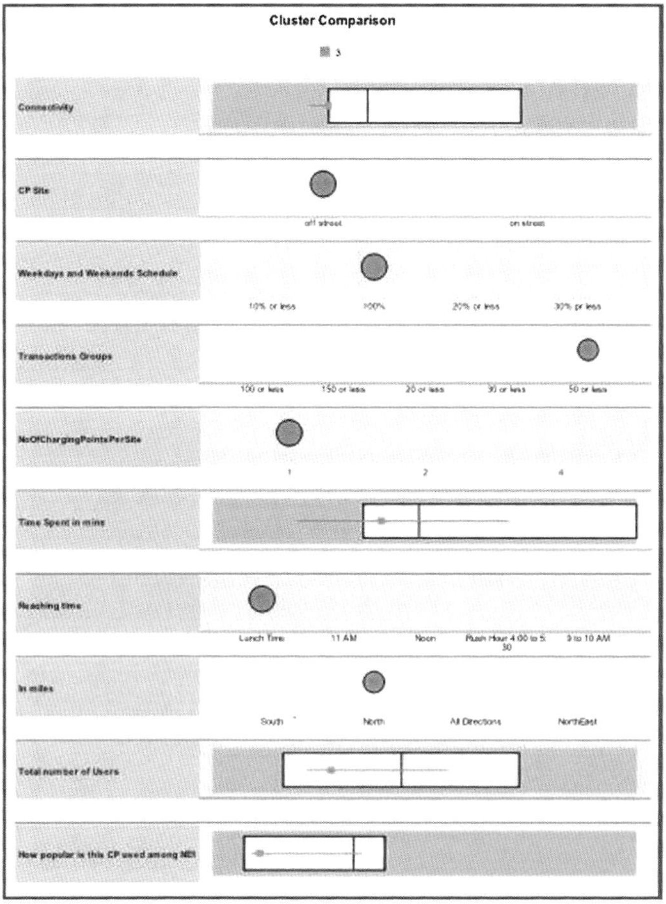

Fig. 9 Cluster 4_"on the go"-SPSS

The study focuses on the impact of the spatial attributes and aspects of the system on its usability at a micro-scale. It analyses the individual charging profiles of a selected (anonymous to the researcher) group of inhabitants of the Newcastle metropolitan area who are frequently using the recharging facilities of NE1. A data reduction analysis was conducted via dimensional analysis technique, TwoStep clustering. The paper investigates how and to what extent the spatial configuration of the network impacts the charging patterns of EV users (Figs. 7, 8, 9 and 10).

From a planning perspective, the planners and policy makers would need to have a clear indicative description of the recharging facilities design characteristic and configuration that provides design key elements as well as guidelines for what to expect to get in terms of business need. Among the four clusters, the second, "The

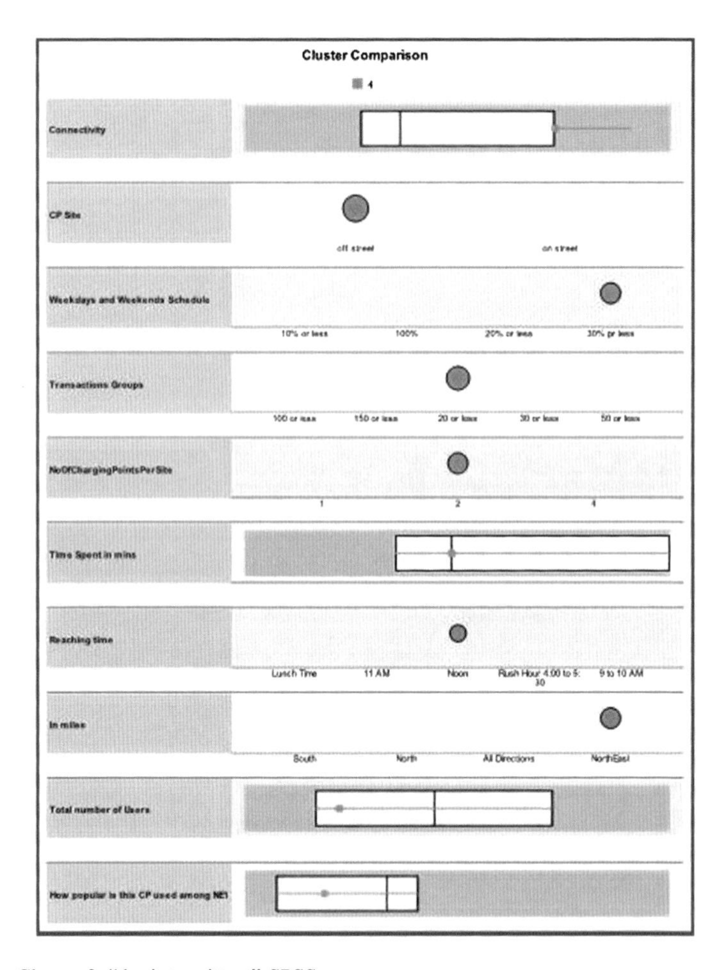

Fig. 10 Cluster 3_"the intermittent"-SPSS

Comfy", Fig. 8, is the chosen one to be replicated. Its design and spatial features are to be desired when designing for recharging facilities. With regard to the process of assisting in the planning of future EV system, the study's outcomes and recommendations are to maximise the EV system with a nature of "The Comfy" cluster. The "Comfy" cluster is a group of charging facilities that generate profit. This cluster set-up meets the business needs of the EV system as it hits the highest number of transactions made by a comparatively very large number of users. It is an accessible off-street facility with a significant value of connectivity, and users tend to have a long charge period (4–6 h), which reflects suitability. However, a cautionary note to add is that perhaps this also reflects that they mainly use it for free parking, soon not to be on offer in this mode any longer.

References

Burgess, M., Harris, M., Walsh, C., Carroll, S., Mansbridge, S., & King, N. (2013). Longitudinal assessment of the viability of electric vehicles for daily use. *Paper presented at the 4th Hybrid and Electric Vehicles Conference 2013*, London, UK.

Barros, A., De Silva, P., & Holanda, F. (2007). Exploratory study of space syntax as a traffic assignment tool. *Proceedings, 6th International Space Syntax SymposiumSpace Syntax Symposium* (pp. 1–14). Istanbul: Space Syntax Symposium.

Caccam, M., & Refran, J. (2012). *Clustering analysis*. Retrieved from http://www.slideshare.net/jewelmrefran/cluster-analysis-15529464.

Chen, T. D., Kockleman, K. M., & Khan, M. (2013). The electric vehicle charging station location problem: A parking-based assignment method for Seattle. *The 92nd Annual Meeting of the Transportation Research Board in Washington DC.*

Chiu, T., Fang, D., Chen, J., Wang, Y., & Jeris, C. (2001). A robust and scalable clustering algorithm for mixed type attributes in large database. In *The 7th ACM SIGKDD International Conference on Knowledge Discovery and Data Mining* (pp. 263–268). New York: ACM.

Christensen, L., Norrelund, A. V., & Olsen, A. (2010). *Travel behaviour of potential electric vehicle drivers. The need for charging*. Denmark.

De Haan, P., Mueller, M. G., & Peters, A. (2006). Does the hybrid Toyota Prius lead to rebound effects? Analysis of size and number of cars previously owned by Swiss Prius buyers. *Ecological Economics, 58*(3), 592–605.

Dong, J., Liu, C., & Lin, Z. (2014). Charging infrastructure planning for promoting battery electric vehicles: An activity-based approach using multiday travel data. *Transportation Research Part C, 38*, 44–55.

Egbue, O., & Long, S. (2012). Barriers to widespread adoption of electric vehicles: An analysis of consumer attitudes and perceptions. *Energy Policy, 48*, 717–729.

Elbanhawy, E. (2014). Space-time analysis of electric vehicles advocates. Forthcoming in: *proceedings of ITEC-IEEE.* Michigan.

Elbanhawy, E., & Dalton, R. (2013). A syntactic approach to electric mobility in metropolitan areas NE 1 district core. *9th International Symposium on Spatial Data HandlingSpace Syntax Sympouim,* Korea.

ElBanhawy, E., Dalton, R., Thompson, E., & Kotter, R. (2012). Heuristic approach for investigating the integration of electric mobility charging infrastructure in metropolitan areas: An agent-based modelling simulations. *2nd International Symposium on Environment Friendly Energies and Applications (EFEA)* (pp. 74–86). Newcastle Upon Tyne, UK. Retrieved from https://ieeexplore.ieee.org/stamp/stamp.jsp?arnumber=06294081.

ElBanhawy, E., Dalton, R., & Nassar, K. (2013). Integrating space-syntax and discrete-event simulation for E-mobility analysis. *American Society of Civil Engineers, 1*(91), 934–945. Retrieved from https://ascelibrary.org/doi/abs/10.1061/9780784412909.091.

ElBanhawy, E., & Nassar, K. (2013). A movable charging unit for green mobility. *ISPRS—International Archives of the Photogrammetry, Remote Sensing and Spatial Information Sciences, XL-4/W1*, 77–82.

FLO, R. (2014). *Electric vehicles charge points*. Retrieved from http://www.retroflo.com/free-electric-vehicle-chargers/domestic-ev-charge-points.

Franke, T., & Krems, J. F. (2013). What drives range preferences in electric vehicle users? *Transport Policy, 30*, 56–62.

Franke, T., Gunthrt, M., Trantow, M., Krems, J., Vilimek, R., & Keinath, A. (2014) Examining user-range interaction in battery electric vehicles-a field study approach. *The 5th International Conference on applied human factors and ergonomics AHFE*, Krakow, Poland.

Ge, S., Feng, L., & Liu, H. (2011). The planning of electric vehicles charging station based on grid partition method. *IEEE*, 2726–2730.

Gil, J., Tobari, E., Lemlij, M., Rose, A., & Penn, A. (2009). The differentiating behaviour of shoppers clustering of individual movement traces in a supermarket. *Proceedings of the 7th International Space Syntax Symposium*. Stockholm: KTH.

He, F., Wu, D., Yin, Y., & Guan, Y. (2013). Optimal deployment of public charging stations for plug-in hybrid electic vehicles. *Transportation Research, Part B, 47*, 87–101.

Heffner, R. R., Kurani, K. S., & Turrentine, T. (2007). Symbolism and the adoption of fuel-cell vehicles. *The World Electric Vehicle Association Journal* (Vol. 1). Retrieved from http://escholarship.org/uc/item/5934t20f.

Hillier, B., & Hanson, J. (1984). *Social logic of space* (p. 275). Cambridge: University of Cambridge.

Hwang, S. W., Jin Kweon, S., & Ventura, J. (2013). Optimal location of discretionary alternative-fuel stations on a tree-network. *10th Annual College of Engineering Research Symposium, April 2, 2013* (p. 7). State College.

Klein, J. (2007). *Why people really buy hybrids*? Retrieved from https://www.toplinestrategy.com.

Lam, A. Y. S., Leung, Y.-W., & Chu, X. (2014). Electric vehicle charging station placement: formulation, complexity, and solutions. *arXiv, 2* (1310.6925), 1–10.

Larson, J., Bradlow, E., & Fader, P. (2005). An exploratory look at supermarket shopping paths. *International Journal of Research and Marketing, 22*(4), 395–414.

Lindblad, L. (2012). *Deployment methods for electric vehicle infrastructure*. Uppsala University. Retrieved from https://www.diva-portal.org/smash/get/diva2:538789/FULLTEXT01.pdf.

Liu, Z., Wen, F., & Ledwich, G. (2013). Optimal planning of electric-vehicle charging stations in distribution systems. *IEEE Transactions on Power Systems, 28*, 102–110.

Lumsden, M. (2012). *Successfully implementing a plug-in electric vehicle infrastructure. A technical roadmap for local authorities and their strategic partners limited* (pp. 1–112). London. Retrieved from www.ietstandards.com/evreport.

Milligan, G. (1996). Clustering validation: results and implications for applied analyses. *Clustering and Classification* (341–375). River Edge: World Scientific.

Mooi, E., & Sarstedt, M. (2011). Cluster analysis. *A Concise Guide to Market Research* (p. 237). Berlin: Springer.

Namdeo, A., Tiwary, A., & Dziurla, R. (2013). Spatial planning of public charging points using multi-dimensional analysis of early adopters of electric vehicles for a city region. *Technological Forecasting and Social Change*. doi:10.1016/j.techfore.2013.08.032.

NE1. (2009). NE1, Newcastle business improvement district. Retrieved from http://www.newcastlene1ltd.com/.

Nie, Y. M., & Ghamami, M. (2013). A corridor-centric approach to planning electric vehicle charging infrastructure. *Transportation Research Part B: Methodological, 57*, 172–190.

OLEV. (2011). *Making the connection—The plug-in vehicle infrastructure strategy*. Office of Low Emissions Vehicles (OLEV): United Kingdom.

Ozaki, R., & Sevastyanova, K. (2013). Going hybrid: An analysis of consumer purchase motivations. *Energy Policy, 39*, 2217–2227.

Pierre, M., Jemelin, C., & Louvet, N. (2011). Driving an electric vehicle. A sociological analysis on pioneer users. *Energy Efficiency in Transport and Mobility, 4*(4), 511–522.

Robinson, A. P., Blythe, P. T., Bell, M. C., Hübner, Y., & Hill, G. A. (2013). Analysis of electric vehicle driver recharging demand profiles and subsequent impacts on the carbon content of electric vehicle trips. *Energy Policy, 61*, 337–348.

Schaeffer, S. (2007). Graph clustering. *Computer Science Review*, 27–64. Retrieved from https://s3-us-west-2.amazonaws.com/mlsurveys/13.pdf.

Steinhilber, S., Wells, J., & Thankappan, S. (2013). Socio-technical Inertia: Understanding the barriers to electric vehicles. *Energy Policy, 60*, 531–539.

SwitchEV. (2013). *SwitchEV Final Report*. Newcastle Upon Tyne, UK. Retrieved from http://www.futuretransportsystems.co.uk/_diskcache/296-final-report-for-seminar.pdf.

Turan, F. A., Tuzuner, A., & Goren, S. (2012). Modeling electric mobility in Germany: A policy analysis perspective. *Proceedings of the 2012 International Conference on Industrial Engineering and Operations Management Istanbul*, Turkey.

Turner, A. (2004). *Depthmap 4: A researcher's handbook*. London, UK: Barlett School, University College London.

Wirges, J., & Linder, K. (2012). Modelling the development of the regional charging infrastructure for electric vehicles in time and space. *European Journal of Transport and Infrastructure Research, 4*(12), 391–416.

Zabala, E., Lopez, J., Arxuaga, A., Zamalloa, M., Pardo, D., & Izagirre, J. (2012). A whole approach for the electric vehicle infrastructure in the Basque Country. *Electric Vehicle Symposium (EVS 26)*.

An Architecture Vision for an Open Service Cloud for the Smart Car

Matthias Deindl, Marco Roscher and Martin Birkmeier

Abstract Project Oscar (Open Service Cloud for the Smart Car, see: http://www.fir.rwth-aachen.de/en/research/research-projects/osc-ar-01-me1203), a 3-year collaborative project running from January 2012 to December 2014, introduces an open platform to reduce heterogeneity in tracking and fleet management systems as well as provides electric vehicle (EV) data to the smart grid. This includes various interfaces providing access to smart car data as well as interfaces for integrating services. Thus actors, developers, systems and various other components are connected. The basis for the open ICT innovation platform is the "Open Service Cloud" (OSC), offering generation of added value to a multitude of actors involved. Contrary to most existing solutions, the OSC is open to third parties, thus establishing a platform serving as gateway for additional services and applications. The Oscar architecture enables vehicles to closely interact with the OSC via wireless ICT solutions and the integration of in-car systems. Additional driver interaction is implemented through an "In-Car"-Tablet, enabling a framework for third-party applications. The server acts as an integrator between smart cars, smart traffic and a smart grid. This allows the development of a smart charging algorithm (SCA), which also provides an energy demand forecast for the whole EV fleet. Based on data provided by the OSC this not only benefits the supplier in the short term, but also allows accurate long-term developments of grid infrastructure.

Keywords ICT · Real-time vehicle data collection · Open service cloud · Cloud services · Information broker · Big data · In-vehicle apps · Vehicle data analysis ·

M. Deindl (✉) · M. Roscher · M. Birkmeier
FIR e.V. an der RWTH Aachen, Institute for Industrial Management at RWTH Aachen University, Campus Boulevard 55, 52074 Aachen, Germany
e-mail: matthias.deindl@fir.rwth-aachen.de

M. Roscher
e-mail: marco.roscher@fir.rwth-aachen.de

M. Birkmeier
e-mail: martin.birkmeier@fir.rwth-aachen.de

W. Leal Filho and R. Kotter (eds.), *E-Mobility in Europe*,
Green Energy and Technology, DOI 10.1007/978-3-319-13194-8_15

Fleet data analysis · Integrating additional sensors · Third-party services · Third-party applications · Vehicle-to-grid · V2G · Electric vehicle integration · Smart grid

1 Introduction and Motivation

Electric vehicles (EVs), if not the dominant form of transport for the future, could at least play an important role in the future devolpment of utility vehicles and the transport of passengers and goods. Furthermore, with the introduction of smart grids, EVs act as an enabler for an increasing share of renewable generation (Mühlenhoff 2010). With the focus shifting away from the car as a status symbol and more towards the mobility provided itself, businesses such as car sharing or rental agencies are on the rise (Proff et al. 2012). Especially in an urban short distance environment, EV mobility exhibits many advantages over today's cars.

Therefore, solutions for effective use and operation of EVs have to be developed. Apart from car sharing and fleet management applications, this can also incorporate many advantages for traffic monitoring and control as well as electricity grid efficiency and load planning. Therefore, the development has to create an interface between smart cars, smart traffic and smart grids (see Fig. 1).

Minimising costs while maximising efficiency and durability is a key challenge. The shift towards increased use of EVs leads to a need for additional information- and communication technologies (ICT) which is not yet prevalent in todays' cars. Parameters

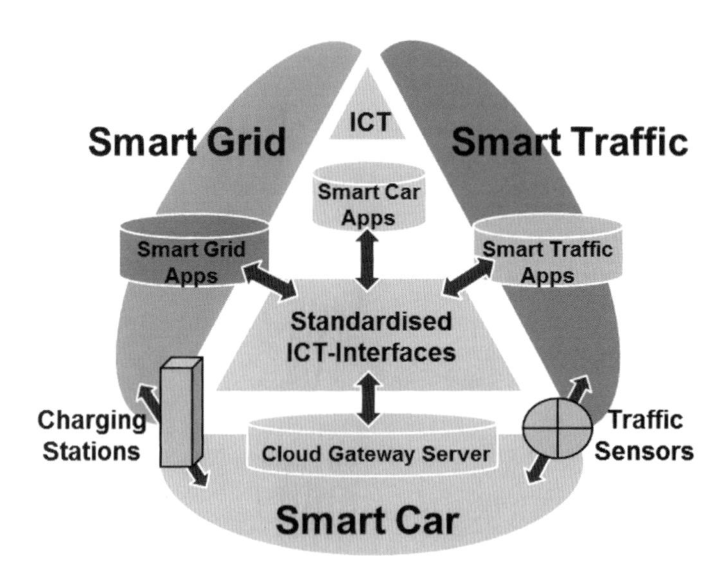

Fig. 1 Project Architecture connecting smart car, smart grid and smart traffic

not relevant today in internal combustion vehicles, such as battery condition, weather and outside temperature have to be taken into account, while existing functionality such as range calculation increases in importance and has to be improved. Additionally, the search for and navigation to available charging stations as well as the application of smart charging algorithms is becoming a vital part of in-car services.

ICT solutions can improve consumer acceptance of EVs and offer a large potential for added value. Especially when employing open structures and standards, third parties as well as consumers can benefit immensely from data exchanged with their car.

Smart traffic monitoring and control allows quicker reactions and a more suitable response to emergency situations. Simultaneously, traffic flow and congestion data is available much more quickly and can even be predicted in advance. This in turn can display warnings to drivers and impact navigation routing though an in-car tablet. Smart grid functionality in turn allows advanced predictions about electricity consumption and local grid loads, allowing the provider to adjust local prices and generation capacity accordingly, decreasing the need for grid extensions (Appelrath et al. 2012). Furthermore, the inclusion of vehicle-to-grid (V2G) functionality provides added value to EV owners while simultaneously enabling increased renewable generation and grid stability (Engel 2010). Finally, the conjunction between smart grids and smart cars provides users with access to reliable information about parking spaces and charging stations and can even enable advance booking, integrated directly into the navigation process. Thus, it reduces the need for blanket coverage of charging stations otherwise necessary without the use of intelligent cloud solutions. Thus, a link to emerging smart grids (Appelrath et al. 2012) and the incorporation of information about charging station status and power is considered particularly important for efficient and flexible use of EVs (see Fig. 1). The high initial cost of the underlying computing architecture and EVs can be mitigated through partnerships and sharing of existing or newly deployed systems with others. This also allows possible usage of a bigger data set to further improve the quality of services derived from data collected through the OSC. With an increase in the number of users the quality of service offered increases as well, because the information gathered from several sources is more reliable, less error-prone and more diversified.

Among system users, car manufacturers, car sharing agencies, charging infrastructure providers and fleet managers in particular profit through the collection of real-time car data. Car manufacturers for example can get a better understanding about the costumers' behaviour and needs to foster brand loyalty by offering tailor-made added services. Most importantly, collection of real-time fleet status allows much more efficient tracking of locations, routes, bookings, vehicle health status, and thus also scheduling of orders. This conduces to a higher efficiency in vehicle and employee utilisation as well as increased flexibility coupled with overall time and cost savings. At the same time, on-board computers and car connectivity in general will further develop, adjusting to emerging needs identified through the use of this system. Certain benefits can be directly passed on to all drivers in particular. High vehicle asset costs are spread out over a large number of users while operating costs are lowered especially for occasional drivers. Additionally, improved parking, high flexibility and easy access are paired with the lowered climate impact of highly utilised vehicles if the overall number of vehicles thereby falls or draws more on renewable energy. Furthermore, a cloud connection allows automatic customization of car settings such as seat, mirror or radio presets. Problems arising from certain business models, e.g. allowing customers one-way transfers, are mitigated by intelligent tracking and scheduling of vehicles and drivers. On the one hand, return routes can be effectively discounted and on the other hand employees having to return cars themselves can be more efficiently deployed (Schäfer and Fricke 2013).

2 Architecture Goal

The Open Service Cloud (OSC) provides a multitude of interfaces enabling access to the data and systems of a smart car as well as integration into the infrastructure of the emerging smart electricity grid. It connects several actors, developers, systems and various other components while providing interfaces for the integration of services. Collection and communication of in-car data occurs in real time. The OSC handles data for single vehicles or whole car fleets, and can provide data-based services to third-party applications. These can also be offered to the driver in the form of apps for handheld mobile devices within the car. The integration of smart grid data allows the use of intelligent charging algorithms to lower operating costs, while simultaneously providing maximum availability and an adequate battery charge adapted to individual driving needs (Roscher et al. 2013). An appropriate security framework takes care of enabling a selection of non-driving security-related control features. Assuming further integration, all this can serve as a basis for a vehicle application store.

A core issue of the OSC development is the implementation as a secure and trusted platform. The vehicle owner has to be provided with complete control over the data which is collected and transmitted from his vehicle in an easily understandable and transparent way. This includes explicitly the ability to determine to which extent data is shared with third parties (data privacy and protection). This is very important due to the fact that not anonymised movement profiles contain very sensitive information about the driver and his behaviour. Thus, differentiated access rights for selected data streams can be granted to individual services that are chosen for use. Preferably, a simple overview shows transparently which data is selected, and how and for which services it is used. Selected data can be anonymised before being made available by the user to benefit services that require usage of data from multiple vehicles, such as traffic analytics, forecasts, energy demand analytics or estimate use of charging stations and parking spaces.

Energy demand, charging station information and projections in particular are a vital part in the successful implementation of smart grid technology and widespread charging station rollout. Making use of variable pricing schemes, the OSC is able to predict ideal charging times to minimise cost and provide additional services to the grid, e.g. use of the EV battery as energy storage unit for load balancing purposes (vehicle to grid). Widespread implementation can help balance the grid as a whole and on a local or regionalised scale, thus reducing the need for investment in the distribution grid. While raising overall electricity consumption, the increasing use of EVs in this way will help reduce fluctuation of residual power needs, enable increased renewable generation and thus lower overall electricity prices (Mühlenhoff 2010). In this way, the OSC acts as an enabling service for major infrastructural change in the electricity grid. Furthermore, inclusion of weather and temperature data allows for the development of advanced range calculation methods that take into account the massive increase in energy consumption as well as lowered battery performance in cold temperatures.

Incentives to promote sharing of data can be either in monetary or non-monetary form, e.g. by an improvement of in-service quality. This will be most pronounced in the use of charging algorithms and scheduling, which work best when supplied with in depth driving profiles. Monetary incentives in turn can be orchestrated by the OSC by reducing service fees or offering discounts for third-party services such as use of charging stations in return to contribution of vehicle data.

In contrast to the majority of solutions provided by Original Equipment Manufacturers (OEMs), the OSC is open to third-party access. A major part of the OSC vision includes the establishment of a future vehicle platform serving as a gateway for additional services and applications. Features usually provided exclusively by big corporations due to their high implementation cost are made available to third parties and small manufacturers through the use of standardised interfaces. Initially developed to focus on an EV perspective, the system can easily be adapted for multiple types of vehicles by making use of controller area network bus (CAN bus) standard and the on-board diagnostic (OBD) system. Both of these are already in place in nearly all cars produced today.

An open innovation campaign integrates third parties from the very beginning to ensure usability and functionality of the OSC. By making use of closely integrated networks and known partners, open innovation integrates their point of view and their requirements in the innovation process. This new scientific approach stands in contrast to the conventional approach of "closed innovation", characterised by companies using ideas and technical skills available only from within their own domain, other than through trusted (systems) suppliers. Furthermore, several interviews, surveys and workshops were conducted (Krenge and Roscher 2013) to integrate the additional perspectives of users and customers. Finally, other user groups were detected and identified by performing a "netnography"-search (Kozinets 2002). Inviting these additional third-party users to workshops and field tests offered further input and feedback regarding OSC usability.

3 State of the Art

Already, there are many smart phone applications dealing with driving tasks and fleet management, sometimes individualised for customers to make complex data tracking and providing easier. Provided by an OEM or fleet service provider, these applications are closed systems. Some services are purchased and retailed under a different company's brand name (Kidder 1997; Briggs and Lampton 2013).

Most in-car apps focus on giving directions or navigating to destinations via Global Positioning Systems (GPS). Real-time information about traffic and congestion as well as nearby incidents and radar traps can be provided to the customer. These applications also might prove very useful for logistics networks, fleet management and taxis, offering economical and expeditious route planning, combined with effective error reports to result in more qualitative driving. By connecting in-car apps to a network running logistical planning systems, rerouting recommendations can be included in real time to enable the driver to reach his destination faster. Additionally, driving habits of users can be identified and analysed to enable more precise fuel consumption information and thus include recommendations to fill up, e.g. when and where prices are predicted to be particularly low (Briggs and Lampton 2013). The usage of parking applications enables drivers to find parking spaces more quickly and even make reservations and advance payments to ensure punctuality at their destination. Furthermore, the cost of car ownership can be decreased by controlling car resources such as fuel efficiency monitoring.

These closed or partially closed systems are intended for use with a specific analyser and are usually developed by the analyser manufacturer (OEM) or a third-party supplier. They do not offer interconnectivity or flexibility, especially in connection to a wide array of possible EV charging providers (Callaghan 2013).

An important issue in modern logistics is vehicle tracking to improve delivery operation. This includes real-time tracking and decision making for incident management providing reactions to changed requirements. Various solutions to static vehicle routing problems (VRP) are already available. Using these solutions, enables the creation of daily or weekly delivery schedules for logistics enterprises by specifying goods carried per vehicle and destination parameters using a determined route while minimising cost and maximising customer satisfaction. However, static schedules are very fragile to incidents and disturbances. Breakdowns, congestion, ramp overloads and unforeseen reverse logistics requests can quickly lead to a disruption (Automotive News Europe 2006).

Tracking the operating vehicles in the field offers a solution. Real-time data about location, technical status and freight of vehicles can transform the static system into a dynamic one, able to instantly react to changes. Location data can be collected through GPS, while technical data can be read from the OBD or CAN bus system. Since CAN bus rarely provides such information, additional sensors and microcontrollers have to be integrated to control and track special systems such as cooling, hydraulics and tire pressure. Most of the time, these systems communicate via proprietary devices without integration into the vehicles' ICT.

Various wireless technologies enabling the transmission of data to a server have emerged during the last decade (General Packet Radio Service—GRPS, Universal Mobile Telecommunications System—UMTS, Long-Term Evolution—LTE, etc.). A real-time fleet management system (RTFMS) which consists of a logistic information service (LIS), the vehicle-borne mobile data terminal (MDT) and the mobile communication infrastructure has been described by See (2007), This connects a typical vehicle routing server to data collected from vehicles and enables the MDT to provide its data to the LIS in real time (See 2007). Fleet management systems typically include software to wirelessly receive data from the fleet's vehicles. Thus, it is possible to monitor an individual vehicle's positions in geographical information systems (GIS) while tracking freight via RFID (Prasanna and Hemalatha 2012). Furthermore, the software performs plausibility checks to determine, e.g. whether the vehicle followed the predetermined route and made all scheduled stops on time. Calculations also include possible delays due to incidents on the route as well as average consumption and expected fuel cost (Vivaldini et al. 2012). While already important in keeping costs reasonable today, consumption calculation will become even more important with further EV deployment because of low overall range, long recharge times and considerable impact of temperature and weather.

Available solutions are provided as commercial software and hardware systems that need to be individually adapted to every single logistics enterprise, causing major expenses. Logistics enterprises in turn depend on selected software providers and their data analysis and logics. Changes in requirements may lead to the implementation of an entirely new system or the costly integration into an existing system environment. Enterprises unwilling to carry the very high costs associated with such an approach find themselves in a lock-in situation, giving up the major advantage of additional information (Walby 2003).

In terms of smart grid infrastructure and implementation of intelligent charging stations, a lot of development still has to be done. Currently, vehicle-to-grid (V2G) functionality is hindered especially by insufficient standardisation of communication and control between the grid and the vehicle. Intelligent metering on both sides still has to be implemented through the use of standards and norms. Furthermore, grid operators and power suppliers have to establish suitable market models and tariff systems to reward participants adequately (Yilmaz and Krein 2013).

4 Open Service Cloud Architecture

Smart grid functionality in regard to the OSC usability has to include certain elements that are laid out in the following section, beginning with the most essential ones. Furthermore, requirements and integration possibilities for OSC-connected EVs are presented and condensed into development needs.

Operating an EV connected to the OSC while not being connected to a smart grid or supplied with variable pricing means giving up a large cost-saving potential. So the most essential functionality for OSC-integration into the smart grid has to be variable pricing schemes coupled with smart metres and the necessary infrastructure to support both. Thus, algorithms included in the OSC are able to predict and schedule active charging times according to given user requirements and derived driving profiles. Especially for home use, this has to incorporate the capability to either switch the corresponding power outlet on and off automatically or to connect to the car itself through the OSC to remotely control charging at that end. Both possibilities have to include an option for manual override, so user flexibility is not impeded (Glanzer et al. 2011). Planning and scheduling algorithms can easily be derived from household appliance algorithms already in use in smart grid field trials such as SmartWatts (2012). Since the requirements for both application domains are essentially the same (i.e. need to turn on power for a certain amount of time and be ready by a certain time), they only have to be adapted to make use of or be included in the OSC services.

A further development in smart grid technology is the inclusion of EVs as active electricity storage for load balancing use—vehicle-to-grid (V2G). When planning and scheduling allow for it, the battery is then made available to the smart grid as power storage. Coupled with variable pricing schemes, this allows further revenue generation for owners and service providers through charging EV batteries at times of low prices and re-injecting the electricity at peak hours (Zahedi 2012).

For the smart grid itself, OSC data can be used to consider regional developments and requirements. To keep variable pricing from simply shifting peak loads occurring from EV charging to different times, individual pricing to spread out the charging over a bigger timeframe, in order to flatten the overall load peak, can be feasible. The next step in pricing development would be dynamic prices coupled to e.g. current EEX rates and current renewable generation as well as grid status and regional constraints (e.g. the need to locally balance a sub-grid because of limited distribution capacity) in real time.

To achieve all this, the electricity grid has to make use of advanced ICT infrastructure in substations and consumer connections, while integrating data collected through the OSC. Advanced prediction systems and scheduling algorithms have to be implemented to generate or predict prices in advance, thus enabling the planning of charging times in the first place. The necessary tariff system also has to be developed and implemented to encourage user interaction and make scheduling of charging and V2G functionality feasible (Yilmaz and Krein 2013).

The Oscar architecture, as laid out in Fig. 2, enables vehicles to closely interact with the OSC while interaction with the driver is realised through an in-car tablet with a special framework for third-party applications (DSC).

The OSC provides data access and feedback for business and private use as well as interfaces and services for third-party services. In this way, additional sources of information and information services, such as weather or traffic data, can be integrated and in turn be provided for vehicles, services and apps. The architecture is based on scalable massive parallel processing data storage for big data volumes.

The integration of large numbers of EVs along with rising shares of volatile renewable electricity generation leads to new challenges for future electrical

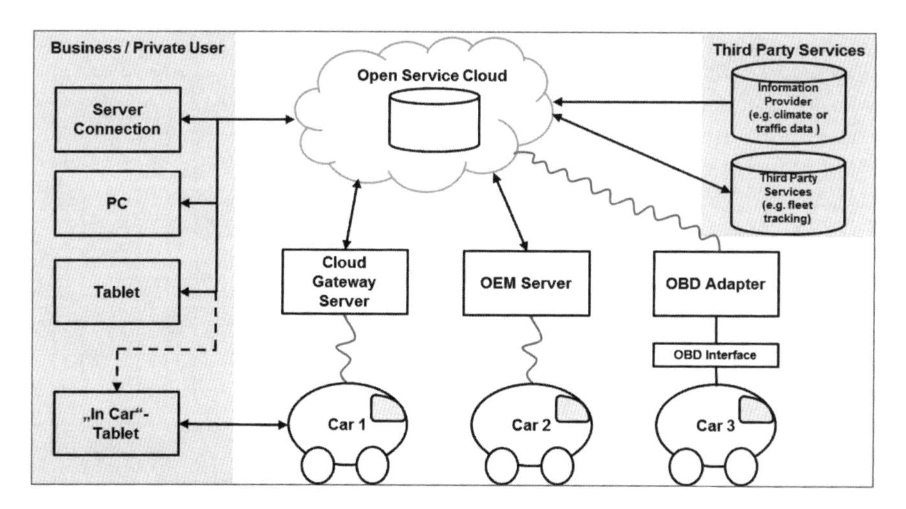

Fig. 2 Principle of the open service cloud

systems (Proff et al. 2013). The current hierarchical network structure has to be revised and developed to include high amounts of automation and ICT.

Figure 2 shows interfaces provided by Oscar while differentiating three different types of vehicles. The newly developed communication module in Car 1 allows the integration of CAN bus. Thus, not only reading of data streams but also a feedback of control signals is made possible. Control possibilities have to be kept strictly limited; however, due to the high risk of this process. The in-car tablet included in this communication module runs a special framework (DSC) providing a secure platform for in-car apps.

Car 2 makes use of its already existing OEM infrastructure, which sends and receives CAN bus data but additionally also provides selected data to the OSC. Limited systems do not allow any in-car apps or control features to be available, but could be retrofitted with the necessary technology and interfaces. This type of connectivity allows the inclusion of vehicles already fitted with OEM solutions.

Finally, Car 3 is only connected through an OBD adapter. In this way, any existing car with OBD connectivity could retroactively be enabled to provide data for the OSC. While control functionality in this case is not available, the car and its driver can still profit from OSC data and services as well as in-car apps.

The OSC itself integrates third-party services and services being offered to third parties, such as data push of certain data aggregations. The owner of the OSC acts as an information provider and manages both the OSC itself as well as its services.

Due to its original focus on EVs, the Oscar server was built as an integrator between smart cars, smart traffic and smart grid (see Fig. 1). Interfaces provided by the OSC thus target all possible stakeholders and user groups. The grid provider can monitor vehicle movements and varying demand for electrical power, while manufacturers can perform diagnostics and further development using vehicle and incident data. Additionally, routing and scheduling by fleet managers can be improved in real time and car-sharing agencies can provide additional services without constraints imposed by OEM platforms. Finally, individual drivers are able to track detailed data on their car and drive while making use of a multitude of services provided through third parties.

Major issues that have to be faced are privacy and security. Complete control, transparency and the possibility to anonymise data have to be provided with ease of use to the data owner. As already mentioned earlier, this includes a transparency overview detailing data use and privacy options along with the possible outcome for OSC data and services received in return. It has to be clear which data is used by whom, for which purpose(s) it is used, and which services and providers get access along the way. This has to include the possibility to define detailed access rights for individual data and services, along with the possibility of anonymising data before making it available.

As a first step to the tracking, managing and planning of fleets, data have to be collected from vehicle hardware and transmitted to the OSC in a standardised way, enabling further analysis of data by fleet managers or vehicle users. Oscar includes three methods to collect vehicle data: OBD adapters, existing OEM infrastructure and the implementation of an innate data collection structure (Fig. 2).

First, the OBD adapter offers an easy way to collect data from nearly any vehicle built after 2001. Engine and emission behaviour are collected and can be read by an OBD interface before being sent to the OSC server. After being stored in a standardised way, the data are made available for further analysis. Since this data collection is, contrary to CAN bus, not bidirectional, no control signals can be fed back to the vehicle. Additional connectivity, such as a smart phone or in-car-tablet, has to be provided to enable interactive systems instead of mere monitoring functionality.

Using an existing OEM infrastructure to collect data from vehicle hardware and transform it to the standardised format to be sent to the OSC is a further possibility. This enables fleet managers that have already invested in a data-collecting infrastructure to keep existing systems in their fleet's vehicles active while still benefitting from services and opportunities provided by the OSC. Again, smart phone or tablet integration is advisable to feed data back to the vehicles and their users, in case the existing infrastructure architecture design does not provide this functionality.

Finally, the most easily controllable and modifiable solution makes use of an individually developed data collection structure. This can be aligned to the vehicle, as it is done for the Streetscooter electric vehicle in the Oscar project.[1] Table 1 provides an overview over what kind of data is planned to be collected in this case. Apart from information about charging and movement status, vehicle speed and mileage, outside temperature, detailed battery data and GPS coordinates are also transmitted. This information is supplied with a unique vehicle ID and a timestamp. The data transmitted allows easy tracking of vehicles through GPS coordinates and detailed collection of traffic information derived from vehicle speed

[1] http://www.streetscooter.eu/.

Table 1 Streetscooter data description

Signal name	Description	Unit	Sensitivity	Range
Cabin preconditioning	Flag if cabin preconditioning is activated	On/Off	–	–
Drive mode	Identificator for currently selected drive mode	0 = Off 1 = Neutral 2 = Drive 3 = Reverse 4 = Other	–	Other: special drive modes such as economy, sport, etc
Total distance travelled	Vehicle mileage	km	1 km	0–999,999
Vehicle speed	Current speed as displayed in the speedometer	km/h	1 km/h	0–300
Ambient temperature	Temperature of outside environment	°C	1 °C	−80–175
Ignition status	Ignition on/off	On/Off	–	0, 1
Battery state of charge	Relative battery charge level	%	1 %	0–100
Battery voltage	Voltage currently supplied by battery pack	V	1 V	0–400
Battery current	Current currently supplied by battery pack	A	0.1 A	−3000.0–3000.0
GPS longitude	Longitudinal GPS coordinates of vehicle	°	10^{-6} °	−180–180
GPS latitude	GPS latitude	°	10^{-6} °	−90–90
GPS height	GPS height	m	0.1 m	−1000–8000

and matching with external services and GIS data. Furthermore, battery information provided allows conclusions to be generated and provided on actual battery state of health from the state of charge, voltage and current data. For this conclusion additional knowledge is necessary. For example, to calculate the state of charge, the transmitted "Battery State of Charge" in per cent has to be multiplied with the battery capacity, which is available for use. Parts of the capacity are not accessible due to safety issues and construction constraints. Such details have to be provided for each vehicle and are stored in the vehicle's profiles. The technical insight can be limited according to concerns of the OEM.

Coupled with the environmental temperature and driver profiles, advanced algorithms allow detailed information to be provided about range under current conditions. Finally, a preconditioning flag enables the consideration of additional energy use through heating or air conditioning which improves the planning of future charging processes and range calculations.

Through the cloud gateway server, data is provided to the OSC for further analysis. Moreover, a tablet PC connected to the car's hardware offers direct access to information gained by data analysis in the OSC. This way, gathering and analysing information in real time based on OSC data enables instant driver and vehicle support. Therefore, the system is perfectly adaptable to the needs of real-time fleet management.

4.1 Data Profiles

The OSC itself is vehicle independent while still being part of the data collection process. It collects and stores data gained from information providers' servers, such as weather data, traffic data (density, congestion, speed, traffic lights) and infrastructure data (road works, road blocks). Since project Oscar includes various stakeholders with individually different requirements, data in the OSC has to be stored in different individualised data profiles. At least six different profiles are currently intended, containing all data with importance for fleet tracking and real-time fleet managing analysis (Fig. 3). However, more profiles can be implemented when necessary to facilitate the integration of further services and third-party stakeholders.

In the first profile, weather data from a third-party information provider such as the DWD (Deutscher Wetterdienst—German Meteorological Service) is collected. The second profile gathers traffic data, again from a third-party information provider like the MDM (Mobilitätsdatenmarktplatz—Marketplace for Mobility Data[2]). This data includes street conditions, traffic conditions as well as traffic signal states. The MDM aggregates traffic information from cities and municipalities that run their own traffic information servers, but independently from one another without being part of a larger network. Profile 3 focuses on the user, their driving and the resulting impact. This also includes vehicle and battery health (in case of an EV) as well as average energy consumption. The next profile stores route-data linked to typical itineraries taken, particularities and characteristics. Specific vehicle data such as battery or fuel level, battery and engine health as well as typical distances travelled are stored in profile 5. Finally, profile 6 provides data of charging stations. This contains the occupation state and power drawn and is collected in Germany by the LISY framework of the Ladenetz project.[3]

All profiles are updated in real time, with every update triggering aggregations and computations which update associated key values of the profiles. To supply additional services, further individualised data profiles can be added when the need arises.

Towards the goal of transparency to the user in the Oscar project, it is essential to give the user the possibility of reconstructing data collected and associated information (i.e. what information is generated from which data). Furthermore, the user

[2]http://www.mdm-portal.de/.

[3]www.ladenetz.de.

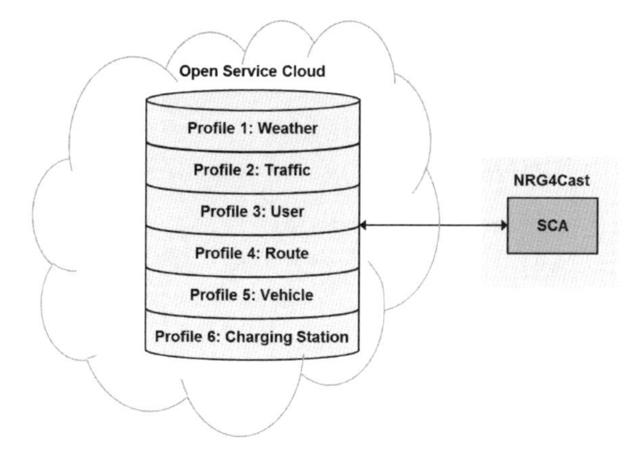

Fig. 3 Example for data profiles in the OSC

will have the ability to exclude data they consider too sensible to share, restrict access individually or anonymise certain data first. Understanding and administrating their own data is a key goal for users, towards which Oscar provides statistics of data usage and processing. Rewards for users providing a lot of possibly sensible information are in place to encourage other users to freely submit their data as well. This way, a large, significant data pool as a base for advanced processing and analysis can be created.

The platform for in-car applications in Oscar is the DSC Framework. It offers a development kit for both Android and Apple smart phones and tablets while providing a secure operating framework for subordinated applications. Exclusive access to the CAN bus for in-car apps is provided with data being instantly retrieved and control signals being able to be fed back. Security in the DSC framework is provided for different aspects. Applications and functionalities are restricted and checked to avoid distraction or unexpected outcomes. Most importantly, unacceptable access to the CAN bus is excluded to prevent disturbances in car usage, damage to hardware and risk to passengers. The framework acts as wrapper for the apps and signals exchanged with the CAN bus.

Control features of in-car applications are strictly limited as well, due to security issues. Depending on the regional legal situation and the actors provided by a car, the range of possible control features is set. Features not related to car safety or use should be possible to control. Examples include climate control as well as seat and mirror settings.

5 Conclusion

In a first step, motivations and goals of the OSC development have been laid out. Connecting cars, especially EVs and whole fleets, to a unified cloud architecture has been argued to offer a wide variety of services providing added value. Existing services can be improved while new services are made possible through extensive interconnection and information exchange between third-party providers. Business models that make use of the emerging architecture and information systems will create added value to drivers and other stakeholders as well as other third-party developers.

The current state of the art has been analysed and the development opportunity for the OSC identified. Existing systems neither offer the necessary amount of flexibility, nor do they include access for third-party service providers. From this, possible uses of an open cloud architecture have been developed.

The underlying architecture is designed to connect cars, users and service providers to allow open use of collected data while still retaining high levels of data security and anonymity. Possible uses for the inclusion of smart grid data have been presented, offering a large new field of possibilities for EV owners. Exploiting variable pricing schemes in smart grids through advanced charging algorithms and V2G functionality combined with prediction and scheduling systems provided via the OSC will offer substantial added value to the owner. The corresponding requirements of smart grid development have been identified and illustrated in detail. This shows the high impact of OSC as an enabling factor for smart electricity grid development and increased renewable generation as well as for EV inclusion and usage. Furthermore, data collection systems and details such as types of data collected have been described, showing high usability and interoperability as well as the potential to achieve significant added value.

Finally, flexible data profiles being used in the OSC for third-party connectivity have been designed. Tailored to specific services and applications, these can be easily adapted and individualised to emerging services. Thus, a maximum of flexibility and usability is facilitated.

References

Appelrath, H. -J., Beenken, P., Bischofs, L., & Uslar, M. (2012). IT-Architekturentwicklung im smart grid. In *Perspektiven für eine sichere markt- und standardbasierte Integration erneuerbarer Energien*. Berlin, Heidelberg: Springer.

Automotive News Europe. (2006). Lean and Green: Volkswagen adding fuel-efficient BlueMotion versions to volume car lines. AutoWeek.com, 3 April 2006. Retrieved January 26, 2010.

Briggs, J., & Lampton, C. (2013). Top 10 iPhone apps for your car. Retrieved May 23, 2013, from http://www.howstuffworks.com/gadgets/automotive/5-iphone-car-apps.htm#page=4.

Callaghan, J. V. (2013). Guidance for industry and FDA staff, replacement reagent and instrument family policy. Document issued on December 11, 2003, U.S. Department of Health and Human Services Food and Drug Administration Center for Devices and Radiological Health Office of In Vitro Diagnostic Device Evaluation and Safety. Rockville, MD. Retrieved from http://www.fda.gov/downloads/MedicalDevices/DeviceRegulationandGuidance/GuidanceDocuments/ucm071465.pdf.

Engel, T. (2010). *Mess- und Abrechnungskonzepte für den Elektro-Auto-Kraftstoff*. Conference papers—VDE Publishing House. *E-Mobility: Technologien—Infrastruktur—Märkte. Conference: VDE-Kongress. Leipzig, 11.08.2010*. Retrieved July 30, 2014, from https://www.vde-verlag.de/proceedings-en/453304071.html.

Glanzer, G., Sivaraman, T., Buffalo, J. I., Kohl, M, & Berger, H. (2011). *Cost-efficient integration of electric vehicles with the power grid by means of smart charging strategies and integrated on-board chargers*. Paper at 10th International Conference on Environment and Electrical Engineering (EEEIC), Rome.
Kidder, T. (1997). *The soul of a new machine*. New York: Random House Inc.
Kozinets, R. V. (2002). The field behind the screen: Using netnography for marketing research in online communities. *Journal of Marketing Research*, *39*(1), 61–72.
Krenge, J., & Roscher, M. (2013). *Beschreibungsmodell für IKT-Geschäftsmodelle in der Elektromobilität*. INFORMATIK 2012—GI-Edition—Lecture Notes in Informatics (LNI). Koblenz: Bonner Köllen Verlag.
Mühlenhoff, J. (2010). *Erneuerbare Elektromobilit* (Renews Spezial, 30). Retrieved July 30, 2014, from http://opus.kobv.de/zlb/volltexte/2013/20543/pdf/30_Renews_Spezial_Erneuerbare_Elektromobilitaet_april10_online.pdf.
Prasanna, K. R., & Hemalatha, M. (2012). RFID GPS and SGM based logistics vehicle load balancing and tracking mechanism. *Procedia Engineering, 30*, 726–729.
Proff, H., Schőnharting, J., & Schramm, D. (Eds.). (2012). *Zukünftige Entwicklungen in der Mobilität. Betriebswirtschaftliche und technische Aspekte*. Wiesbaden: Gabler Verlag.
Proff, H., Pascha, W., & Schőnharting, J. (Eds.). (2013). *Schritte in die künftige Mobilität. Technische und betriebswirtschaftliche Aspekte*. Wiesbaden: Springer Gabler.
Roscher, M., Fluhr, J., & Lutz, T. (2013). *Optimized integration of electric vehicles with lithium iron phosphate batteries into the regulation service market of smart grids*. SMARTGREENS 2013—2nd International Conference on Smart Grids and Green IT Systems, pp. 88–92, Aachen: SCITEPRESS—Science and Technology Publications.
Schäfer, K., & Fricke, V. (2013). Cloud-computing in der Automobilindustrie nimmt Fahrt auf. *ATZagenda, 2*(1), 52–55. doi:10.1007/s40357-013-0016-0.
See, W. (2007). Wireless technologies for logistic distribution process. *Journal of Manufacturing Technology Management, 18*(7), 876–888.
SmartWatts (2012). Smart Watts. Retrieved March 18, 2014, from http://www.smartwatts.de/fileadmin/smartwatts/mediapool/downloads/SmartWatts_allgemeine_Praesentation_121113.pdf.
Vivaldini, M., Pires, S. R. I., & de Souza, F. B. (2012). Improving logistics services through the technology used in fleet management. *Jistem, 9*(3), 541–562.
Walby, S. (2003). *Complexity theory, globalisation and diversity*. Paper at the Conference of the British Sociological Association, University of York. Retrieved from http://www.leeds.ac.uk/sociology/people/swdocs/.
Yilmaz, M., & Krein, P. T. (2013). Review of the impact of vehicle-to-grid technologies on distribution systems and utility interfaces. *IEEE Transactions on Power Electronics, 28*(12), 5673–5689.
Zahedi, A. (2012). *Electric Vehicle as distributed energy storage resource for future smart grid*. Paper at the 22nd Australasian Universities Power Engineering Conference (AUPEC), Bali.

Inductive Charging Comfortable and Nonvisible Charging Stations for Urbanised Areas

Steffen Kümmell and Michael Hillgärtner

Abstract For a wide acceptance of E-Mobility, a well-developed charging infrastructure is needed. Conductive charging stations, which are today's state of the art, are of limited suitability for urbanised areas, since they cause a significant diversification in townscape. Furthermore, they might be destroyed by vandalism. Besides for those urbanistic reasons, inductive charging stations are a much more comfortable alternative, especially in urbanised areas. The usage of conductive charging stations requires more or less bulky charging cables. The handling of those standardised charging cables, especially during poor weather conditions, might cause inconvenience, such as dirty clothing etc. Wireless charging does not require visible and vandalism vulnerable charge sticks. No wired connection between charging station and vehicle is needed, which enable the placement below the surface of parking spaces or other points of interest. Inductive charging seems to be the optimal alternative for E-Mobility, as a high power transfer can be realised with a manageable technical and financial effort. For a well-accepted and working public charging infrastructure in urbanised areas it is essential that the infrastructure fits the vehicles' needs. Hence, a well-adjusted standardisation of the charging infrastructure is essential. This is carried out by several IEC (*I*nternational *E*lectrotechnical *C*ommission) and national standardisation committees. To ensure an optimised technical solution for future's inductive charging infrastructures, several field tests had been carried out and are planned in near future.

Keywords E-mobility · Inductive charging · Charging stations · Urban areas · Germany

S. Kümmell
IAV GmbH, Carnotstrasse 1, 10587 Berlin, Germany

M. Hillgärtner (✉)
Department of Electrical Engineering and Information Technology, University of Applied Sciences Aachen (FH Aachen), Eupener Str. 70, 52066 Aachen, Germany
e-mail: hillgaertner@fh-aachen.de

W. Leal Filho and R. Kotter (eds.), *E-Mobility in Europe*,
Green Energy and Technology, DOI 10.1007/978-3-319-13194-8_16

1 Introduction: E-Mobility Status and Requirements

The automotive world is working hard on electrifying its products. Electrification ranges from mild hybrid to plug-in-hybrids (PHEV). The latter are the link to the pure battery electric vehicles (BEV) which have only an electric drivetrain, such as the fuel cell electric vehicles (FCEV). Although all of these types are a kind of electric mobility in the general understanding, E-Mobility only means vehicles with an electric propulsion system and an electric range sufficient for daily operation. Those kinds of vehicles have typically in common that they can be charged from the grid, except for the fuel cell electric vehicles.

The number of models available to the customer is increasing but not yet comparable to the variety of models with internal combustion engine. Nevertheless, electric vehicles (EVs) have become very important already for some markets, such as Norway, the Netherlands or specific places in the US. Coupled with incentives with reduced costs or special permissions to ease everyday life these places have in common that a charging infrastructure exists or is developing (Mock and Yang 2014; Wietschel et al. 2013). For PHEVs, a public charging infrastructure might be expendable although fuel economy will be worse but for BEVs it is mandatory to achieve comparable travel capabilities like conventional vehicles. Therefore, Tesla Motors instal their supercharger infrastructure in EV hotspots and along transport axes (Green Car Congress 2012; Mein Elektroauto 2012).

Tesla Motors has also presented a battery swap solution for their Model S for even faster "recharge". This concept may be familiar from Better Place, which has gone into administration. This former start-up turned E-Mobility provider was one of the greatest hopes for E-Mobility but is to this date one of the biggest failures. Tesla may be more successful with this concept as they do not sell E-Mobility as a service but in form of a desirable product. But superchargers, as well as battery swap or future concepts such as charging on the move, are means to extend the range on special occasions. Most of the time EVs are charged at home or at work (PlugInsights 2013; Shepard 2013).

Conductive charging solutions are state of the art (Klinger-Deiseroth 2013; Cars 21 2013). At the end of 2013, in Germany 4,400 public charging stations (BDEW—Bundesverband der Energie und Wasserwirtschaft e.V. 2014) were available for 12.156 BEVs (KBA—Kraftfahrtbundesamt 2014). Beside the public charging stations, private solutions such as wall boxes near by the private parking spaces of the BEV's owner are installed in an undisclosed dimension. In any case, the vehicle has to be connected to this charging infrastructure with a charging cable to start the recharge process (see Fig. 1). This procedure is time consuming and can be uncomfortable, e.g. in bad weather conditions in open areas. Also charging stations and cables in urban areas constitute an obstacle for pedestrians. Public charging stations are likely to become a subject of vandalism, similar to the case of parking metres or telephone booths.

Fig. 1 Conductive charging of electric carsharing in Brussels and CHAdeMO charging station (*Photos by* IAV)

Therefore, wireless charging systems with minimal necessary user interaction and no parts sticking out of the street would be the ideal charging solution for urban places.

2 Wireless Charging

As discussed before and depicted in Fig. 1, conductive charging stations cause a significant diversification in townscape. Modern wireless charging stations such as the inductive charging station in Fig. 2 are a suitable alternative.

In the rest of this chapter, different wireless charging technologies are presented. Inductive charging with an air gap induction is derived as suitable for automotive usage. Therefore, the fundamental working principles of this are explained.

2.1 Wireless Charging Options

Electrical energy is always fed by electromagnetic fields. Even for the normally used cables, it can be showed with Maxwell's equations that energy transport is carried out by electric and magnetic fields caused by the voltage and the current of the lines of the electrical circuit (Kraus 1992). Energy transport by electromagnetic fields is getting more obviously in real-life applications such as the microwave oven, in which the heating energy is induced into the food by a 2.45 GHz electromagnetic field.

Fig. 2 Inductive charging system for buses at main train station, Brunswick, Germany (*Photos by* IAV)

The aim of wireless charging stations for automotive usage should always be the realisation of a comfortable, competitive and efficient energy transfer between the station and the vehicle. The charging process shall start without user-action such as plugging a charging cable into a socket. Hence, a transfer over a certain air gap is always needed and the electromagnetic fields have to be guided without causing significant losses. This might be possible by directed fields which are known, e.g. from laser or beam radio. Beside the safety issues caused by the power transported, those techniques are not convincing from an economic point.

The usage of the electric field for power transfer between charging station and vehicle seems to be an interesting option (Kline et al. 2011), because the elements needed for power transfer and electromagnetic shielding are lighter and cheaper than those for inductive charging. The maximum amount of power which can be transformed via an air gap is regulated mostly by two aspects: the capacity of the capacitor and the used frequency. Due to EMC (*E*lectro*m*agnetic *C*ompatibility) and radio broadcast regulations, it is not possible to use reasonable high frequencies, so that increasing the power up to the needed minimal level of 2.0 kW is only possible by a significant capacity. The capacity is impacted by the active area as well as the size of the air gap. The bigger the area and the smaller the air gap, the larger is the capacity. If utilising the undercarriage, a large air gap is the result of the vehicle's chassis clearance. Of course it would be possible to reduce the air gap to a minimum by the usage of mechanical moving facilities, but those moving parts are service intensive and hence not suitable for the usage in urban public areas. The maximum surface area is given by the size of the vehicle's underfloor. If utilising the tyres, the area is limited by their footprint. With these given boundary conditions, the power possibly transferred by the electrical field is limited and hence currently not the best choice.

2.2 Inductive Charging Basic Principle

Inductive charging is based on the functional principle of a transformer with air gap (Kürschner 2010; Prasanth 2012). The principle layout is provided in Fig. 3.

The air gap causes a stray field, which can be influenced by well-designed iron circuits and coils. Nevertheless, the air gap causes stray fields which do not contribute to the power transfer. They can be represented by stray inductances L_σ (see Fig. 4) in the electrical equivalent diagram. To reduce the impact of these, the transformer is operated in resonance by applying additional capacitors and the

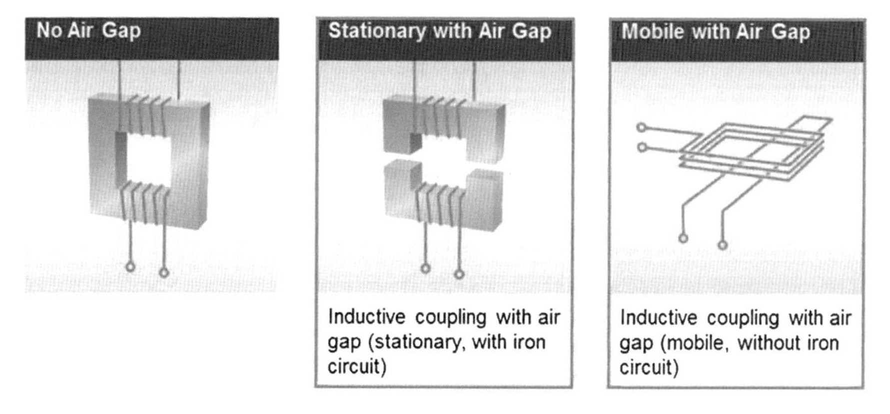

Fig. 3 Inductive circuits for wireless charging (framed) (IAV)

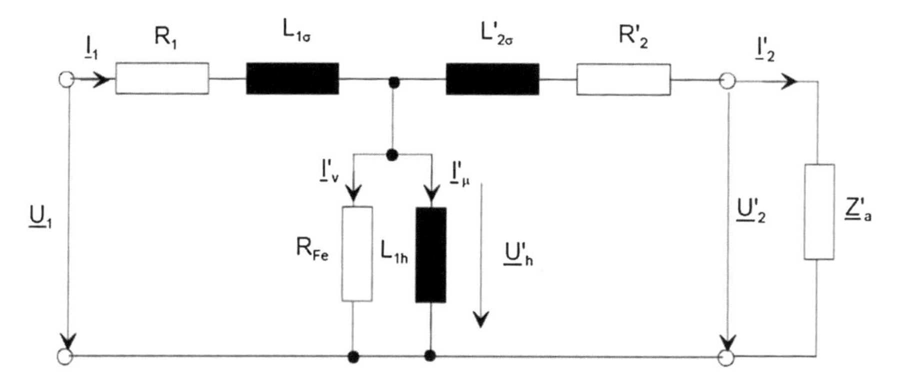

Fig. 4 Equivalent electrical diagram of a transformer

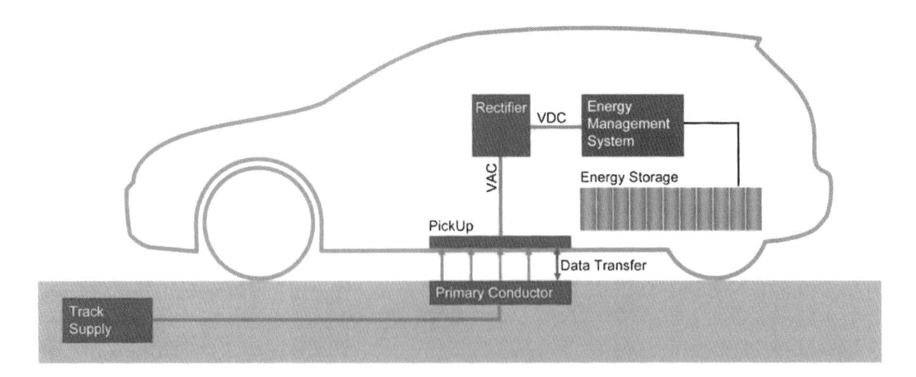

Fig. 5 Basic elements of an inductive charging system (IAV)

usage of matched frequencies. Further information on the resonance operation mode of inductive charging systems can be found, e.g. in (Kürschner 2010).

The maximum, transferable power is again directly proportional to the area which is covered by the transformer building primary (e.g. ground floor) and secondary coil (e.g. vehicle underfloor, see Fig. 5). This surface is restricted by the dimensions of the vehicles in which the secondary coil shall be integrated. For best performance, primary and secondary coils should have a maximum coverage on each other.

In contrast to the capacitive charging, the distance does not have a direct influence to the maximum transferable power but on the stray inductances and hence on the overall efficiency of the system. This can be optimised by the resonance operating mode.

The maximum, transferable power is furthermore influenced by the magnetic field strength. In contrast to the capacitive charging, this can not only be influenced by the applied current but also by the amount of windings of the used coils. Hence it is possible to transfer the power needed. Today's experimental setups suitable for passenger vehicle application have shown power transfers of up to 22 kW with high efficiency (up to 95 %) (Itasse 2013).

If significant power is transferred through the air gap between road surface and vehicle's underfloor, it has to be ensured that no safety regulations are failed. In contrast to conductive charging, the protection of the public against the risk of an electrical shock has to be discussed much less intensively with wireless charging systems because the user cannot get in contact with potentially harmful voltages—no open, electrical contacts are present. For conductive charging systems, high safety regulations have also be respected to ensure the electrical safety even for wet weather conditions. Furthermore, the risk of current in a fault being present has to be assured against for conductive systems (e.g. by intensive usage of fault-current circuit breaker). Due to the floating-potential construction of the wireless charging system, this is not existing with this alternative technology.

On the other hand, protection mechanisms have to be found for the high magnetic field strengths needed during inductive charging. For the protection of the general public, the valid regulations for the exposure to electromagnetic fields (ICNIRP 1998) have to be respected in all areas around the vehicle, which are accessible during the charging process. Investigations of the authors on several prototype systems have shown that the limits are kept as long as the vehicle is placed correctly. Furthermore, it has to be ensured that a contamination of the active zone, e.g. by metalised parts or living objects, cause a prompt reduction of the field strength to ensure safety during the charging process. Dependent on the manufacturer, detection of metalised parts is implemented in the systems, because they cause a detectable change of the magnetic field. The constant monitoring of the active air gap zone in realised prototype stations is carried out by LIDAR (*Li*ght *D*etection *a*nd *R*anging) or RADAR (*Ra*dio *D*etection *a*nd *R*anging) systems (see Fig. 6), so that a save usage in public areas can be ensured today. The usage of less cost-intensive and nonvisible solutions is part of several, current research projects.

As depicted in Fig. 5, an inductive charging system consists of two separated parts. The primary part, which consists mainly of the primary coil of the transformer and the supplying power electrics, can be completely installed below the road surface, because no direct user interaction is needed. Hence a nonvisible, vandalism safe construction for the usage in urban areas can be realised. The communication needed between vehicle and charging station (e.g. placement information,

Fig. 6 Prototype inductive charging station for the usage in general public. The active power transfer zone (marked in *blue*) is monitored by a LIDAR system (marked *yellow*) (*Photo by* IAV) (color figure online)

clarification of the maximum power transfer rate and accounting information) is fulfilled with a wireless connection. Beside the usage of well-established wireless standards like WLAN or cellular mobile applications, it is also possible to use the magnetic field generated for the power transfer. The secondary part inside the vehicle consists of the pick-up coil and a rectifier with control mechanism. The transferred energy can be directly fed into the vehicles energy management system which makes it easy to realise the standard conductive without extra effort.

These technical boundary conditions make inductive charging the most powerful wireless charging system for E-Mobility purposes.

2.3 Inductive Charging as the Optimal Solution for E-Mobility Purpose

From an automotive point of view, every on-board system should in general be as small and as light as possible. Also, cost should be as low as possible with the maximum of the desired effect, e.g. efficiency of the power transfer over an air gap. It is mandatory that the system is compliant with all the relevant safety standards, e.g. for electromagnetic compatibility. For wireless charging the real-life trade-off is best for inductive charging (see above). With a transfer power of 3–6 kW, a wireless system is capable of charging, e.g. the 18.8 kWh battery in the BMW i3 in 3.5–7 h from empty to full like a conductive charging system in Germany. At this power level, a wireless charging system can be made small and light enough to fit, e.g. behind the number plate or at the underbody of the vehicle (Schrieber 2010). The position behind the number plate has the advantage that the air gap of the wireless charging system can be minimised by driving against the correspondent infrastructure side, e.g. in a private garage. Also, foreign objects are less likely to be between both sides of the system. On the other hand, an active adjustment mechanism is required for interoperability to charge all kinds of vehicles with, e.g., different heights. Also the vertical column to mount this adjustment-damper mechanism causes the same problems for urban areas as conventional charging stations.

Most vehicle projects as well as standardisation are focussed on the mechanical integration of the secondary side at the undercarriage although the longitudinal position differs from front, middle to the back of the vehicle. An integration at the under chassis has the advantage that if positioning is necessary it could be done by the vehicle. Different vehicle heights can be accounted for in development with a suitable air gap tolerance. For private areas such as the garage at home, an installation on top of the floor is possible as well as an installation flush with the surface in public places. Additionally, the vehicle serves as a cover of the transfer area which means that accessibility for humans during the charging process is limited which is beneficial for EMC compliance (see above).

If inductive charging systems should be an alternative for urban E-Mobility, they shall be as flexible as possible, such that vehicles equipped with an inductive changing system can be charged on any installed primary side.

Due to different vehicles' chassis clearance, the distribution of the stray fields will vary and hence the resonant frequency which has to be used for optimised efficiency will also change from car to car. Hence, the operating frequency of the system needs to be variable in a certain region to ensure usability over a wide range of vehicles. Furthermore, the efficiency is also directly influenced by the coverage of primary and secondary coil—they must have a coverage of more than 90 % which makes it necessary to match primary and secondary coil's size. These two examples shall indicate the need for a well-defined standardisation of inductive charging systems. National and international standardisation committees are working hard on such standards but so far only draft standards are available.

3 Charging System Status

3.1 Conductive Charging

In comparison, the standardisation of conductive charging has progressed further. In the international standards IEC 61851 and IEC 62196, requirements and connection types for conductive AC (*A*lternating *C*urrent) and DC (*D*irect *C*urrent) charging are described. Due to slightly different requirements and pre-standard developments, there are different charging solutions for different areas of the world described. On top, there are national standards like CHAdeMO or proprietary solutions like Tesla's supercharger which are not described in international standards but are a fact on the streets in form of charging stations and more importantly in form of electric vehicles with these charging connections. These charging options are further extended by national household grid connections which will be at least a fall-back solution in the foreseeable future. In a long-term sight, consolidation of charging options by the market is likely but for the early adoption phase of E-Mobility it is important not to scare any customers off by missing charging infrastructure for their vehicle.

But even if a suitable conductive charging station is available it is not necessarily possible to charge your car. Some charging stations require a key card to identify the user and initiate the charging process. Others are activated after a phone call by the utility company. At best, the charging process starts after you plug-in your vehicle either by authorisation of the vehicle or a mobile metre in the charging cable (ubitricity 2013). The experienced EV user will know vehicles and the compatible charging infrastructure eventually. But especially for inexperienced users, e.g. in EV carsharing, charging the vehicle is an obstacle for adoption or at least a nuisance (Beyer 2012) because of

- missing information concerning the charging process
- difficulties plugging in the charging cable

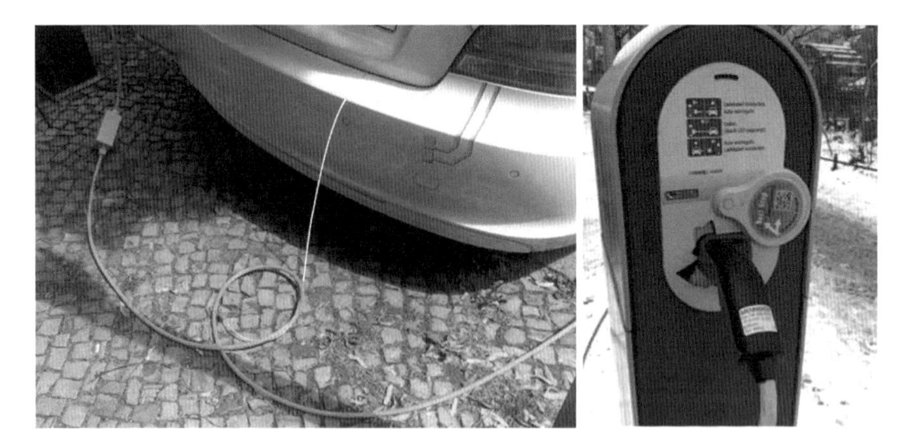

Fig. 7 Cable lock in electric carsharing and damaged charging station (*Photos by* IAV)

- uncertainty if charging cable belongs to vehicle or charging station
- no storage place for the charging cable in the vehicle.

For some of these complaints, a simple solution like a cable lock is available but others will occur if users have to deal with damaged charging stations (see Fig. 7).

With inductive charging, there is the chance to learn from the conductive standard and infrastructure development. With a uniform international standard additional costs accruing for the automotive manufacturers and hence customers can be avoided. Wireless charging technology for EVs itself clears up most of the complaints of users with conductive charging stations.

3.2 Inductive Charging

With an inductive charging infrastructure provided that is effective there is no need for an action on part of the user to charge the relevant vehicle. Charging cables for everyday usage would not have to be carried on-board of the vehicle. Cables to charge in case of an emergency could be reduced, equivalent to a spare tyre or a tyre repair kit with a dedicated place in the vehicle which does not have to be easily accessible on an everyday basis. The inductive charging infrastructure can be integrated into urban areas, submerged completely beneath ground. Therefore, protection against vandalism is better than with conductive charging systems.

The electromagnetic field to transfer the power between primary and secondary side is not influenced by weather phenomena like rain or snow. This means, inversely, that there is also no risk for users in this weather to soil, e.g. their hands with wet charging cables because there are none. This has another advantage—without the physical plug-in process there is no wear on either the charging plug or the socket.

The vehicle will probably be more frequently connected to the grid as there is no annoying part for the user to fulfil. This could have a positive effect on the battery lifetime or opens up opportunities for smart grid applications. At least it will minimise the risk of human errors for the charging process—unintentionally or on purpose.

This is why inductive charging over an air gap was already incorporated in carsharing tests in France in the 1990s (Nietschke et al. 2011). With the growing interest in E-Mobility, another series of passenger vehicle projects for inductive charging were conducted by Conductix, Indion and W-Charge till September 2011 (Conductix Wampfler AG/Daimler AG 2011; BMW AG/Siemens AG 2011; Audi Electronics Venture GmbH et al. 2011) which confirmed the potential of this technology for passenger vehicles. Some automotive manufacturers like the Volkswagen Group have made announcements to bring inductive charging for electrified vehicles in the near future (Mörer-Funk 2014; Lane 2014).

Currently, there are also research and development projects worldwide for inductive charging for commercial vehicles, e.g. buses in public transport. Recharging incrementally at bus stops like in Genoa and Turin (Italy) could prolong the range of pure electric buses enough to maintain service in bad weather conditions or scattered charging failures to the overnight charge at the bus depot with minimal on-board energy storage capacity (Gordon 2009; Spiegel 2012). The latter could even be further reduced with dynamic charging routes like in Guma (South Korea) where vehicles with appropriate inductive charging systems are supplied with enough energy to accelerate and maintain constant speed while they are on the move (Rovito 2014). Also a combination of stationary and dynamic charging is possible like in Augsburg (Köbel 2013). Such a technology applied to highways with the necessary power requirements could lead to an E-Mobility with the same range and even more comfort than today's fuel-based mobility.

4 Conclusion

Conductive charging are the current state-of-the-art solutions. But as obvious these are to charge an electric vehicle, wireless inductive charging would have advantages in urban areas as shown above—it is possible to integrate them invisible in townscape. Furthermore, they offer much higher comfort to the user by a reduced vandalism risk. Challenges of this technology can be resolved with the appropriate safety systems. For a widespread adoption in mass production serial cars, an international standard has to be finalised such that solutions of different manufactures are compatible to each other. The standardisation process is accompanied by several prototype field tests in which the improved user comfort and confidence with electric vehicles are underlined.

References

Audi Electronics Venture GmbH, Fraunhofer IWES, Paul Vahle GmbH & Co. KG, Volkswagen AG. (2011, Oktober). W-Charge. Abschlussbericht zum Verbundvorhaben Kontaktloses Laden von Elektrofahrzeugen im Rahmen des FuE-Programms "Förderung von Forschung und Entwicklung im Bereich der Elektromobilität".

BDEW - Bundesverband der Energie und Wasserwirtschaft e.V. (2014, May 11). Retrieved from Zahl der Ladepunkte nimmt weiter zu: http://www.bdew.de/internet.nsf/id/20131028-pi-zahl-der-ladepunkte-nimmt-weiter-zu-de.

Beyer, H. (2012, March 12). Elektrifizierte Mietsysteme der Deutschen Bahn. Berlin.

BMW AG/Siemens AG. (2011, Oktober). Indion. Abschlussbericht zum Verbundvorhaben Kontaktloses Laden von batterieelektrischen Fahrzeugen im Rahmen des FuE-Programms "Förderung von Forschung und Entwicklung im Bereich der Elektromobilität". München.

Cars 21. (2013, Jan 28). Cars 21. Retrieved from EU proposes minimum of 8 million EV charging points by 2020: http://www.cars21.com/news/view/5171.

Conductix Wampfler AG/Daimler AG. (2011, Oktober). Conductix. Abschlussbericht zum Verbundvorhaben Kabelloses Laden von Elektrofahrzeugen im Rahmen des FuE-Programms "Förderung von Forschung und Entwicklung im Bereich der Elektromobilität". Weil am Rhein.

Gordon, J. (2009, July). Bus breakthrough. Electric and Hybrid Vehicle Technology, 63–64.

Green Car Congress. (2012, Sept 25). Green car congress. Retrieved from http://www.greencarcongress.com/2012/09/tesla-20120925.html.

ICNIRP. (1998). Guidelines for limiting expisire to time-varying electric, magnetic, and electromagnetic fields. *Health Physics, 74*(4), 494–522.

Itasse, S. (2013, July 05). Automobil Industrie Vogel. Retrieved from Automobil Industrie Vogel: http://www.automobil-industrie.vogel.de/index.cfm?pid=11360&pk=4.

KBA - Kraftfahrtbundesamt. (2014, May 11). Bestandsbarometer 2014. Retrieved from http://www.kba.de/cln_031/nn_125398/DE/Statistik/Fahrzeuge/Bestand/2014_b_bestandsbarometer_teil2_absolut.html.

Kline, M., Izyumin, I., Boser, B., & Sanders, S. (2011, March 06). Capacitive power transfer for contactless charging. Applied Power Electronics Conference and Exposition (APEC), Twenty-Sixth Annual IEEE, pp. 1398–1404.

Klinger-Deiseroth, C. (2013, Q2). Vielfältiges Laden. Mobility 2.0, pp. 36–39.

Köbel, C. (2013, Oct 29). Primove inductive charging for electric mobility. ETEV 2013 Nürnberg.

Kraus, J. D. (1992). *Electromagnetics* (4th ed.). New York: McGraw-Hill.

Kürschner, D. (2010). *Methodischer Entwurf tolernazbeghfteter induktiver Energieübertragungs systeme*. Aachen: Shaker Verlag.

Lane, R. (2014, March 26). Ecomento. Retrieved from Wireless charging for Volkswagen electric cars from 2017: http://ecomento.com/2014/03/26/wireless-charging-for-volkswagen-electric-cars-from-2017/.

Mein Elektroauto. (2012, Nov 11). Mein Elektroauto. Retrieved from Niederländer nutzen nicht das ganze Potenzial ihrer Elektroautos bzw. Plug-In Hybridautos: http://www.mein-elektroauto.com/2012/11/niederlander-nutzen-nicht-das-ganze-potenzial-ihrer-elektroautos-bzw-plug-in-hybridautos/6515/.

Mock, P., & Yang, Z. (2014, May 10). The international council on clean transportation. Retrieved from http://www.theicct.org/sites/default/files/publications/ICCT_EV-fiscal-incentives_20140506.pdf.

Mörer-Funk, A. (2014, April 10). Auto Motor und Sport. Retrieved from Audi bringt reines E-Auto ab 2017: http://www.auto-motor-und-sport.de/news/auto-motor-und-sport-kongress-2014-audi-bringt-reines-e-auto-ab-2017–8305777.html.

Nietschke, W., Fickel, F., & Kümmell, S. (2011, April). Inductive energy transfer for electric vehicles, ATZ, pp. 22–27.

PlugInsights. (2013, Dezember). SmartGridNews. Retrieved from http://www.smartgridnews.com/artman/uploads/2/PlugInsights_U.S._PEV_CHARGING_STUDY_2013_media_copy.pdf.

Prasanth, V. (2012). *Wireless power transfer for E-Mobility*. Delft: Delft University of Technology.

Rovito, M. (2014, May 01). Charged EVS. Retrieved from OLEV technologies' dynamic wireless inductive system charges vehicles while in motion: http://chargedevs.com/features/olev-technologies-dynamic-wireless-inductive-system-charges-vehicles-while-in-motion/.

Schrieber, H. (2010, July 02). Autobild.de. Retrieved from autobild.de: http://www.autobild.de/artikel/schriebers-stromkasten-teil-86_1206801.html.

Shepard, S. (2013, June 27). Forbes. Retrieved from Tesla tries out battery swapping: http://www.forbes.com/sites/pikeresearch/2013/06/27/tesla-tries-out-battery-swapping/.

Spiegel Online. (2012, June 09). Spiegel Online. Retrieved from Induktives Ladesystem für E-Busse: Kraft ohne Kabel: http://www.spiegel.de/auto/aktuell/induktives-ladesystem-fuer-e-busse-a-837696.html.

ubitricity. (2013, Sept 05). Forum Elektromobilität e.V. Retrieved from Forum Elektromobilität e. V.: http://www.forum-elektromobilitaet.de/assets/mime/c20985878e8c3b4dd381320e3e3b1ddf/130905_WS_Ladeinfrastruktur_ubitricity.pdf.

Wietschel, M., Plötz, P., Kühn, A., & Gnann, T. (2013, Sept). Fraunhofer ISI. Retrieved from Fraunhofer-Institut für System—und Innovationsforschung ISI: http://www.isi.fraunhofer.de/isi-media/docs/e/de/publikationen/Fraunhofer-ISI-Markthochlaufszenarien-Elektrofahrzeuge-Zusammenfassung.pdf?WSESSIONID=41ab7e1629229f2c88656727a7ecd5a6.

Information and Communication Technology for Integrated Mobility Concepts Such as E-Carsharing

Michael Rahier, Thomas Ritz and Ramona Wallenborn

Abstract During the past decade attitude towards sharing things has changed extremely. Not just personal data is shared (e.g. in social networks) but also mobility. Together with the increased ecological awareness of the recent years, new mobility concepts have evolved. E-carsharing has become a symbol for these changes of attitude. The management of a shared car fleet, the energy management of electric mobility and the management of various carsharing users with individual likes and dislikes are just some of the major challenges of e-carsharing. Weaving it into integrated mobility concepts, this raises complexity even further. These challenges can only be overcome by an appropriate amount of well-shaped information available at the right place and time. In order to gather, process and share the required information, fleet cars have to be equipped with modern information and communication technology (ICT) and become so-called fully connected cars. Ensuring the usability of these ICT systems is another challenge that is often neglected, even though it is usability that makes carsharing comfortable, attractive and supports users' new attitudes. By means of an integrated and consistent concept for human-machine interaction (HMI), the usability of such systems can be raised tremendously.

Keywords Information and communication technology · Fully connected car · E-carsharing · Mobility management · Integrated mobility · Human–machine interaction · Usability

M. Rahier (✉) · T. Ritz · R. Wallenborn
Mobile Media and Communication Lab at the University of Applied Sciences Aachen, Eupener Straße 70, 52066 Aachen, Germany
e-mail: rahier@fh-aachen.de

T. Ritz
e-mail: ritz@fh-aachen.de

R. Wallenborn
e-mail: wallenborn@fh-aachen.de

W. Leal Filho and R. Kotter (eds.), *E-Mobility in Europe*,
Green Energy and Technology, DOI 10.1007/978-3-319-13194-8_17

1 Introduction: Social Change of Attitude Towards Sustainable Shared Mobility

The rising success of social networks in the past decade has changed the common attitude towards sharing of personal data and sharing in general. Being part of a network hence changed the information and communication behaviour of the population as well and led to high demands on flexibility and usability (Initiative D21 2013). From transport and mobility service providers it is expected to become part of this network and to provide the same amount of flexibility and usability.

Referring to cars as the most common means of transport, the load on public roads has increased continuously in the last years. But now there is a change of mobility behaviour. The car has lost its value as a status symbol and is rather seen as a pure means of transport. Even now, especially young metropolitans use different mobility services because they can usually travel faster by bike or bus than by car (Arnold et al. 2010; Becker 2012). Additionally, topics like the rising number of vehicles, increasing carbon dioxide levels and sound emissions within cities as well as limited fossil fuels promote research activities and development in the field of electric mobility (Hirte and Nitzsche 2013, BUMB 2009). Besides technical innovations such as automotive design and energy management, electric mobility provides the opportunity to create new mobility concepts and systems, such as e-carsharing.

E-carsharing is an example for an innovative mobility system and means an organized and shared usage of electric cars. Unlike car rental, registered customers have access to different types of vehicles at distribution stations throughout the city and round the clock. E-carsharing is a mobility-on-demand service and can be part of an integrated mobility service among taxi, bike, private car and public transport. It cannot reach the full range of self-determination (such as a private car) nor can it replace means of transport to far-off destinations in time and space (e.g. bus and train) (Saretzki and Krämer 2009). It started as a station bounded service like for example cambio CarSharing (Cambio 2008), but in the past few years free floating concepts entered the market. Big automobile manufacturers were the first ones that entered the market with these new carsharing approaches, for example Daimler with *car2go* and BMW with *DriveNow* (Car2go 2014; DriveNow 2014).

Carsharing is an ideal facilitator of electric mobility services. Many disadvantages of electric cars such as high costs due to battery technology, low speed limit and short range between 100 and 150 km in comparison to conventionally powered vehicles can be compensated with integrating electric cars into carsharing services (BMBF 2014; Handelsblatt 2012). The costs can be allocated to all customers and regarding the matter of range, carsharing vehicles are mainly used for city driving with short distances anyway. Baum et al. 2012 provide an elaborate evaluation of the potential and limitations of e-carsharing. A research project (Ec2go 2012) about an electric mobility model for carsharing illustrates that in most cases people drive less than 25 km with small carsharing vehicles (study from 10/2009 to 9/2010, see

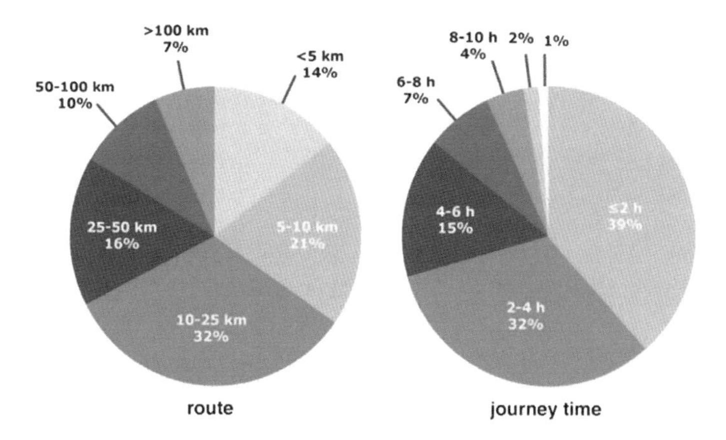

Fig. 1 Analysis route and journey time of compact cars of a carsharing fleet in Aachen

Fig. 1). Only 7 % of the trips were longer than 100 km. An appendant customer survey showed that about 75 % of the trips could have been covered with an electric car (Anthrakidis et al. 2013).

2 Challenges of E-Carsharing and Integrated Mobility Concepts

With e-carsharing there are various challenges to overcome. On the one hand the shared mobility services need to address mainstream needs and expectations, but on the other hand users increasingly request individual adaption of services to fulfil their own specific needs (Zukunftsinstitut 2014). For the purpose of carsharing, electric vehicles must be available for and adaptable to a large user group. Depending on the fleet size and the business model the task of managing availability is not easy, as Baum et al. (2012) showed. If, for example, the business model does not obligate users to bring their car to a certain station after the trip, availability in a certain area cannot be guaranteed. Cars can be spread over the complete city area, not necessarily close to the next user with transport requirement. Hence, it is important to create an appropriate motivation in order to enable an equal distribution of cars. Even if a car can be assigned to a user successfully, the next challenge is to adapt this car as much as possible to this user and his likes and dislikes, which has to be known first.

Drivers prefer to have their individual preferences in a carsharing vehicle like they have in their private car, for example the seat position, mirror adjustment, as well as music and navigation targets. Another characteristic of carsharing drivers is that they predominantly drive occasionally instead of taking the car every day. Furthermore, most people have never driven an electric car before, which makes it necessary to provide advice to drivers with advanced driver assistance systems (Anthrakidis et al. 2013; Arnold et al. 2010). In addition, the low level of acceptance of electric vehicles in the population represents an enormous challenge for society. Even though future mobility is associated with electric cars, hardly anybody in Germany would buy an electric car nowadays (about 81 %), because according to public opinion electric mobility restricts the personal mobility behaviour. In line with this point of view, very few people have driven an electric car yet (Arnold et al. 2010).

Another important challenge typical for electric mobility is intelligent energy management. The flexibility that is demanded by users contrasts with the limited range of electric cars due to insufficient battery technology. Therefore, the management of the scarce resource energy is not a trivial task and has to be considered when providing an electric mobility service. Integrated mobility services that combine more than one means of transport are a possible solution for the current weakness of electric mobility, but have to be managed in a sensible manner.

A major part of these challenges can be overcome by provision of suitable information to the corresponding task, wherever this task may be executed. Besides the right location, this information must also be available at the right time and in a format that is easy to process. The provision of this kind of information is called information logistics and, as the name implies, information has to be transported like on a road network (Haftor and Kajtazi 2009). For establishment of the required information network the concerned means of transport have to be connected to their environment (other means of transport, mobility providers, users, infrastructure, etc.) by modern information and communication technology (ICT). This technology also enables the collection and processing of information (Lešková 2013). Hence, with respect to e-carsharing, electric fleet cars have to become so-called fully connected cars.

3 The Fully Connected Car and the Resulting Information Flood of Mobility Data

Networking describes a conjunction of different components. We are already linked by various mobility networks such as car, bus and train. These physical connections allow people to meet friends or business partners at any place and at almost any time. Beyond the physical connection a virtual connection via the Internet and wireless networks also interconnects people. Due to increasing mobile Internet coverage, data and information are available ubiquitously. Continuous communication with friends via social networks is thus possible, keeping users always connected to each other. Current technologies such as cloud computing support and accelerate this trend. The previously described areas of the physical and virtual world need to be more interconnected and integrated into the organization of life. Day by day, individual mobility is organized independently and afresh. This way of organization is neither efficient for the individual nor for the environment.

By means of ICT every means of transport can be woven into the networked environment in order to share data with other means of transport, the infrastructure, users and mobility providers. It is thus possible to turn them from closed information cells to smart mobile devices that can both provide internal data and process external data at the same time. The huge amount of information inside the network can be used to ease the common mobility planning, making it efficient and capable of being integrated into the individual organization of life. Managing this network is a complex task that can only be handled by ICT systems as well. First attempts to organize mobility networks can be seen for example in timetable information systems.

The cross-linking of the above-mentioned stakeholders could be divided into three levels, depending on the spatial extension of communication: micro-level, meso-level and macro-level (see Fig. 2). Each level has its own transmission technology and transmission protocol, which makes translation between levels necessary.

At the micro level, the control units of vehicles are connected by in-vehicle networks, e.g. the Controller Area Network (CAN). Via this network they exchange vehicle specific information like speed, lighting, steering angles, seat and mirror settings among other things. Although the format of this information is standardized, its interpretation is left to the vehicle OEM (Zimmermann and Schmidgall 2011). Only information that is needed by vehicle workshops for diagnostic purposes is standardized. This standard is called on-board diagnostics (OBD) and was defined in the ISO norms 9141, 11519, 11898, 14230, 15765 and 15031 (WGSoft 2014). It was first applied to vehicle models in 2001 and is accessible via a hardware interface in the vehicle interior, mainly in the footwell on the driver side.

On the next level, the meso level, internal data is shared with the environment, which is called vehicle-to-x (V2X) communication. V2X is split into three subtypes: vehicle-to-device (V2D), which is often referred to as vehicle-to-driver, vehicle-to-vehicle (V2V) and vehicle-to-infrastructure (V2I) (RITA 2014). From a technical point of view, possible communication partners can be either different vehicles, mobile devices such as smartphones, tablets and smart watches or infrastructure elements such as traffic lights, railroad gates or traffic control systems. V2X communication can be used for warnings due to accidents, traffic jams, and danger spots or for information on available parking space. It requires a wireless connection, for example, Wi-Fi (IEEE 802.11p), Infrared or Bluetooth (Zimmermann and Schmidgall 2011).

With mobile data networks like Universal Mobile Telecommunication System (UMTS) and Long Term Evolution (LTE) as well as cloud computing, information is ubiquitously available. This is a requirement for the macro level, which allows access to all the data of all mobile devices and vehicles. Questions about current traffic situations on the entire route or on the utilization of a vehicle fleet can be answered. In addition, charging stations and alternative mobility options can be considered when planning the route.

Due to the extensive cross-linking of the vehicle with its environment and communication within this network enormous amounts of data are produced, which

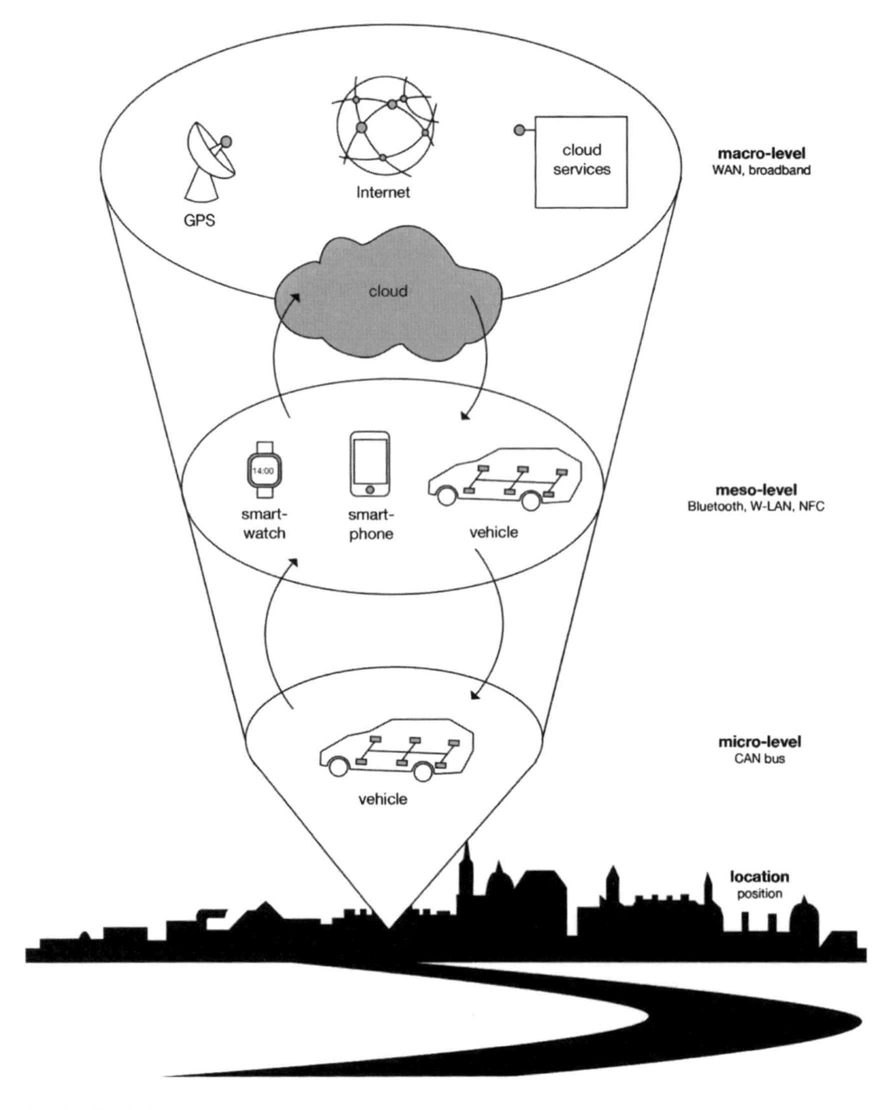

Fig. 2 The fully connected car

need to be coped with by dedicated data technologies and ICT systems. Mobility data sets are large, heterogeneous and frequently changing. Data technologies that are able to handle this kind of data are called big data technologies (GI 2013) and consist of dedicated database and network technologies, visualization techniques as well as processing algorithms (e.g. pattern recognition). These technologies help to process the large number of real-time vehicle data and to recognize typical driving situations. In addition, typical usage patterns of drivers and charging processes of the vehicle fleet can be determined in the area of fleet management.

Besides vehicle-specific information also custom data can be exchanged (e.g. for personalization). In preference profiles driver-specific data such as music, contacts, appointments, emails, etc., are bundled. If these profiles are stored in the cloud, they can be used on all the user's devices for a consistent multi-modal personalization (Behrend et al. 2011). For example, publicly available vehicles like carsharing vehicles will be shared with many other users, so that the vehicle must be adjusted to the user's needs first before the trip starts. With ICT, individual settings can be saved and automatically restored before driving. This requires a holistic infrastructure and architecture model (Ericsson 2013).

4 Overcoming Complexity with ICT

Information and communication technology covers both hardware and software components. Relevant components for use in the area of e-carsharing are explained in the following sections. Their site of operation helps to group them into backend, infrastructure, user and in-car components. All components are mentioned and briefly described in the following, while the focus of this paper is on in-car components.

Backend: E-carsharing requires a considerable amount of administration of users, cars and bookings in corresponding databases. Appropriate databases mainly differ in structure and performance. While in the past relational databases were state of the art, NoSQL databases are catching up nowadays (DBEngine 2014) due to their performance advantage in data analysis. As part of integrated mobility concepts or in the context of macro-level cross-linking, also data exchange with foreign systems, such as public transport or public authorities, has to be considered. Interfaces for this exchange are predominantly implemented as Representational State Transfer (REST) services due to their flexible and scalable architecture (Richardson and Ruby 2011). In addition to administration, an easy-to-use user interface for registration, car localization and booking is needed. A dedicated website can act as this interface, serving also promotional purposes. An underlying REST service enables mobile access with applications across platforms. Both databases and websites can be hosted on ordinary server systems, either local or remote, as part of cloud computing, to outsource server maintenance and to enhance scalability and reliability. For advanced analysis of large mobility data sets, big data technologies are the key enablers. These technologies include, among others, distributed systems, supporting a divide and conquer approach on several servers, each equipped with multicore processor(s), huge amounts of RAM and solid-state disks (SSD) (Parashar et al. 2013). As an example for a corresponding database engine serves *Elasticsearch*, a key-value database that is distributable to several physical server nodes and capable of REST service integration. Its basic version is available for free and has neither data visualization plug-ins nor customer support (Elasticsearch 2014).

Infrastructure: In the context of meso-level cross-linking, V2I communication will become an important factor to optimize routing in terms of raising the road throughput and lowering the individual consumption. Furthermore, traffic security is improved due to mutual warning of hazardous areas. Traffic lights, construction sites, ramp metres and other infrastructure elements will be equipped with so-called road side units (RSU) enabling them to communicate with both road users and public authorities. The current state of research intends IEEE 802.11p for being the transmission standard for this communication (RITA 2014). With regard to concrete software components, no details are published yet.

User: Still on meso level, the user is already integrated in the network in the form of V2D communication. By means of mobile devices, like smartphones, tablets and smartwatches, personalization can be taken from favourite music to a higher level, making it possible to carry individual preferences like seat and mirror settings from one car to another. The current transmission standards for a connection with the car range from USB over Bluetooth to IEEE 802.11n, depending on the operation purpose. A famous example for user integration is BMW*'s ConnectedDrive*. It supports both the exchange of preference profiles with a dedicated USB stick and a Bluetooth connection to a mobile device for social networking, mobile working and entertainment (BMW 2013). However, the basic requirement for the integration of mobile devices into a car is to support their leading operating systems, i.e. Android, iOS, Blackberry OS and Windows Phone (Huffingtonpost 2014), which is a non-trivial task.

In-car: The focus of carsharing booking is on the car chosen by the user. This car can act as some kind of hardware interface between the user and the carsharing organization. Already equipped with high-tech components that, by now, only serve driving purposes, it can be turned into a fully connected car, sharing information with its environment. On the micro level, control units use a vehicle bus system like CAN to exchange vehicle-specific information whose complexity ranges from simple scalar values to driver assistance systems. Integrating a custom device in this bus system will, under consideration of security aspects (Hartung et al. 2014), enable further personalization and organization of e-carsharing vehicles (e.g. restoring custom seat and mirror settings).

Each car with a construction year newer than 2007 has a built-in OBD-interface, primarily for diagnostic reasons (Zimmermann and Schmidgall 2011). Via this interface standardized vehicle information like speed and diagnostic trouble codes (DTC) can be retrieved. A full list of supported values, so-called Parameter IDs (PIDs), can be found in the SAE norms J1979 and J2012. Depending on the required level of detail, it is also possible to access the CAN bus directly from the OBD-interface pins 6 and 14 (SAE J1962). By means of a wireless OBD/CAN interface like *OBDLink MX Bluetooth* (ScanTool 2014) even seat and mirror settings can be read and written by a connected device.

Besides OBD and CAN information, it is also essential to track the position of a car. This is done by means of a Global Positioning System (GPS) receiver that transfers position data to a connected device in the form of National Marine Electronic Association (NMEA) protocol data sets (NMEA 2013). In the current standard GPS III the accuracy of GPS is about two metres (Eckstein 2014). The recent development of GALILEO, a new European positioning standard with dedicated satellites, will raise accuracy of positioning to values below one metre. GALILEO is designed to be supported by current hardware after a simple firmware update (Eckstein 2014). Like OBD/CAN interfaces, GPS receivers exist in wired and wireless versions.

In order to gather and process both vehicle and position data, as well as sharing data with the environment, a powerful device with appropriate connectivity is needed. This device can be either a mobile device, like a smartphone or tablet, a data logger or a Car-PC, i.e. a built-in computer for in-vehicle usage.

The main advantages of a mobile device are availability and upgradeability since most users carry a recent device anyway and are steadily connected to the Internet via high-bandwidth transmission technologies like UMTS for software updates and data exchange. The main disadvantages are the usability while driving and the inconsistency of hardware features and operating systems, which makes it difficult to test functionality in advance.

A data logger is a dedicated device for bundling and logging (vehicle-specific) data either locally on a flash memory (e.g. Secure Digital (SD) card) or remotely by sending information to a server via General Packet Radio Service (GPRS) or, most often, UMTS. Data loggers can also be equipped with a Wireless Local Area Network (WLAN) or Bluetooth adapter in order to act as a V2D-bridge between the car and a mobile device. With this approach, a mobile device only needs a single wireless connection to the logger for access to all relevant data items (Kuck 2014).

In order to eliminate the dependency on the user's device, its hardware features and operating system, an in-car Personal Computer (Car-PC) has to be used. Another advantage is the extendibility with regard to input devices and networking hardware. The type of Car-PC is influenced by the need for connectivity (physical connection ports), the need for performance (construction technique and cooling), the available space and weight. Examples for ready-to-use Car-PCs are the *FleetPC*-models by CarTFT.com. They are fanless, intended for 12 V operation, have an Intel multicore processor, an integrated CAN interface can be equipped with various mini-PCIe extension cards (e.g. UMTS, GPS, WLAN, etc.) and external antennas (CarTFT 2014).

The user interface has to be in sight and within reach of the driver, ensuring a safe usage with the least possible distraction from the actual driving task. Input devices, such as a touch screen or a rotary pushbutton like the *iDrive Touch Controller* (BMW 2014), help to improve usability, making on-trip interaction safer and more ergonomic than with mobile devices. Since touch screens are sometimes difficult to use during vibrations and carry with a considerable amount of distraction, the current trend is to replace them with a combination of a simple display and a rotary pushbutton or touchpad respectively. The latter help to keep the eyes on the road while interacting with the system, even more when providing haptic feedback (Continental 2013). The ideal mounting location for the display is the centre console, right at the usual location of the radio, while the transmission tunnel is the ideal place for the input device. With regard to networking hardware, V2V communication requires a dedicated IEEE 802.11p Wi-Fi adapter that is not yet applicable to mobile devices. Either it is part of the data logger or the Car-PC. IEEE 802.11p only standardizes the physical (PHY) and Media Access Control (MAC) layer of the Open Systems Interconnection (OSI) stack. For higher layers there are three different standardization approaches: CALM (Communications Access for Land Mobiles) by the ISO Technical Committee 204, WAVE (Wireless Access in Vehicular Environment) by the IEEE and ITS (Intelligent Transport Systems) by the European Telecommunications Standards Institute (ETSI). At the time of this paper, no uniform standard is determined (Santi 2012).

5 Hiding Complexity Behind Intuitiveness; Human–Machine Interaction Requirements for E-Carsharing

Overcoming the various challenges of carsharing and electric mobility is not only a technical issue. A simple and user-friendly handling is essential for the user's acceptance. The human–machine interaction (HMI) describes the interaction between a human being and a machine or a system using interactive applications. Computer and mobile devices like smartphones and tablets as well as in-car systems are represented by the term system. Typical for an interactive system is a bilateral communication like a dialogue between user and machine. It is characterized by alternation between the user's actions and those of the system (Heinecke 2012).

Most of the time the driver's optical channel is busy with the task of driving. During the ride 80–90 % of the information is perceived visually. The driving situation itself and the environment are stressful for the driver and due to the mass of information he is confronted with in the car, he must keep up a high level of attention (Bruder and Didier 2009). According to the DEKRA Road Safety Report 2011 nearly 70 % of German accidents happen in city areas due to frequently changing and difficult traffic situations (DEKRA 2013).

In addition to that, many functions of contemporary cars are already too complex and distract the driver (Schaal and Schmidt 2011). Specially performed analyses demonstrated that in most cases the behaviour of integrated in-car systems differs from what people are used to from their mobile devices (e.g. rotary knobs in the vehicle (Audi 2014) and touch gestures on mobile devices). This leads to conflicts for the users in terms of interaction patterns, because the expectations of the customers are not met. A consistent compliance of interaction patterns, colours, symbols and information structure between in-car systems and mobile applications would improve the operational safety. Hence, the result of a then called integrated HMI (see Fig. 3) is that the driver can focus on his driving task instead of being distracted by operating the system.

For in-car systems, the European Statement of Principles (ESoP) established by the Task Force HMI in 1998 with the latest update in 2008 describe interaction

Fig. 3 Integrated HMI for e-carsharing applications

paradigms in the vehicle for creation of a usable and secure system. These guidelines contain instructions for design, installation, presentation of information, interaction and system behaviour for systems the user interacts with while driving the car. In general, the distraction caused by using these systems should be minimized with the help of ICT. It is not allowed to present information which stimulates a dangerous behaviour. A uniform design (colours, symbols, metaphors, etc.) provides easy understanding and guidance. In addition, the operation of the system with only one hand must be possible (European Union 2008). As a basis for these guidelines the human ergonomics should be considered in system design. In order to provide the user a pleasant arrangement of control elements and a sufficient impact area of the buttons, the touch-sensitive area of the surface should be at least equal to the size of the middle finger's width of a 95-percentile man. The body size of a European 95-percentile man is at 1.88 m (Jürgens et al. 1999). This corresponds to a size of about 2 cm. In addition, the button should have a touch-insensitive frame to avoid incorrect operation of at least 5 mm on all sides (Pfeil 2005; Norm ISO 15536, Part 1 2008).

Own developments and analyses showed that the display brightness must be taken into account as well (Ritz et al. 2014). Too bright displays can glare the driver during the ride, especially when driving at night. Darker displays with white letters provide better readability and are less straining for the eyes than white screens with black writing. For a fast assimilation of information there must be enough contrast between the single elements on the display. Furthermore, the font requires a different size for use in a car than for use on a mobile phone. While driving, words and hints need to be readable easily and quickly. Therefore, it is recommended to take a larger font. Additionally, buttons, icons and notes must be presented in sufficient sizes because of the greater distance between the interface and the driver. For better visibility of the symbols, the diagonal surface should have the size of an average finger or touch surface of an adult human.

Furthermore, there is a standard for human–machine interaction called ISO 9241. The standard consists of different parts including dialogue principles in part

110 and ergonomics of human–system interaction in part 210. In addition to the interaction standard, the usability standard is also important to ensure a usable and user friendly application independent if it is an application for in-car systems or mobile devices like smartphones (Norm ISO 9241, Part 110 2008, Part 210 2010).

6 Application Examples for E-Carsharing

E-carsharing includes topics like finding and booking free cars, driver assistance systems, fleet and charging management, as well as follow-up mobility, which becomes more and more important. With an intelligent human–machine interface electric mobility will be made available for everyone. For example, vacant electric cars can be searched and booked via a mobile application. Electric cars are located via GPS and displayed in this app on a digital map including the battery's state of charge and a predicted reach. The customer can book free cars spontaneously. An integrated navigation for pedestrians can then navigate him to the chosen car. The booking can also be cancelled optionally. Furthermore, an overview of all previous trips is given to the customer.

After a successful booking process, the driver has to check the car for damages before driving. A dedicated app for damage reports gives the driver an overview of all damages, their position and a short description. If there is a new damage, the user can add it graphically and the updated report will be sent right to the provider. If more information is needed an automatic phone call to the provider will be set up. There are two ICT-based alternatives for damage reporting: either via the mobile app or the car's driver assistance system (see Fig. 4). With the mobile app it is also possible to take pictures from the damage and include them in the report.

Fig. 4 Damage report mobile app (*left*) and driver assistance system (*right*)

Inside the car, electric mobility can be made manageable and usable for everyone by using a driver assistance system. It allows customers to configure individual preference profiles for the vehicle (e.g. setting of the preferred seat and mirror positions) and informs the driver about specific e-car characteristics (e.g. no engine noise and fuel consumption). By analyzing vehicle data that are sent via the CAN bus, driving situations can be identified in which an electric car behaves differently than a conventional car. The system can then support the driver with specific visual and auditory notifications. For example, when the battery level is low, the customer is encouraged to load the car.

Considering carsharing as a part of multimodal mobility, the system can provide information about other transport systems (departure, arrival, price, etc.), including booking facility. Since e-carsharing is considered as a supplement and not a substitute for public transport, it should be provided to the customer as a holistic mobility option for a low-carbon and environmentally conscious driving to the final destination.

7 Conclusion: ICT Is the Key Enabler for Future Smart Mobility

E-carsharing, as integrated and intermodal mobility concept, follows the trend of sharing and increased environmental awareness. Both the challenges of mobility planning itself and the challenges of electric mobility can be reduced by using information and communication technologies. Thus, an intelligent electric car can be an important factor in future smart mobility. High demands on flexibility and usability of shared mobility concepts can be met by information logistics providing an appropriate amount of information in the right place, time and form. For this purpose, ordinary means of transport have to be fully connected with their environment, building up a network that provides a basis for extensive exchange of data. Equipped with a Car-PC, tablet or smartphone, the car itself becomes a mobile device in a network between mobility providers and users. Information and communication technology is the key enabler for processing, distribution and presentation of the shared data. With easy-to-use mobile applications like smartphone apps or driver assistance systems, e-carsharing becomes manageable for everyone. In addition, preference profiles can be used for personalizing shared e-cars, making the actual sharing transparent to the user.

As individual mobility is changing from buying own vehicles to an efficient combination of various mobility services like public transport and e-carsharing, ICT becomes the only means to handle the resulting planning complexity. Furthermore, by compensating the weaknesses of electric mobility, an essential building block of future mobility, it offers the possibility to increase its acceptance in society and thereby promotes sustainable mobility.

References

Anthrakidis, A., Jahn, R., Ritz, T., Schöttler, M., Wallenborn, R., & Warmke, G. (2013). *Urbanes eCarSharing in einer vernetzten Gesellschaft* (71p, 80pp). Steinbeis-Edition.

Arnold, H., Kuhnert, F., Kurtz, R., & Bauer, W. (2010). Elektromobilität—Herausforderungen für Industrie und öffentliche Hand. *Fraunhofer Institut für Arbeitswissenschaft und Organisation (IAO)*. Retrieved May 5, 2014, from http://www.iao.fraunhofer.de/images/downloads/elektromobilitaet.pdf. 51p, 11p, 11p.

Audi (2014). *Audi connect*. Retrieved March 27, 2014, from http://www.audi.com/content/com/brand/en/vorsprung_durch_technik/content/2014/03/connect.html.

Baum, H., Heinicke, B., & Mennecke, C. (2012). Carsharing als alternative Nutzungsform für Elektromobilität. *Zeitschrift für Verkehrswissenschaft, 83*(2), 63–109. Univ. Köln. Retrieved from http://z-f-v.de/fileadmin/archiv/Heft%202012-2%20Volltext.pdf.

Becker, J. (2012). *Das Auto wird zur Nebensache*. Retrieved March 26, 2014, from http://www.sueddeutsche.de/auto/autohaeuser-der-zukunft-das-auto-wird-zur-nebensache-1.1436667.

Behrend, T., Wiebe, E., London, J., & Johnson, E. (2011). Cloud computing adoption and usage in community colleges. *Behavior and Information Technology, 30*(2), 231–240.

BMBF (2014). Elektromobilität: Das Auto neu denken. *Bundesministerium für Bildung und Forschung*. Retrieved March 26, 2014, from http://www.bmbf.de/de/14706.php.

BMW (2013). *BMW connected drive*. Retrieved March 27, 2014, from http://www.bmw.com/com/en/insights/technology/connecteddrive/2013/index.html.

BMW (2014). *iDrive touch controller*. Retrieved May 2, 2014, from http://www.bmw.de/de/footer/publications-links/technology-guide/idrive-touch-controller.html.

Bruder, R., & Didier, M. (2009). Gestaltung von Mensch-Maschine-Schnittstellen. In H. Winner, S. Hakuli, G. Wolf (Eds.), *Handbuch Fahrerassistenzsysteme* (314–324p). Vieweg+Teubner/GWC.

BUMB (2009). Vorteile elektrischer Antriebe. *Bundesministerium für Umwelt, Naturschutz, Bau und Reaktorsicherheit*. Retrieved March 26, 2014, from http://www.bmub.bund.de/themen/luft-laerm-verkehr/verkehr/elektromobilitaet/vorteile-elektrischer-antriebe/.

Cambio (2008). *Cambio website*. Retrieved March 26, 2014, from http://www.cambio-carsharing.de/.

Car2go (2014). *Car2go website*. Retrieved March 26, 2014, from https://www.car2go.com.

CarTFT (2014). *FleetPC-Series*. Retrieved May 2, 2014, from http://www.cartft.com/catalog/gl/126.

Continental (2013). *Study: Next generation touchpad with haptic feedback makes control tasks easier and safer*. Retrieved April 21, 2014, from http://www.continental-corporation.com/www/pressportal_com_en/themes/press_releases/3_automotive_group/interior/press_releases/pr_2013_11_08_touchpad_en.html.

DBEngine (2014). *DB-Engines ranking*. Retrieved May 2, 2014, from http://db-engines.com/en/ranking.

DEKRA (2013). *Verkehrssicherheitsreport 2013 Landstraßen: Strategien zur Unfallvermeidung auf den Straßen Europas* (11p.). Retrieved March 27, 2014, from http://www.dekra.de/Share/Blaetterberichte/VSR_2013_DE/blaetterkatalog/index.html.

DriveNow (2014). *DriveNow website*. Retrieved March 26, 2014, from https://de.drive-now.com/.

Ec2go (2012). *Elektromobilität & CarSharing—Das Projekt ec2go erforscht ein Mobilitätskonzept der Zukunft*. Retrieved May 19, 2014, from http://www.ec2go.de/.

Eckstein, L. (2014). *Aktive Fahrzeugsicherheit und Fahrerassistenz* (178p). Institute of Automotive Engineering, RWTH Aachen University.

Elasticsearch (2014). *Elasticsearch website*. Retrieved April 21, 2014, from http://www.elasticsearch.org/.

Ericsson (2013). *Connected vehicle cloud* (p. 3). Retrieved May 19, 2014, from http://archive.ericsson.net/service/internet/picov/get?DocNo=28701-FGD101192&Lang=EN&HighestFree=Y.

Europäische Union (2008). *Empfehlung der Kommission über sichere und effiziente bordeigene Informations- und Kommunikationssysteme: Neufassung des Euopäischen Grundsatzkkatalogs zur Mensch-Maschine-Schnittstelle*.

GI (2013). Gesellschaft für Informatik. *Informatiklexikon*. Retrieved April 14, 2014, from http://www.gi.de/nc/service/informatiklexikon/detailansicht/article/big-data.html.

Haftor, D. M., & Kajtazi, M. (2009). *What is information logistics* (20pp). Sweden: Linnaeus University.

Handelsblatt (2012). *Die Vor- und Nachteile des E-Autos*. Retrieved March 26, 2014, from http://www.handelsblatt.com/auto/nachrichten/elektromobilitaet-die-vor-und-nachteile-des-e-autos/7205944.html#image.

Hartung, F., Hillgärtner, M., Schmitz, G., Schuba, M., Adolphs, F., Hoffend, J., & Theis, J. (2014). *IT-Sicherheit im Automobil*. Technical Paper, University of Applied Sciences Aachen.

Heinecke, A. M. (2012). *Mensch-Computer-Interaktion—Basiswissen für Entwickler und Gestalter* (99p). Springer.

Hirte G., & Nitzsche, E. (2013). Evaluating policies to achieve emission goals in urban road transport. *Zeitschrift für Verkehrswissenschaft, 84*(2), 112–137. TU Dresden.

Huffingtonpost (2014). Android leads in operating system share. Retrieved May 2, 2014, from http://www.huffingtonpost.com/michael-r-levin/android-leads-in-operatin_b_5200491.html.

Initiative D21 (2013). *D21-Digital-Index* (52pp). Retrieved March 26, 2014, from http://www.initiatived21.de/wp-content/uploads/2013/05/digialindex_03.pdf.

Jürgens, H. W., Matzdorff, I., & Windberg, J. (1999). *Arbeitswissenschaftliche Erkenntnisse—Forschungsergebnisse für die Praxis—Internationale anthropometrische Daten*. Retrieved May 19, 2014, from http://www.Users/ramona/Downloads/AWE108.pdf.

Kuck, A. (2014). *Entwicklung eines CAN auf W-LAN Gateways für die Realisierung einer Restreichweiteanzeige in einem Elektrofahrzeug (Bachelor Thesis)*. University of Applied Sciences Aachen.

Lešková, A. (2013). *Services In automotive business based on car connectivity possibilities* (p. 28). Technical Paper, Technical University of Košice.

NMEA (2013). *NMEA 0183-Datensätze*. Retrieved May 2, 2014, from http://www.nmea.de/nmea0183datensaetze.html.

Norm ISO 15536, Part 1 (2008). *Ergonomics—Computer manikins and body templates: General Requirements*.

Norm ISO 9241, Part 110 (2008). *Ergonomics of human system: Dialogue principles*.

Norm ISO 9241, Part 210 (2010). *Ergonomics of human system: Human-centered design for interactive systems*.

Parashar, M., Bui, H., Jin, T., Sun, Q., & Zhang, F. (2013). *Addressing big data challenges in Simulation-based Science* (4p). Rutgers Discovery Informatics Institute. Rutgers University New Jersey. Retrieved May 5, 2014, from http://www.pppl.gov/sites/pppl/files/parashar-cdse-bigdata-pppl-01-14.pptx.pdf.

Pfeil, U. (2005).*Informationsdesign im Fahrzeug. Entwürfe und Prototypen von Bedien- und Anzeigekonzepten eines Fahrerinformationssystems unter Berücksichtigung ergonomischer Richtlinien* (Bachelor Thesis). Hochschule der Medien Stuttgart.

Richardson, L., & Ruby, S. (2011). *RESTful Web Services* (49p, 299p). O'Reilly.

RITA—US Department of Transportation: Research and Innovative Technology Administration (2014). *Connected vehicle applications*. Retrieved April 21, 2014, from http://www.its.dot.gov/research/v2v.htm.

Ritz, T., Siekmann, K., & Wallenborn, R. (2014). *Anforderungen an die Gestaltung multimodaler Mobilitätsanwendungen* (eBusiness-Lotse Aachen). Retrieved May 19, 2014, from http://www.ebusiness-lotse-aachen.de/sites/default/files/mediathek/20140304_eBL_Broschüre_Gestaltungsanforderungen_multimodale_Mobilit%C3%A4t_FHAachen.pdf.

Santi, P. (2012). *Mobility Models for next generation wireless networks* (145p). Wiley.

Saretzki, U., & Krämer, C. (2009). *Das ‚öffentliche' Auto. Kooperationen zwischen ÖV +CarSharing* (13p). VDM.

ScanTool.net LLC (2014). *OBDLink MX Bluetooth Scan Tool*. Retrieved April 22, 2014, from http://www.scantool.net/obdlink-mx.html.

Schaal, K. M., & Schmidt, J. (2011). *Innovationen: Bediensysteme, Ablenkungsmanöver*. Retrieved March 27, 2014, from http://www.ace-online.de/ace-lenkrad/test-und-technik/blickwinkel-736.html.

WGSoft (2014). *OBD-2 Allgemeines, technische Informationen*. Retrieved April 21, 2014, from http://www.obd-2.de/obd-2-allgemeine-infos.html.

Zimmermann, W., & Schmidgall, R. (2011). *Bussysteme in der Fahrzeugtechnik—Protokolle, Standards und Softwarearchitektur* (45p, 417p, 173pp). Vieweg+Teubner.

Zukunftsinstitut (2014). *Megatrends—Die großen Treiber der Gesellschaft*. Retrieved March 26, 2014, from http://www.zukunftsinstitut.de/megatrends.

Thermal Management in E-Carsharing Vehicles—Preconditioning Concepts of Passenger Compartments

Daniel Busse, Thomas Esch and Roman Muntaniol

Abstract The issue of thermal management in electric vehicles includes the topics of drivetrain cooling and heating, interior temperature, vehicle body conditioning and safety. In addition to the need to ensure optimal thermal operating conditions of the drivetrain components (drive motor, battery and electrical components), thermal comfort must be provided for the passengers. Thermal comfort is defined as the feeling which expresses the satisfaction of the passengers with the ambient conditions in the compartment. The influencing factors on thermal comfort are the temperature and humidity as well as the speed of the indoor air and the clothing and the activity of the passengers, in addition to the thermal radiation and the temperatures of the interior surfaces. The generation and the maintenance of free visibility (ice- and moisture-free windows) count just as important as on-demand heating and cooling of the entire vehicle. A Carsharing climate concept of the innovative ec2go vehicle stipulates and allows for only seating areas used by passengers to be thermally conditioned in a close-to-body manner. To enable this, a particular feature has been added to the preconditioning of the Carsharing electric vehicle during the electric charging phase at the parking station.

Keywords Carsharing · Thermal management · Thermal comfort · Electrical vehicle · Passenger compartment · Preconditioning of vehicle interior · Heat to passenger · Heat to seat

D. Busse (✉) · T. Esch · R. Muntaniol
Institute of Applied Thermodynamics and Combustion Technology, Aachen University of Applied Sciences, Hohenstaufenallee 6, 52064 Aachen, Germany
e-mail: busse@fh-aachen.de

T. Esch
e-mail: esch@fh-aachen.de

W. Leal Filho and R. Kotter (eds.), *E-Mobility in Europe*,
Green Energy and Technology, DOI 10.1007/978-3-319-13194-8_18

1 Introduction

The recent development and market introduction of battery electric or hybrid vehicles as an alternative to traditional gasoline- or diesel-powered vehicles is being driven by politics and also by the automotive industry now for several years. After an initial euphoria and first successes in cities and regions with strict environmental regulations and generous funding programmes, the current discussion is more characterized by a balance between benefits and costs of a system change in automotive drive concepts (AMS 2011). In addition to numerous other issues, the identification of appropriate initial markets is crucial to give all parties the opportunity to develop sensible vehicles, usage models and support frameworks.

The focus of this paper is on the subject of Carsharing fleets for the integration of electric mobility: Concepts of cars with shared use offer interested customers the opportunity to test vehicles without having to bear the high cost. In addition, it may allow a faster penetration of electric drive technology to speed and economy of electric vehicles can be improved by minimizing downtimes. Furthermore, Carsharing has developed to a very dynamic market in recent years and it suggests itself to a change in the mobility behaviour of broad social layers. Signs of a trend away from the car as a ‘status symbol’ for the flexible use of vehicles and transportation to their suitability for the currently planned trip purpose can be recognized and supported by various studies (ISI 2011).

For electric vehicles the air conditioning and thermal management of the vehicle represents a major challenge (Ackermann 2011). Classic solutions which are used in conventional vehicles with internal combustion engines such as belt-driven refrigeration compressors, electric windows and seat heaters can only be used very partially in electric vehicles due to the electrical charging of the traction battery, as draining the electric battery for these uses would be counterproductive for the driving range (Ackermann et al. 2013). Therefore, innovations in the field of temperature and climate management are essential for electric vehicles especially in Carsharing applications.

2 The Thermomanagement Concept

The Carsharing climate concept stipulates and allows for only seating areas used by passengers to be conditioned. Heating of the whole interior is not always energetically sensible. Tight surface heating compensates for the lowered interior temperature. The basic idea of the close-to-body conditioning (Heat to Passenger) saves energy by reducing the average interior temperature of the vehicle and increases the comfort by a demand-controlled individual seat heating. The use of surface heating (Heat to Seat) includes seat and backrest, floor mats, armrest, steering wheel, headrest, seat belt, even slices. According to the usage profile, the response time of the heating elements is very fast. The concept of close-to-body temperature control may also include the concept of eClothing, a climatic seat belt and air headrest. Vacuum insulation panels are used in the vehicle body elements on the basis of high-performance foams from polyurethane, which provide an up to seven times better thermal performance (heat, cold) compared to conventional polymer insulating materials. Infrared-reflective films in the glazing reduce the heating up of the vehicle interior, and infrared-reflective coatings and pigments in exterior paint and interior reflect heat radiation from the sun and light. Transparently, coloured photo cells generate enough electricity for a continuous air flow in the interior of the Carsharing vehicle at high sunlight. The climate centre is displaced from the dashboard to the proximal areas of the occupants. The comfort can be produced by simple controls and customized comfort elements can thus be taken into account.

3 Benchmarking of Thermal Management Parameters of Electric Vehicle

For subsequent validation of simulation models, a detailed benchmark of the university's research vehicle, a Mitsubishi i-MiEV, was performed. In addition to the electrical energy flows in the drivetrain and in the auxiliaries, the thermodynamic behaviour of the drivetrain as well as the vehicle compartment was analysed.

Temperature was recorded at eight different measuring points: four temperature measurement points were set up in the engine cooling circuit (the drive motor, the power electronics, the battery charger and the DC/DC converter), two in the heating circuit, one ambient temperature measuring point and one in the passenger compartment. Electric current was measured at three measuring points.

In addition, the battery voltage of the vehicle was recorded and evaluated.

In the high voltage (HV) electrical system, the voltage of the drive battery and the electrical power to the interior heating, air conditioning compressor and power train were also documented. In addition to the thermal behaviour of the drivetrain at different load profiles, heating and cooling curves of the interior were recorded (see Fig. 1). The data sets were evaluated and used to validate the thermal management model.

4 Carsharing Specific Thermal Management Requirements

To ascertain requirements and user profiles for Carsharing of relevance to actual usage, evaluation data of Carsharing provider *Cambio Mobility Services* in Aachen were used. Their (as of 2014) 9 e-vehicles in a fleet of about 100 vehicles are in use. From the usage data provided, information such as average distance travelled, time of booking, etc. were provided. Furthermore, a user survey among approximately 700 participants was conducted amongst Carsharing customers. These data were

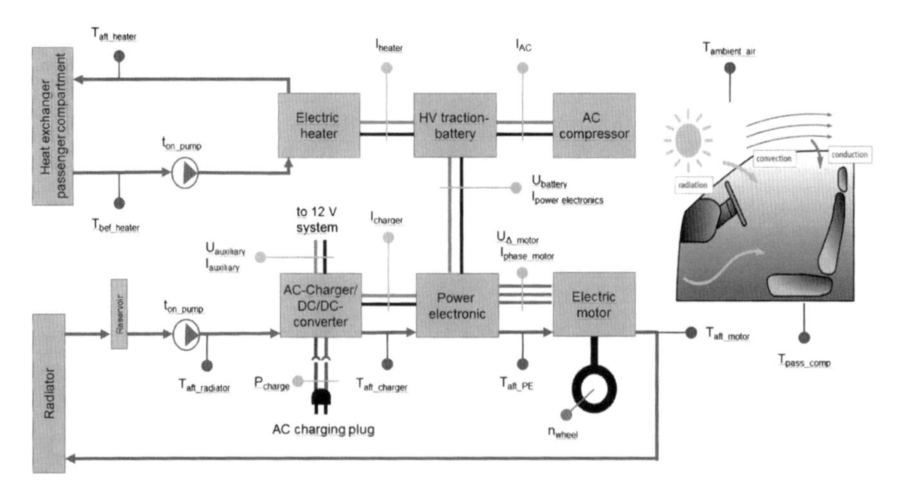

Fig. 1 Measuring positions during benchmark investigations of the Mitsubishi i-MiEV vehicle

used to interpret and evaluate the Carsharing specific potentials. The survey results and the user data of Carsharing customers analysed were then evaluated at the vehicle level with respect to thermal management and drivetrain with respect to the user requirements. The following identified key data parameters have been taken as a decisive basis in the conceptual design:

- in terms of *distance travelled per user*, in 58 % of cases this ranged from 0 to 20 km
- the *average distance travelled per user* was 13.9 km
- in terms of *frequency of use of Carsharing*, 62 % of customers used it less often than three times per month
- the *average use time/vehicle booking duration per user* was 2.5 h
- in terms of *predictability of bookings*: only 8 % of trips of each ride are known only less than 1 h before use
- in terms of highway usage, 58 % of users said that they seldom to never go on the highway with a *Carsharing vehicle*.

4.1 Load Profiles of Drivetrain

In order to establish the longitudinal dynamics model for the subsequent concept evaluation, a requirement profile was defined as a vehicle profile. These requirements are similar to the performance of the university's research vehicle and shall comply with requirements representing a small zappy town car for Carsharing use. From these data profiles, a requirement design cycle has been generated to test the parameterized models (see Table 1).

Table 1 Requirements of vehicle performance for Carsharing purpose

Acceleration		Elasticity	
0–30 km/h	3 s	30–50 km/h	2.5 s
0–50 km/h	5 s	50–70 km/h	4 s
0–100 km/h	15 s	70–100 km/h	10 s
v_{max}	120 km/h	100–120 km/h	10 s
Gradeability		*Body parameter/weights*	
4 %	100 km/h	Tare weight	900 kg
20 %	30 km/h	Gross vehicle weight	1260 kg
30 %	20 km/h	c_W	0.35
35 % (max)	1 km/h	Front surface	2.5 m^2

For the longitudinal dynamic evaluation of the concepts the New European Drive Cycle (NEDC) was used as reference. With its 11.007 km length an NEDC cycle corresponds approximately to the average travel distance of a Carsharing user per booking. Furthermore, for a more detailed validation of the longitudinal dynamic model a real reference route through Aachen (Aachen City Cycle, ACC) was defined. Various benchmark measurements for different traffic situations have been conducted with the university's research vehicle (see Table 2).

4.2 Load Profile Vehicle Interior

Since for E-mobility climatic conditions have a major impact on vehicle performance (range), temperature profiles were sought to evaluate, design and optimize the thermal system performance. From the idea of a Carsharing vehicle for the use in urban areas and to analyse the potential for optimization with regard to the specific usage requirements, it was decided to use the weather data of Aachen, representing a typical central European city. With respect to the thermal requirements the daily values of temperature of the German Meteorological Service (Deutscher Wetterdienst, DWD) weather station 10501 Aachen (located at 50°47′N006°05′E) were evaluated over the time period between 1991 and 2010. From the DWD source, the following values were specified for each month in the mentioned period: Minimum temperature, mean daily minimum temperature, mean temperature and mean maximum temperature, as well as maximum temperature. The day referred to here is a 24-h calendar day without distinction between day and night. From this

Table 2 Comparison of used driving cycles

Cycle	Driving distance (km)	Driving time (s)	Average velocity (km/h)	v_{max} (km/h)
NEDC	11.007	1180	33.6	120
ACC	24.2	Appr. 3600	28–31	100

temperature data three design points for use in warm ambient temperatures (*summer*, S1–S3) were defined, as well as three for cold ambient temperatures (*winter*, W1–W3). For the design point *summer* the values of the period 01.04–30.09 were considered, and for *winter* accordingly 01.10–31.03. The mean daily maximum temperature, averaged over the summer months of April to October, is defined as the first design point for the summer (S1), and the second design point (S2) is the mean of the average daily maximum of the months of April to October. The absolute daily maximum has been determined as the highest design limit for this time period. For summer design points the thermal requirements on warm summer days, for example at noon, should be reflected. For the winter design points this was translated accordingly (see Table 3).

In addition to the expected temperature, solar radiation is important. For central European latitudes such as Aachen one can expect a solar radiation of 600–1000 W/m^2 on a summer day with clear skies (see Table 4).

4.3 Simulation Tools and Procedures

To determine the purpose of analysis and the optimization potentials, vehicle sub-models have been constructed. As a simulation tool, in this case GT-Suite by Gamma Technologies was chosen. With its programme modules GT-Power and GT-Cool it offers, among others, options, the possibility of simulating the vehicle's

Table 3 Evaluation of the weather data of the Aachen DWD weather station (from 1991 to 2010)

Point	Name	Evaluated by	Temperature value (°C)
S1	Average summer day (01.04–30.09)	Average of median daily maximum summer days	20
S2	Average summer extremes (01.04–30.09)	Average of the maximum temperature summer days	28
S3	Absolute extreme summer (01.04–30.09)	Extremum 1991–2010	37
W1	Average winter day (01.10–31.03)	Average of median daily maximum winter days	0
W2	Average winter extremes (01.10–31.03)	Average of the maximum temperature winter days	−4
W3	Absolute extreme winter (01.10–31.03)	Extremum 1991–2010	−16

Table 4 Solar radiation for central Europe (http://www.wetterstation-bremen-nord.de/index.php?inhalt_mitte=content/solar.inc.php)

Season	Heavy overcast to cloudy foggy (W/m^2)	Light to medium clouds (W/m$^{2)}$	Clear to slightly diffuse sky (W/m$^{2)}$
Summer	100–300	300–600	600–1000
Winter	50–150	150–300	300–500

longitudinal dynamics as well as the thermodynamic behaviour of different vehicle systems (engine, interior, cooling …). Thus, a longitudinal dynamic model of the drivetrain, and a model of the vehicle interior have been built and validated.

4.4 Vehicle Cabin

With the thermal model of the passenger compartment the required heating or cooling capacity for achieving the comfortable temperature environment at predetermined conditions is detected. The model simulates the driving environment in all material respects and consists of eight interrelated concentrated point masses. All the doors of the vehicle are represented by a single mass, as are all door panels, all side windows and the entire interior. Further individual masses represent the roof, the windscreen and the rear window. Besides the weight, the components are characterized by their surface described by thermal conductivity, by emissivity and absorption coefficient and by transmittance (only for glass). For the heat exchange between the vehicle components and the indoor air or the ambient air, the heat transfer is taken into account by convection. For the sake of simplification, a constant heat transfer coefficient is defined between an interior side and environment side heat transfer. The heat conduction within the individual masses is taken into account by a constant thermal conductivity. View factors for the roof, doors and windows are also considered.

4.5 Interior Ventilation

The layout of the heating and ventilation system of the model was largely inspired by one of the conventional vehicles. Both a fresh air mode and a recirculation mode can be simulated. During the standing and charging phase of the vehicle a recirculation mode of the ventilation system effects to save energy and to accelerate the warming or cooling of the interior is applied. In the subsequent preparation phase and during the trip a fresh air mode takes place. Both in recirculation mode and fresh air mode, the air current operates a fan into the passenger compartment with a constant flow rate of 150 m^3/h. The function of the air speed of the vehicle is neglected in order to simplify the simulation model. The graph (Fig. 2) shows the cabin model with the air circuit and the control elements in the simulation environment GT-Cool.

In *Temperature Control* all subcomponents are combined, regulating the heating or cooling the indoor air and the air delivery. The change of air circulation mode to fresh air operation is dependent on the duration of the individual phases of operation. The heating or cooling of the vehicle interior air comfort temperature and maintaining the comfort temperature can be controlled by a PID (proportional-integral-derivative) controller. The PID controller determines this function of the

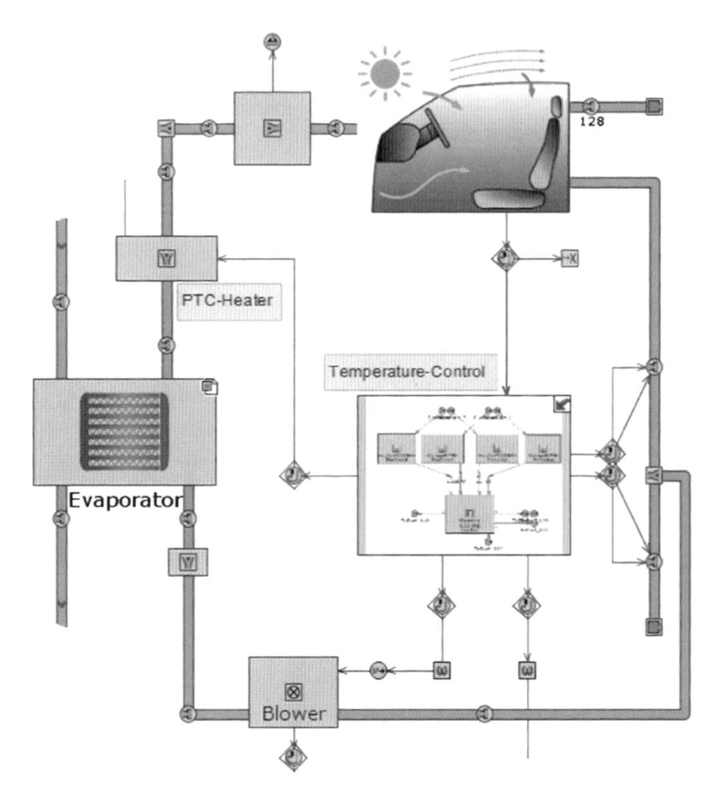

Fig. 2 Model of interior ventilation and heating in GT-Suite environment

instantaneous average indoor temperature, the required heating power of the *PTC Heaters* or the required speed of the compressor. For heating the indoor air, a PTC (positive temperature coefficient) heater is used, which is mounted in the air circulation downstream of the *Evaporator*. The heat thus passes directly through the ventilation system (*Blower*) into the interior.

4.6 Cooling Circuit

The cooling circuit model consists of the components of an electric compressor, a condenser, a TX valve (TXV), an evaporator, a receiver/dryer and refrigerant pipes. A refrigerant R134a is used. The selected compressor is a positive displacement compressor (positive displacement type). The simulation model of the compressor is based on map data. The map comprises a function of the compressor speed-reduced to the reduced mass flow rate, the pressure ratio of the refrigerant and the efficiency of the compressor. The compressor speed is controlled according to the case study of PID controllers depending on the indoor temperature in different

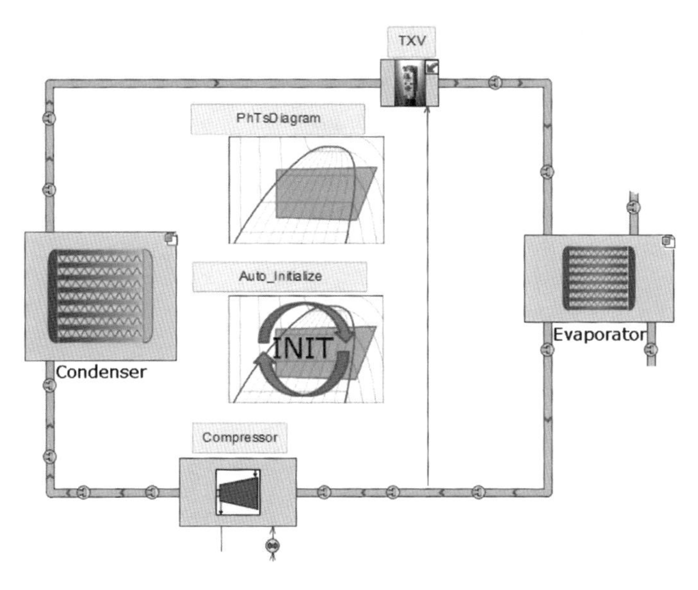

Fig. 3 Model of the cooling circuit of the passenger compartment

speed ranges. The used amount of energy to drive the compressor, depending on the load point, is taken from the battery. For the modelling of the condenser, the software tool COOL3D was used. With the help of COOL3D the condenser was initially mapped three-dimensionally, and then generated as a discrete one-dimensional model. Similarly, the evaporator was imaged. This makes a detailed illustration of the condenser possible. The discretized 1D model of the condenser is integrated into the overall model. The TXV was depicted as a four-quadrant model which reproduces the real behaviour of the thermal expansion valve. Figure 3 shows the coolant circuit in the simulation environment of GT-Cool.

4.7 Experimental Validation of the Models

The vehicle longitudinal and thermal models were validated by means of measurements on the research vehicle Mitsubishi i-MiEV from the aforementioned benchmark (Eckstein et al. 2011). The consumption of the drivetrain was calibrated by the model with the measurement results from the i-MiEV under laboratory conditions in the NEDC on the chassis dynamometer as well as in real driving in Aachen City Cycle (see Table 5).

The cabin model was calibrated for each temperature point regarding heating and cooling. As a criterion the time-dependent behaviour by heating/cooling curves was compared. Second, the required heating/cooling power in the steady state was adjusted to maintain the temperature constant between model and real vehicle. In

Table 5 Comparison of measured energy consumption of i-MiEV vehicle in different cycles and weight configurations with longitudinal dynamics model

	Aachen City Cycle (sporty driving style)	Aachen City Cycle (restrained driving style)	NEDC
Distance	24.2 km	24.2 km	11.022 km
Time	2809 s	3086 s	1180 s
Average vehicle speed	31 km/h	28.2 km/h	33.6 km/h
ec2go (900 kg and altitude profile)	11.26 kWh/100 km	8.95 kWh/100 km	13.34 kWh/100 km
ec2go (1260 kg and altitude profile)	14.15 kWh/100 km	10.96 kWh/100 km	15.26 kWh/100 km
i-MiEV (1160 kg)	15.69 kWh/100 km	13.45 kWh/100 km	17.55 kWh/100 km
ec2go simulation		*i-MiEV*	
• Energy consumption from battery		• Energy consumption from socket	
• Altitude profile included		• Only to drive required auxiliary loads in service	
• No electrical consumers			

Figs. 4, 5, 6 and 7, measured values on the research vehicle i-MiEV and the simulation results are plotted against each other. The model was adjusted to reflect the different vehicle body geometry and the other components of the selected i-MiEV compared to ec2go design. The Mitsubishi i-MiEV features—in contrast to ec2go concept—a coolant fluid heater. A PTC heating element is warming up a coolant fluid which is pumped to the coolant–air heat exchangers in the interior. The ec2go concept provides a PTC air heating which guides the warmed up air directly into the interior. This difference was not adjusted for in the comparison, and this explains the sluggish response of the heating of the i-MiEV vehicle compared to the ec2go model during the first 270 s.

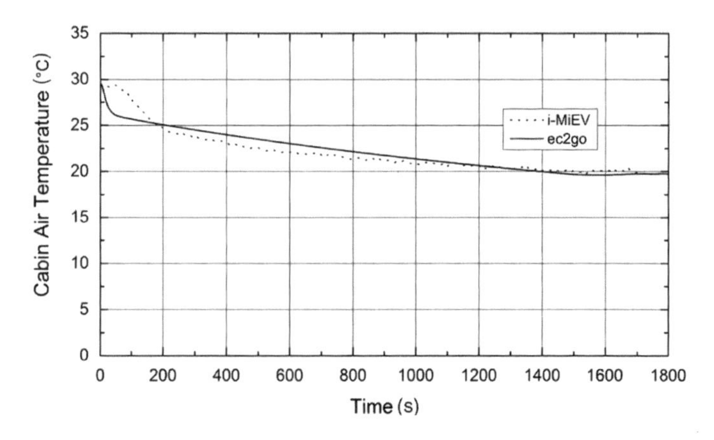

Fig. 4 Compare cooling behaviour of the ec2go model and the i-MiEV research vehicle

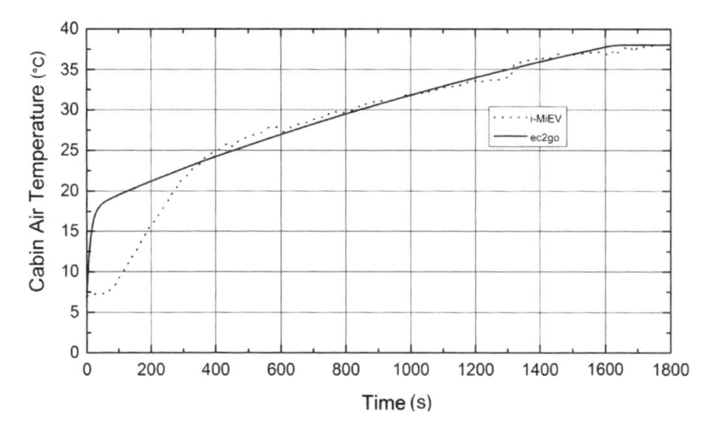

Fig. 5 Compare heating behaviour of the ec2go model (PTC air heater) and the i-MiEV research vehicle (PTC water heater)

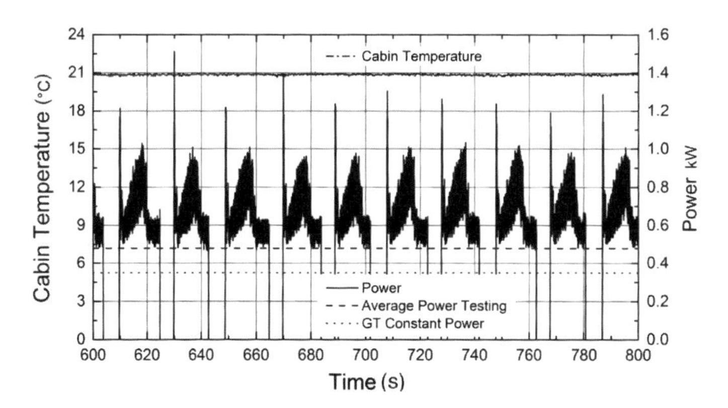

Fig. 6 Compare cabin temperatures and electrical power of the compressor at 30 °C ambient temperature

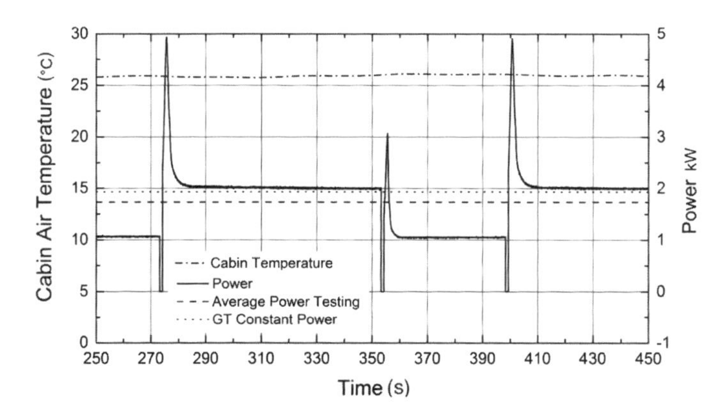

Fig. 7 Compare heat output at 7 °C ambient temperature

4.8 Thermomanagement Interior

The task of the concept is to be as user-oriented and energetically favourable as possible to provide a feeling of satisfaction and also to ensure driving safety aspects such as good visibility. To minimize the influence of climatic conditions on the performance, a thermal management concept was created based on the specific terms of use of a vehicle with exclusive use in a Carsharing application. This concept is to exploit the optimization potential at the expected service as fully as possible for a Carsharing vehicle. The concept features were optimized and evaluated in the following by means of simulations.

4.9 Preconditioning in Electric Vehicle

The electric vehicles are parked in Carsharing stations and connected to the electricity grid to recharge the battery. During charging, the energy is used from the grid to keep the vehicle continuously at a predetermined temperature level of the booking system. The difference between the temperature level during the preconditioning and the comfort temperature for the user is selected so that the vehicle can reach the final heating or the final cooling within the limited preparation phase quickly. If the vehicle is booked, the booking system is calculating the beginning of the final heating and final cooling depending on the prevailing weather. While driving, the internal temperature of the preheated vehicle is only going to keep on comfort temperature. Since the energy for heating/cooling need not be provided by the vehicle battery, the energy storage in the vehicle can be made smaller, or the range of the electric vehicle can be increased.

4.10 Inclusion of Booking Data and Carsharing Station

To underline the brand's trademark ec2go, special Carsharing base stations are possible in the form of solar carports. Affixed on the carports, solar panels provide clean energy for charging the electric vehicle, underscoring at the same time the ecological brand image of Carsharing. Furthermore, the solar carports offer summer protection from direct sunlight on the electric vehicle and its interior surfaces. Sheltered from direct sunlight the average indoor temperature can be kept stable on the outside temperature level only through venting and less AC power will be needed. Out of the Aachen Carsharing benchmark data it is known that in 92 % of all cases the booking of vehicles takes place more than an hour before each ride. In case of a sensibly selected mean indoor temperature for the preconditioning phase, there is thus sufficient time for the final heating or the final cooling of the vehicle to the desired comfort temperature of the respective user.

4.11 *Lowering the Interior Temperature When Using Surface Heating Systems*

The analysis of usage data from Cambio showed an expected predominant use of vehicles for short distances. This means that the heating of the entire vehicle interior is not deemed appropriate in many cases. By the application of a targeted more favourable surface heating, controlled energetically, a comfort feeling appears with the passengers already at a low mean interior temperature in winter. A possible energy requirement of such a system that consists of contact surfaces, heating surfaces of the short range and of the long range, is a total of 700 W at steady state after a few minutes. The temperature of the heating surface is then 37 °C. By means of a control strategy, areas can be switched off with rising interior temperature gradually. This allows for further reducing the energy demand in the steady state per seat to 120 W (Ackermann et al. 2013).

4.12 *Potential Analysis Through Simulation*

Along with the design of electric vehicles and its energy storage systems it must be ensured that there is enough energy available to cool the vehicle interior or sufficient to heat it both in winter and in summer. This is necessary both for ride comfort and for the safety of the driver. Here, the heating or cooling is very energy consuming directly after starting the car. The use of energy for preconditioning and temperature variation of the vehicle interior to the final temperature when charging with the energy from the electrical supply grid brings significant savings on energy consumption whilst driving. Furthermore, the Carsharing user benefits from comfort and safety advantages with each ride. Also, the vehicle is always immediately ready to drive at temperatures below 0 °C, and possibly a removal of snow and ice is not necessary.

4.13 *Potential Analysis of Thermal Preconditioning of the Interior*

In the following, the determined energy consumption is compared after driving the electric vehicle with and without previous conditioning under winter and summer ambient conditions. The total power consumption is the portion that is taken from the vehicle battery. It consists of the drive energy and the heating and cooling energy and is the average specified per 100 km driving distance. The individual energy components also contain the energy levels of the corresponding energy consumer. As winter ambient conditions, the design points defined are the design points W1 and W2, and as summer ambient conditions the design points S1 and S2 are chosen.

The total energy consumption of a not preconditioned vehicle before driving is a reference value for the relative power consumption of a preconditioned vehicle. In the simulations in not preconditioned condition it is ensured that by a standing phase at the beginning of the simulation the vehicle temperatures can adapt to the ambient conditions. For the case of 'summer' the heating by the sun is taken into account, in winter it has been derived from an overcast sky without energy entry by solar irradiation.

Next follows the ride in the driving cycle and the beginning of the vehicle conditioning to comfort temperature. For each ride all the energy required for this is removed from the battery. The simulation of the preconditioned ride is divided into three areas. It also begins with a standing phase in which the vehicle heats up on account of the surroundings terms or cools and at the same time the first preconditioning of the vehicle on the interior temperature T1 occurs. Subsequently, the preparation phase for the final heating or final cooling to comfort temperature T2 and the drive at this constant interior temperature takes place. For all scenarios the comfort temperature of 20 °C is applied for ease of comparability. During the standing and preparatory phase, the required energy is drawn from the grid, so that by the beginning of the cycle the battery is fully charged.

The NEDC driving cycle is used which will be traversed once according to the user analysis. In not preconditioned simulations carried out with the beginning of the cycle, a fresh air mode with a constant volume flow of 150 m^3/h and the conditioning on the comfort temperature of 20 °C is applied. The solar radiation is taken into account in the summer points S1 and S2 with a constant value of 850 W/m^2, and neglected under the winter design points. Furthermore, the standing phase lasts an hour for the winter points, and 2 h for the summer points. The preparation phase lasts 1 h for all design points.

First, the energy consumption of the not preconditioned vehicle is considered for the design point W1. After the end of the standing phase a vehicle temperature of 0 °C is set up. With the start time of the NEDC the blower and PTC heater begin to heat up the vehicle interior by fresh air mode. Before entering into the passenger compartment, the intake of cold air is heated up to 64 °C by the PTC heater. The warm air then exchanges heat with the components of the passenger compartment, and heating them. At the same time the average temperature of the passenger compartment air increases, and reaches after 1178 s 20 °C. When preconditioned, the passenger compartment is brought to comfort temperature in two steps before beginning the NEDC. In the first step, during the standing phase, the average temperature of the passenger compartment air is preconditioned at 10 °C and kept at a constant level. In the second step, with the beginning of the preparation phase, the final heating is at comfort temperature. At the beginning of the cycle, the passenger compartment is fully preheated with its components (see Fig. 8).

The heating of the vehicle not preconditioned (PC-Off) to comfort temperature is compared with the preconditioned vehicle (PC-On) only during the drive. In this case, the PTC heater is operating under maximum power output to speed up the heating process. Only when the temperature of the passenger compartment air of comfort temperature approaches comfort temperature, the power output of the PTC heaters

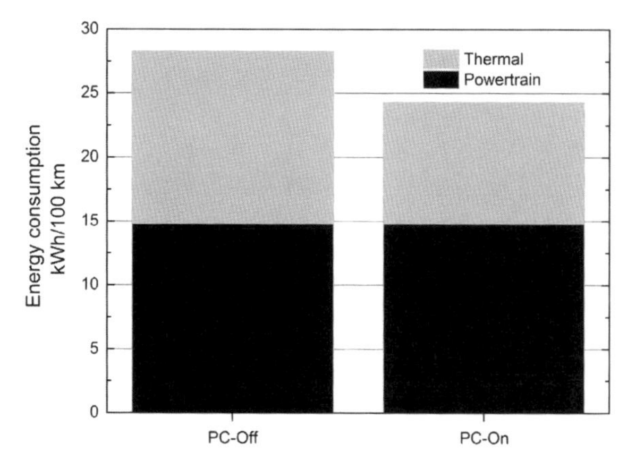

Fig. 8 Total electric energy of a not preconditioned (*PC-Off*) and a preconditioned vehicle (*PC-On*) after driving the NEDC in relation to the energy consumption of the drive at 0 °C outdoor temperature in winter (W1)

decreases. In the preconditioned passenger compartment the PTC heater has the task to keep the temperature constant and thus has a lesser power output. The energy required for heating the vehicle interior represents a non-negligible share of total energy expenditure. Its share is 48.3 %. The energy consumption of the drive—including the electrical load required at a minimum—is 51.7 %. In the case of preconditioning (PC-On) and heating of the vehicle interior through the use of electrical energy from the utility grid, the heating energy consumption in the NEDC can be reduced by up to 27 %. Without heating or cooling the maximum range of the vehicle is 126 km. This range is reduced down to 65.4 km by heating the interior. By preconditioning in winter at 0 °C outdoor temperature, the range increases to 75.2 km.

Other energy savings could be obtained by using potential surface heating in the form of heaters that are mounted in the near and far fields around the driver and passenger. For this purpose, surface heating in the seat and backrest, floor mats, armrest, steering wheel, headrests and seat belt are conceivable. In the winter, heating the cold interior surfaces to the level of the body temperature, the heat output from the body to the surrounding components is reduced and the feeling of comfort is increased. Thereby, the mean interior temperature of the vehicle could be lowered by a few degrees Celsius, without reducing the feeling of comfort. For the preheated interior and an ambient temperature of 0 °C (design point W1), this results in a saving of 0.75 kWh/100 km in the NEDC when a reduction in the average indoor temperature by 2 °C is applied. This corresponds to a constant power output of 250 W over a driving cycle. A possible power consumption of a radiant heating system is 120 W per seat in steady state, depending on the switched surface elements.

For summer ambient conditions, design point S1 and a constant heat radiation of sun are taken into account with a value of 850 W/m^2. In not preconditioning case, the air inside the passenger compartment in the middle is at 31 °C, after standing for 2 h. Then the NEDC starts and also the cooling of the passenger compartment is initiated. The comfort temperature of 20 °C is reached after 700 s, and maintained for the remaining journey time. In the case of the preconditioned ride, it is actively

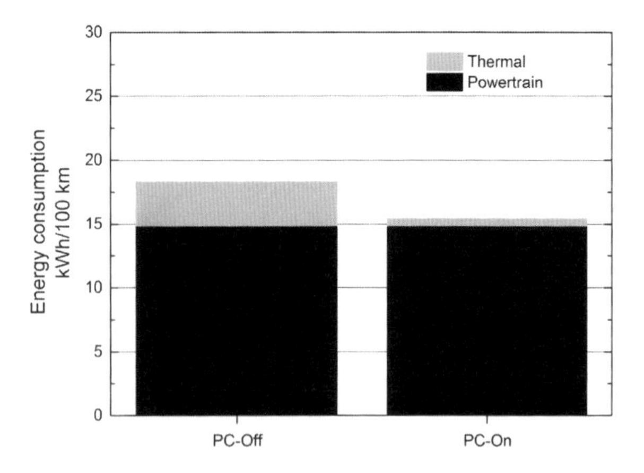

Fig. 9 Total electric energy of a not preconditioned and a preconditioned vehicle after driving the NEDC in relation to the energy consumption of the drive at 20 °C outdoor temperature in summer (S1)

cooled during the standing phase, when the interior temperature of the vehicle exceeds 25 °C. By booking request for the vehicle takes place during the 1 h-long preparatory phase, further cooling the passenger compartment on the comfort temperature of 20 °C takes place (see Fig. 9).

The results obtained by simulation for the overall determined energy consumption for the not preconditioned case splits to 21.2 % in the cooling energy and 78.8 % in pure driving power, including the minimum required auxiliary equipment. The low proportion of electricity consumed for cooling the passenger compartment is to be expected for the design point S1, since the outside temperature is 20 °C and the heat is supplied to the passenger compartment only by the sunlight. A preconditioning reduces energy costs for cooling the passenger compartment significantly by 82.5 %. The range of the vehicle in the not preconditioned case is 100.3 km. By preconditioning the design point S1 the range is increased to 121.7 km.

Under extreme environmental conditions, as defined for the design points W2 and S2, the average consumption of electric power for the heating or cooling of the passenger compartment increases during the passage of an NEDC. During the cooling circuit at the elevated ambient temperature 1.26 kWh/100 km more energy is needed, though the energy increase by heating at lower outdoor temperatures is negligible. By preconditioning the passenger compartment a saving at the design point W2 of 15.6 and 16.1 % at the design point S2 due to heating and cooling energy savings can still can be achieved.

As expected, simulations show that for the sequence of two NEDCs after the other, the average energy consumption of a not preconditioned passenger compartment sinks with fairly long distance. In this case, this is due to the heating or cooling energy. By preconditioning the passenger compartment before driving there are still significant savings in electrical energy at the design points W1 and S1. The simulation results for twice the NEDC show that the energy consumption for heating or cooling the vehicle interior by preconditioning, can be reduced at the design point W1 by 21 % and at the design point S1 by 11 %. The advantage of preconditioning decreases with increasing travel distance.

5 Conclusions

The simulation shows that the direct optimization of thermal management for the vehicle interior to the respective e-Carsharing use has a lot of potential. Due to the electric drive waste heat is hardly available, and if so on a much lower temperature level than in internal combustion engines. Also, existing heating/cooling concepts of vehicle interiors are highly influenced by the internal combustion engine. The simulation models have targeted some potential concepts of E-mobility, and in particular the use of Carsharing. A particular feature has been added to the pre-conditioning of the vehicle during the electrical charging phase in the parking station. There, the vehicle is continuously heated to a bookable value. The required start of final heating or final cooling, depending on the prevailing weather, is adjusted in a customized fashion according to the Carsharing reservation system. Thus, energy is saved by a volume of more than 21 % in winter whilst driving through preheating or precooling at a—for example—solar-powered Carsharing station (maximum operating range improvement by 21.7 % in the summer NEDC load profile). Applying this thermal preconditioning of the vehicle, a reduction of the energy storage capacity is possible. This represents a considerable convenience and also results in safety benefits compared with previous system solutions for personal mobility. In addition to the bookable internal temperature, interior smells and music packages can be individually preordered. For the future, a test of concept ideas would be desirable in a small series by vehicle data analysis and user surveys.

References

Ackermann, J. (2011). *Klimatisierung von Elektrofahrzeugen*, paper at the 3. Deutscher Elektro-Mobil Kongress, IAV GmbH, 6/2011.

Ackermann, J., Brinkkötter, C., & Priesel, M. (2013), *Neue Ansätze zur energieeffizienten Klimatisierung von Elektrofahrzeugen,* ATZ 6/2013 (pp. 481).

AMS. (2011). *Elektroauto-Reichweiten-Vergleich*, Auto Motor Sport, Heft 01/2011 (pp. 142–147).

Eckstein, L., Göbbels, R., & Wohlecker, R. (2011). *Benchmarking des Elektrofahrzeugs Mitsubishi i-MiEV, ATZ* 12/2011 (pp. 964–970).

ISI. (2011). *Integration von Elektrofahrzeugen in Carsharing-Flotten. Simulation anhand realer Fahrprofile*, 5/2011. Karlsruhe: Frauenhofer ISI.

Wetterstation Bremen Nord (2014). Retrieved from http://www.wetterstation-bremen-nord.de/index.php?inhalt_mitte=content/solar.inc.php.

Towards the Integration of Electric Vehicles into the Smart Grid

Ghanim Putrus, Gill Lacey and Edward Bentley

Abstract Electric Vehicles (EVs) have high energy capacity and their anticipated mass deployment can significantly increase the electrical demand on the grid during charging. Simulation results suggest that for every 10 % increase in households operating 3 kW EV chargers in an uncontrolled way, there is a potential increase of peak demand by up to 18 %. Given the limited spare capacity of most existing distribution networks, it is expected that large-scale charging of EVs will lead to potential problems with regard to network capacity and control. This paper presents analysis of these problems and investigates potential means by which the particular features of EV batteries may be used to enable large-scale introduction of EVs without the need for wholesale upgrading of power grids. Smart charging, using a combination of controlled EV charging (G2V) and Vehicle to Grid (V2G), can significantly help. The results presented demonstrate the benefits of smart charging for the grid and consider the impact of grid support on the EV battery lifetime. Various factors that affect capacity degradation of Lithium ion battery (used to power EVs) are analysed and the impacts of G2V and V2G operation on battery capacity loss and lifetime are evaluated. Laboratory test results are provided to quantify the effects of the various degradation factors, and it is shown how these may be ameliorated to allow economic network support using EV batteries without incurring excessive battery degradation in the process.

Keywords Electric vehicle · Smart grid · Li ion battery · Grid constraints · Grid to vehicle · Vehicle to grid · Smart charging

G. Putrus (✉) · G. Lacey · E. Bentley
Faculty of Engineering and Environment, Ellison Building, Northumbria University, City Campus, Newcastle upon Tyne, NE1 8ST, UK
e-mail: ghanim.putrus@northumbria.ac.uk

G. Lacey
e-mail: gillian.lacey@northumbria.ac.uk

E. Bentley
e-mail: edward.bentley@northumbria.ac.uk

W. Leal Filho and R. Kotter (eds.), *E-Mobility in Europe*,
Green Energy and Technology, DOI 10.1007/978-3-319-13194-8_19

1 Introduction

Increasing concern over the effects of climate change resulting from increasing global demand for energy and the persistent reliance on fossil fuels has led world leaders to set a target of a 50 % reduction of greenhouse gas emission by 2050 (the UK has an even more ambitious target of 80 %). In the UK, the contribution to CO_2 emissions from the surface transport sector is some 21 % of the total, leading to recognition of the need to electrify the transport sector to allow the UK to meet its 2050 emission targets. To encourage the uptake of EV and to allow for the expected increase in EV numbers, several countries have put ambitious plans to build the charging infrastructure for EVs (Office for Low Emission Vehicles 2011).

A range of EV is already on the market, chiefly comprising Plug-in Hybrid Electric Vehicles (PHEV) and battery powered EVs; the latter are usually powered by Lithium ion batteries with a capacity of a few tens of kWh (Kampmann et al. 2010). At present the market for EVs is limited, in view of their high price and limited range, but the market is expected to grow with anticipated rises in the price of petrol and advances in battery technology which will result in EV absolute and relative total cost of ownership reduction that will make the EV option more attractive.

Research suggests that uncontrolled charging of EVs can cause problems for the electric power grid due to the associated heavy electrical demand during charging (Putrus et al. 2009). As to whether a national power system is able to support large numbers of EVs, Taylor et al. (2009) found that if 90 % of Australia's peak annual generating capacity is available during off-peak periods, there would be enough energy available within the system to provide charging for EVs to make all existing urban passenger vehicle trips. The impact of the energy requirements of an increased number of EVs on the UK national power grid has been evaluated by a study which concluded that the grid capacity should be adequate for up to 10 % market penetration of EVs (Harris 2009). However, while the supply–demand matching for a region as a whole might be adequate to allow the use of sufficient numbers of charging points to support the EVs, there may be an impact on specific parts of the distribution system, particularly at the Low Voltage (LV) level. Local distribution substations and feeders for different areas may not have enough capacity to handle the increased load created by EV charging.

The impact of EV charging on the grid can be minimised by controlled charging and EVs can even be used to support the grid if their charging schedule is managed appropriately in a concept known as "Grid to Vehicle" (G2V) (Putrus et al. 2009; Jiang et al. 2014). Further, once the transport sector becomes largely electrified, it will be possible to use the energy storage capability of the EVs to mitigate problems arising or anticipated within the national power grid, as well as to provide storage to optimise the use of renewable energy sources (RES). EV batteries have considerable energy storage capacity and controlled charging can allow a schedule whereby they can be charged at a time when the grid has surplus capacity and discharged when the grid has a shortfall in capacity in order to meet peak demands and provide a storage facility for supply/demand matching. In addition, EV batteries can also be used to effectively balance the network frequency, 'shave' peak demand and provide emergency power in case of generation failure (V2G). Examination of this potential forms the subject of this paper.

As to when EVs would be connected to the grid (and thus available for charging and/or V2G), Babrowski et al. (2014) found that vehicle availability at charging facilities in Europe during the day for all countries is at least 24 %. With the additional possibility to charge at work, at least 45 % are constantly available. Results from EV trials (Bates and Leibling 2012) show that vehicles are parked for over 95 % of the time (23 h each day), so there is ample opportunity for them to be plugged in, charged and/or used to support the grid.

The study described in this paper is divided into three parts. The first (Sect. 2) presents the potential impact of EVs on the grid, using simulation results to support the analysis. The second part (Sects. 3–5) describes the means of alleviating the problems arising and presents potential opportunities for using the EV to support the grid supply. The third part (Sect. 6) describes the implication of the latter (EV to support the grid) on the EV battery capacity and lifetime, and is based in part on experimental tests on batteries. Section 6 defines the factors affecting battery degradation and introduces the various degradation mechanisms affecting EV batteries, so that these may be guarded against, allowing the minimum level of degradation to occur whilst the batteries are used to support the grid. Minimising battery degradation will maximise EV battery useful life, allowing the economics of EV operation to be made as favourable as possible. The knowledge of how to minimise battery degradation will allow maximum use to be made of the techniques suggested to maximise EV adoption given the constraints set by the grid. The economic benefits accruing from EV operation in accordance with the findings of this work are also discussed. In this way two of the most important factors militating against large-scale EV introduction, battery degradation and grid capacity, are addressed.

2 Impact of EVs on Existing Power Grid

EVs form a concentrated heavy load on the grid when compared to normal domestic power demands. EVs have high energy capacity and their anticipated mass deployment may lead to uncontrolled loading and a potential increase in peak electrical demand. Serious problems may be created for network operators from heavy charging demand to be met in certain times during the day, uncontrolled 'mobile' loads and seasonal 'migrations' of demand for EV charging.

As explained in Sect. 1, it is likely that the available national generation and grid capacity will be enough to meet the energy requirements of EVs for modest EVs penetration levels. Also, while the national aggregate capacity might be adequate, there are likely to be problems on specific parts of the grid, where local distribution substations and feeders may become overloaded by the increased load created by

EV charging. The following concerns regarding EV charging have been identified (Putrus et al. 2009).

- Uncontrolled loading due to increased deployment of EVs and potential increase in peak demand and overload of substations and feeders.
- Change in voltage profiles and violation of statutory limits.
- Phase imbalance (specific to single phase chargers).
- Reverse power flow (if V2G is adopted).

At the same time, mass deployment of EVs will create a very large energy storage capacity, which when considered as part of a smart grid can provide a valuable support to the grid. In a smart grid, the user will have the opportunity to plug in and charge the battery at will or when the price is right (to allow the possibility of arbitrage, buying power when it is cheap, such as in the middle of the night and reselling at times of peak demand), thus providing energy storage for supply/demand matching. In addition, there will be the need to allow EV operators to earn money by providing ancillary services and network support, e.g. voltage and frequency control using the energy stored in the EV batteries. It will often be possible to charge EVs from available micro-generation such as domestic Photovoltaic (PV) and Combined Heat and Power (CHP), thus charging from renewable energy and leading to efficiency savings due to reduced transmission losses. As a result there will be a need for smart grid interface controllers.

This paper presents the means by which the particular features of EV batteries may be used to overcome the difficulties inherent in the mass deployment of EVs, to enable large-scale introduction of EVs without the need for wholesale upgrading of power distribution systems.

2.1 Simulation Results

A typical LV distribution system is simulated using an Excel-based modelling tool that allows evaluation of the network performance for different operational scenarios in the presence of low carbon technologies, such as EVs and micro-generation (Lacey et al. 2013). The layout of the system is shown in Fig. 1. Typical daily load profiles, shown in Fig. 2, for the UK consumers for both summer and winter seasons were used (Barbier et al. 2007).

As mentioned earlier, EV charging represents a heavy load on the grid and therefore tends to cause overloading of the transformer and feeders as well as high voltage (HV) drops across the distribution system. To analyse this, the distribution system shown in Fig. 1 is considered with maximum (winter) loading conditions and domestic 3 kW EV charging for ~8 h (assuming 24 kWh battery capacity).

The problem facing distribution network operators with the introduction of EVs is that uncontrolled charging will tend to result in people plugging in their EVs when they return home from work at about 6.00 pm, when there is already a peak in demand for power. The problem will become worse as the uptake of EVs increases,

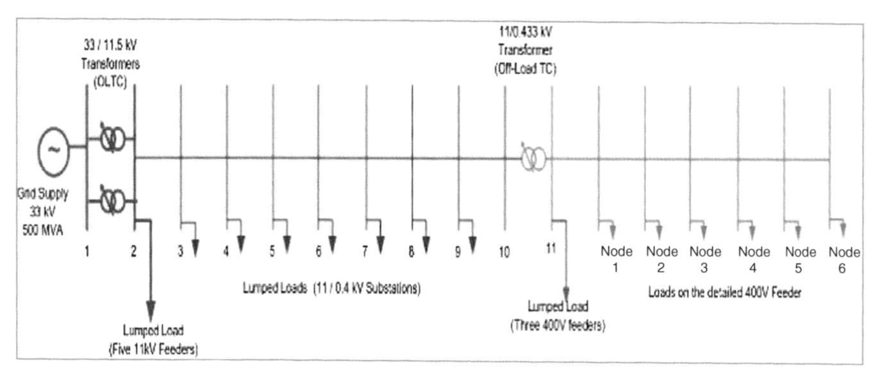

Fig. 1 Distribution network model

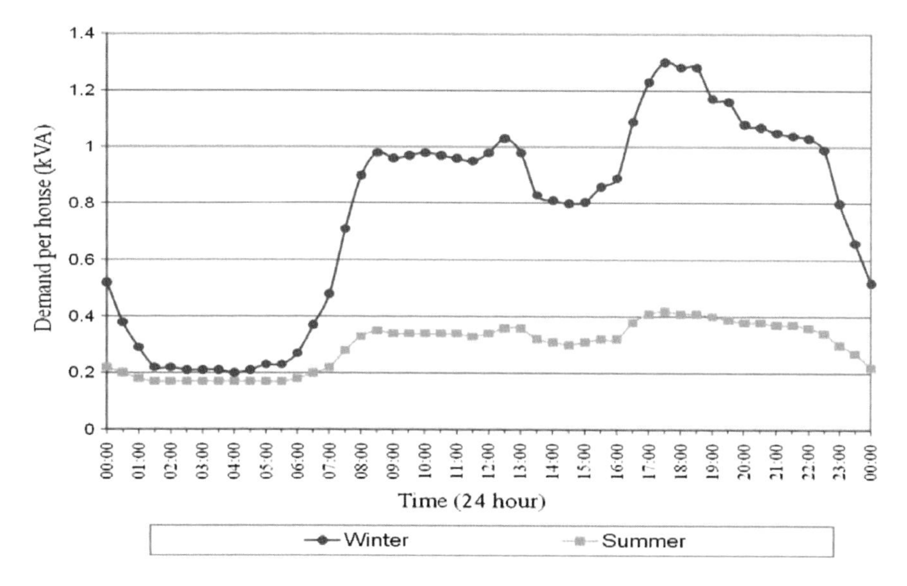

Fig. 2 Typical daily load profile for a domestic load based on ADMD referenced to a nominal 100 consumers and measured at a distribution substation on an outgoing feeder

as shown in Fig. 3 (zero to 30 % of houses having an EV). Peak demand rises by some 18 % for every 10 % increase in houses with an EV. Low voltage substations and feeders do not have a very large degree of spare capacity due to economic constraints, and a problem will be seen to arise at some degree of EV adoption.

The increased loading may also cause the voltage supplied to customers, particularly at the far end of the LV feeder, to fall below the statutory limit. Figure 4 shows the voltage at the far end of the LV feeder (Node 6 in Fig. 1) with different levels of households having a 3 kW EV charger.

Figure 5 shows the voltage profile across the length of the LV feeder (Nodes 1–6 in Fig. 1) for three cases: 'no EV charging' situation and then 20 and 30 % of

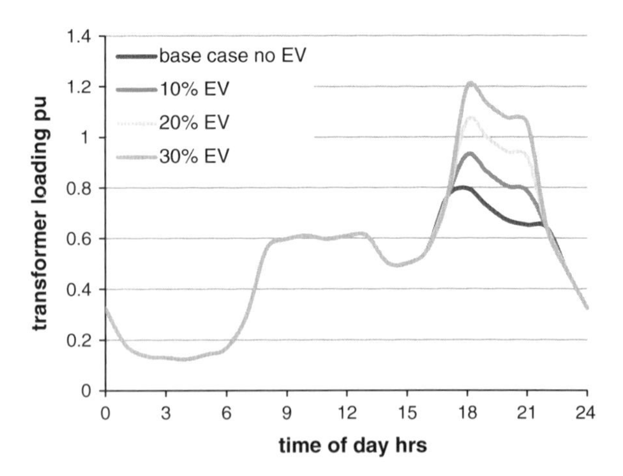

Fig. 3 Transformer loading for EVs with unscheduled 3 kW (uncontrolled) charging

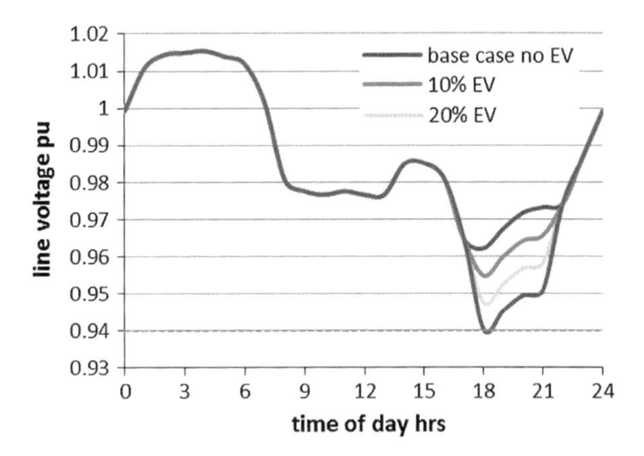

Fig. 4 End of feeder voltage with varying uptake of 3 kW EV chargers

households operating a 3 kW charger at the same time. As can be noted, charging of EVs creates extra loading on the feeders and therefore extra voltage drop. At 20 % level, the system is able to maintain the load voltage within the statutory minimum limit of −6 %, by the operation of the on-load tap changer (OLTC). However, with a 30 % level, the tap changer reaches its limit and the voltages at Node 6 approach the statutory limit.

3 Controlled EV Charging

3.1 Grid to Vehicle (G2V)

Controlled charging, e.g. by using incentives for customers, will reduce daily variations and improve load factor (match network capacity). If successful

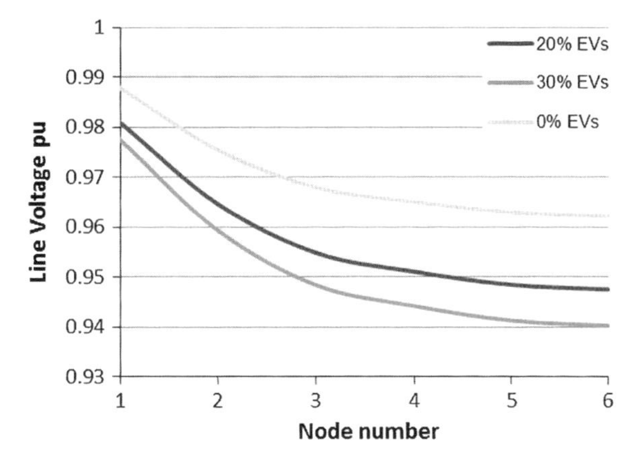

Fig. 5 Voltage profiles for different EV 3 kW charger penetration Levels

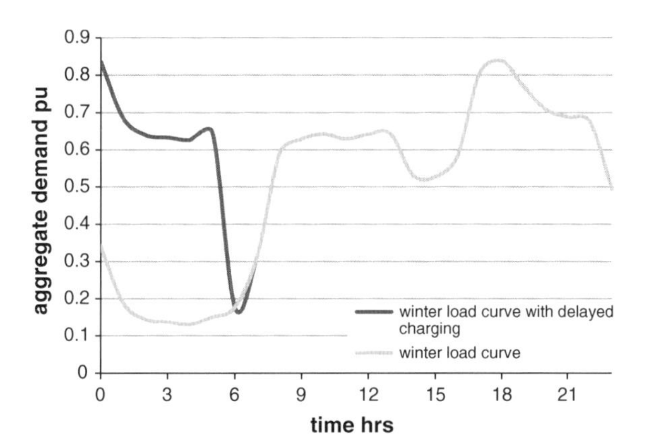

Fig. 6 Winter load curve, showing load levelling effect of delayed 3 kW EV chargers when 30 % of houses have EVs

incentives are introduced, the attractive possibility of levelling the demand load curve over 24 h presents itself. In Fig. 6, it is assumed that 30 % of houses have EVs and these are charged at the optimum time for the grid (after midnight), by staggering/phasing EV charging times. EV charging will then occur at the times when the underlying demand for power is low; so the increase in demand does not exceed the peak level.

Figure 7 shows the effects on the voltage at the far end of the feeder (Node 6 in Fig. 1) caused by delaying charging. As can be seen, the under-voltage caused by uncontrolled charging is eliminated when controlled charging is deployed. This demonstrates the inability of existing distribution systems to support high levels of domestic EV charging whilst maintaining the legal minimum load voltage, unless some form of demand management is adopted.

Barbier et al. (2007) and Sulligoi and Chiandone (2012) reported that with significant renewable energy generation connected to the distribution network, the distribution system may experience a voltage rise, particularly during low demand. G2V may be designed to complement the generation profiles of renewable sources

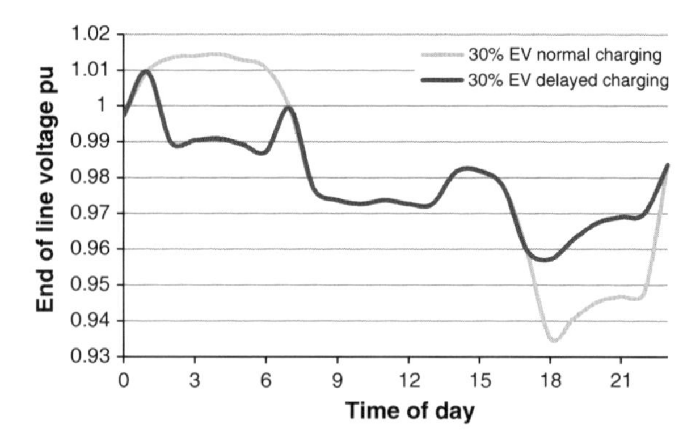

Fig. 7 Showing end of line voltage for 30 % 3 kW EVs with and without delayed charging

and therefore is an ideal approach to charge EVs from renewable energy as well as relieve the grid from extra burden and losses.

3.2 Vehicle to Grid (V2G)

EV batteries have considerable energy storage capacity and controlled charging can allow a schedule whereby they can be charged at a time when the grid has surplus capacity (e.g. surplus renewable energy) and discharged when the grid has a shortfall in capacity (or renewable energy). In addition, EV batteries can also be used to provide supply/demand matching and effectively balance the network frequency and provide emergency power in case of generation failure. In V2G operation, where large numbers of EVs are aggregated and the composite energy stored is able to be used for grid support, the system allows provision of a potentially large-scale power reserve.

The V2G process may be used intelligently to ameliorate EV charging problems. Figure 8 demonstrates how the overloading of transformer caused by the scenario of charging with 30 % of households having EVs at 3 kW may be removed by arranging for the EVs to discharge their stored energy when the system is highly loaded at 6.00 pm, and recharging at a convenient time.

Another problem experienced with LV distribution systems when significant distributed generation (DG) is connected to the system is the potential for over-voltage, particularly when the DG is connected at the end from the supply point (Sulligoi and Chiandone 2012). Figure 9 shows the voltage profile at the far end of the LV feeder in Fig. 1, in the presence of renewable energy generation at a level based on the targets for 2050 (DECC 2010). As can be seen, without the use of controlled charging, the voltage will rise well above the statutory limit of 10 % above the nominal line voltage. This problem may be dealt with by using a suitably timed charging and V2G, as shown in Fig. 9, assuming 40 % of the households have 3 kW EV chargers.

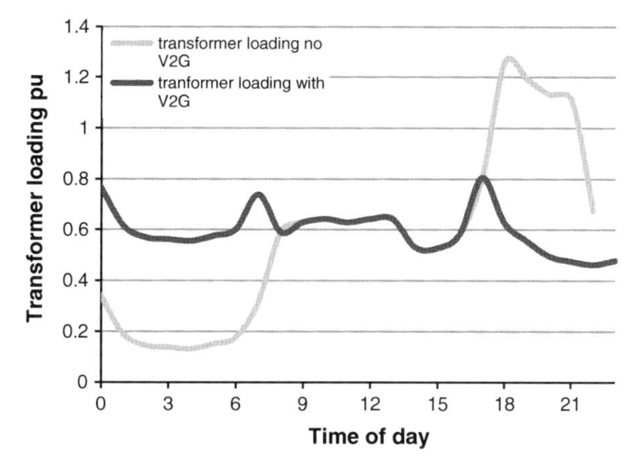

Fig. 8 Transformer overload caused by EV charging, and its mitigation using V2G

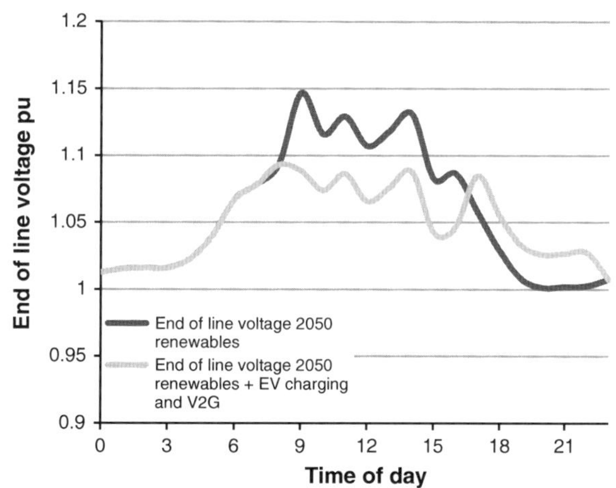

Fig. 9 End of line overvoltage caused by renewables, and its removal using EV charging and V2G

3.3 Smart Charging

Traditional battery chargers operate at nearly constant power for most of the charging time; input power tapering off as the battery is being charged. High-power EV chargers initially charge at constant current and this is then changed to constant voltage before reaching full charge. These chargers do not provide the optimum conditions for protecting the battery and maximising its life span. As will be described in Sect. 6, high charging current may damage EV battery, particularly at low (below 0 °C) and high (above 40 °C) temperatures, and that batteries have their remaining life prolonged by gentle low current charging regimes (Peterson et al. 2010). This shows the need for 'smart chargers' where the charger output (charging rate and time) varies with battery conditions, grid state (available power) and EV user requirements, as shown in Fig. 10.

A smart charger is required to determine the optimal charging current rate by considering the network condition, the battery's state of health (SOH) and state of

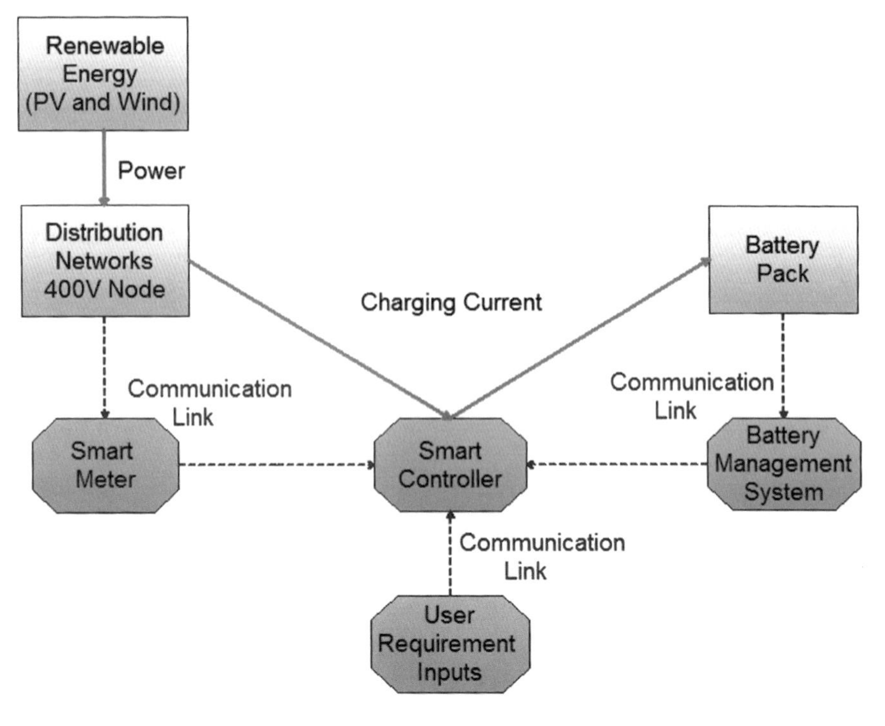

Fig. 10 The concept of EV smart charge controller (Jiang et al. 2014)

charge (SOC) (based on information from the battery management system), and EV user requirements (journey length and charging waiting time) (Jiang et al. 2014). The controller may also respond to direct signals from renewable energy generation or indirect signals (e.g. weather conditions). In this way, smart charging will meet user requirements, maintain battery SOH, support the grid and optimise the use of renewable energy. The rules for optimum smart charging may include the following, given by Jiang et al. (2014):

i. Charge the battery to user specifications (to ensure EV is charged and available for next journey), as long as there are no restrictions from the grid or the battery SOH.
ii. Monitor the grid condition (voltage and thermal limits) and adjust the battery charging current (if needed), in proportion to the deviation from the nominal (rated) limits.
iii. Monitor the battery SOH and adjust the battery charging current (if needed) in order to avoid negative impacts on the battery cycle life.
iv. The priority of each input can be adjusted, depending on the design requirements.

A smart charger which can allow two-way power flow will be needed to provide for V2G and V2H services. Smart charging involves incentivising the EV users to adopt a charge regime which optimises battery health and avoids charging during times of high grid demand whilst still allowing freedom of use for driving.

4 Smart Distribution Grid

A possible configuration of a future smart distribution system is shown in Fig. 11, in which bidirectional communication links and the flow of information are essential. All EV chargers are connected to the household-level smart charge/V2G controller, which is linked to the household smart metre. The metre calculates the net household power supply (e.g. from PV) and demand and sends data continuously to the Medium Voltage (MV) aggregator. The data bit rate can be very low, as it represents a single number sampled perhaps once a minute. In turn, the data received by the local smart metre from the MV aggregator will consist of a signal to control EV charging power demands. If the area served by the MV aggregator is as a whole able to supply all demands, but one particular line is reaching its limits, EV charging on that line alone will be curtailed. If the whole area controlled by the MV aggregator is short of power, all lines will experience a curtailment. The system can be developed to bring on stream V2G power from particular areas of the MV aggregator's control zone where it is needed.

This system allows EV users to charge at differing rates depending for instance on the SOH of the battery and ambient temperature, to maximise EV battery life. In the event that there is a sudden problem with the power availability in the MV system, the control signal can effectively shut down demand for EV charging power.

The MV aggregator under this approach would send data to the HV aggregator, again perhaps minute by minute, advising of the total net power requirement of the MV area. Signals from the HV aggregator will allow the MV aggregator to adjust

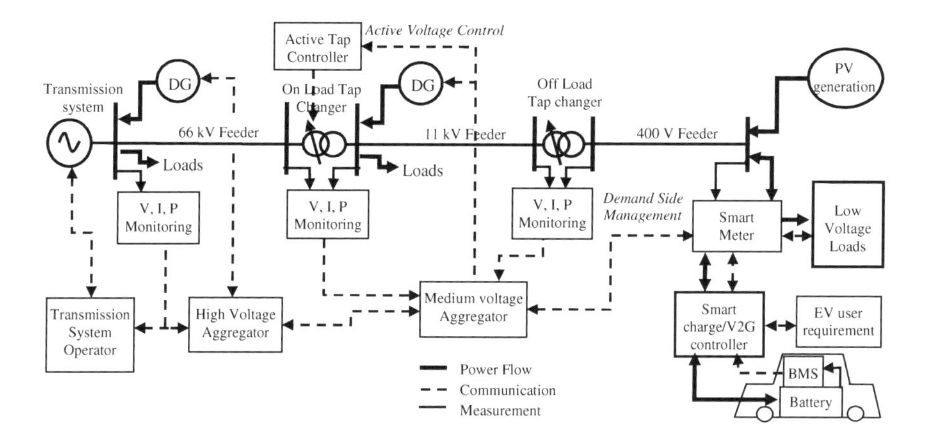

Fig. 11 A smart grid showing medium and low voltage elements of a distribution system incorporating smart charging (adapted from Putrus et al. 2013) (BMS: Battery Management System; DG: Distributed Generation)

supply and demand in their own area, so that overall system balance is achieved. Load flow analysis would be carried out continuously by both HV and MV aggregators to ensure that the areas for which they are responsible operate within the relevant network limits such as transformer and cable loading, and voltage limits.

5 Vehicle to Home (V2H)

Vehicle to Home (V2H) is a small-scale operation of V2G, in which a single EV battery is used to supply power for a single household. The use of V2H is intended to provide power to the home at times of supply failure or during peak demand. This energy may be stored in the EV from the grid or from a local micro-generation during a different time of day. Appliance usage by a single household is not subject to averaging, so the power demand from the single household might resemble that shown in Fig. 12. The average power demand portrayed is moderate, but peaks of 10 kW appear in the load profile.

Potentially, an EV bidirectional charger rated at 7 kW could supply about 70 % of the peak demand, averaging out the load profile so that it would be more readily supplied by, for instance, a PV installation owned by the household. The EV battery is designed to produce peak power outputs greater than 7 kW for short periods, and will not suffer undue damage by being used in this way. On this basis, the grid would merely have to supply the average domestic load rather than the instantaneous demand, rendering the job of the distribution network operator easier. In addition the transmission and distribution losses would disappear, making this option the most efficient as well as the most 'green'.

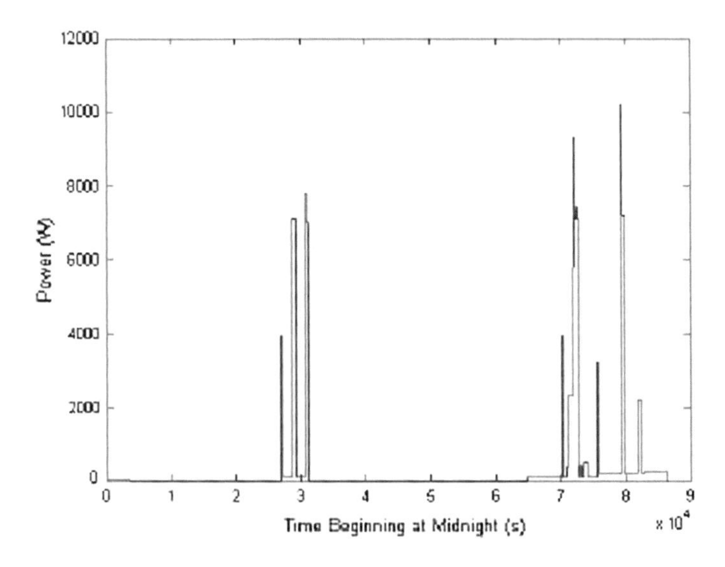

Fig. 12 Instantaneous power demand for a single house (Haines et al. 2009)

The total energy supplied by the battery in V2H will need to be replenished by the grid or by micro-generation to ensure the battery is still charged for driving. The total daily energy demand of a typical house in the UK on average is around 7 kWh in summer and 18 kWh in winter. In the UK, the output of a 3 kW PV installation can provide up to 20 kWh in summer and only about 2 kWh in winter, rendering the household potentially grid independent in the summer.

A further great advantage of V2H is that it can display a smoothing property as far as the grid is concerned. Vehicles are usually parked, and hence available for V2G for around 96 % of the time (Kempton and Tomic 2005). Therefore, the EV battery can provide a good service to the grid or home as well as being able to maintain the requisite average SOC to enable use as a vehicle.

6 EV Battery Degradation Caused by Smart Grid Support

Using EV batteries to balance supply and demand through V2G will result in extra charge transfer through the battery (cycling) and therefore the impact of this on the battery SOH needs to be evaluated to ensure that the effects on the battery (the most expensive part of the EV) are minimal, or even zero. To do this, it is important to define the main parameters that affect battery degradation and then use these factors to model the impact of V2G. As described in Sect. 3.3, by using smart charging of EVs, the battery SOH can be taken into consideration whilst providing support to the grid. In this way, the charger ensures that there will be minimum or no impact on the net charge transfer and therefore the battery SOH. However, there are several factors that affect battery SOH which need be considered, and these are described in detail in this section.

In the following, the process of EV battery degradation and the main factors that contribute to this are presented. The aim is to evaluate whether the use of EV battery to provide grid support (which it is capable of as shown above) will have any impact on the battery SOH; that is, whether the cycling patterns required for V2G or V2H will degrade the battery more quickly than standard (uncontrolled) charging only.

6.1 Methodology for Defining Battery Degradation

The degradation factors are identified and their weights are determined from available literature and from experimental tests conducted by the authors. Test results and research have enabled the life of a lithium ion battery to be predicted with reasonable accuracy using mathematical modelling techniques based on derived coefficients for each of the parameters that affect battery degradation. It should be noted that the results only apply for the type of battery being tested, as the model is derived from experimental tests on a specific battery type. Different batteries, even with the same chemistry, will follow the same trends but the value of each modelling coefficient may be different.

For EV applications, a battery is considered to be at 'end of life' when the fully charged capacity is 80 % of the new value. This is normally described in terms of a reduction in the battery SOH. The SOH at any time represents the percentage of a capacity possessed by a battery at that time when fully charged to the fully charged capacity when the battery is new. So, in this case the SOH will have fallen to 80 %.

6.2 Battery Capacity Loss (Lifetime)

Battery life time is defined by the permanent capacity loss of energy storing capacity, which may be divided into two types: calendar loss and cycle loss. The former is the capacity loss due to the passage of time, whether or not the battery has been in use (charged and discharged). For Li ion batteries, the calendar loss is dependent on temperature and SOC. Degradation tends to slow down when the battery is not in use and results show that a battery maintains its energy storage capacity if stored in a temperature around 5 °C and the SOC is kept low. The battery SOC affects the electrical stress between the electrodes and consequently the battery calendar life (Spotnitz 2003; Lunz et al. 2011a, b).

The use of EV battery to support the grid will result in more cycling (charge/discharge) of the battery. Consequently, concerns have been raised regarding the damage caused to the battery due to this operation and whether the gains for supporting the grid justify the loss in battery life.

The cycle life of a Li ion battery is defined in terms of the capacity loss per cycle due to charge entering and leaving the cell during cycling. The capacity loss is caused by the charge transfer between the electrodes and therefore is dependent on the way the battery is being used during the charging and discharging cycles. Four impact factors affecting battery ageing in terms of SOH have been identified. These factors are the operating temperature of the battery, the average SOC, the Depth of Discharge (DOD) in each cycle and the charging/discharging current flowing into and out of the battery. It is worth mentioning that these factors are interlinked and are determined by the chemistry of the battery as well as the reaction (both chemical and physical) during the charging and discharging process.

6.3 Factors Affecting Battery Degradation

There are four variables at play here; the temperature, the charging rate, the average SOC and the DOD. An attempt is made to analyse them and then to identify the 'sweet spot' where optimum battery SOH allows maximum support for the grid.

i. *Operating Temperature*

Research shows that cycling Li ion batteries at differing temperatures causes varying levels of irreversible battery capacity loss (Kaneko et al. 2013). The effect of temperature on cycle life of Li ion cells is shown in Fig. 13. As shown, the cycle life progressively reduces below 0 °C and above 50 °C. Cycling cells outside a specified range accelerate the capacity loss in the cells and when the temperature approaches 70 °C, a thermal runaway becomes likely. The battery thermal management system must be designed to keep the cell operating within the specified range (usually around room temperature) at all times to avoid premature wear out of the cells. It is worth noting that the cycle life quoted in manufacturers' data sheets is usually based on operation at room temperature ($\sim$20 °C), which may not be realistic for EV applications.

ii. *Charging Rate*

Dubarry et al. (2011) and Ning (2004) showed that battery cycle loss accelerates with charging current rate. Using experimental data and modelling based on electrochemical behaviour, Dubarry et al. (2011) showed that for the first 500 cycles or so the capacity fade is linear. This result is backed up by the results presented in Ning (2004) which also shows an experimentally linear rate with current density, as shown in Fig. 14.

The values are empirical but a correlation can be found using a base of 20 % at 1C rate after 500 cycles; this gives a degradation rate for the 23 kW of 0.0004. Scaling from Fig. 14 gives the battery loss values in Table 1 for different charging rates based on commercial charging stations. To verify these results several more tests were undertaken at different charge rates and all the results obtained appear to show that the lower the charging rate the better the SOH.

iii. *State of Charge*

The SOC of an EV battery is the battery capacity at any time expressed as a percentage of maximum capacity. It is usually determined by integration of the

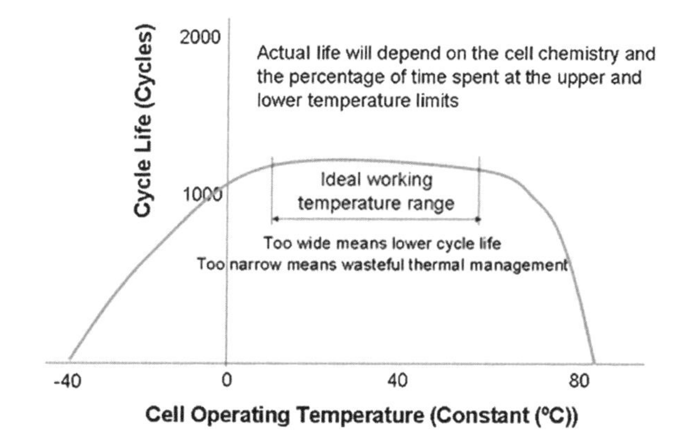

Fig. 13 Variation of battery cycle life (irreversible capacity loss) with temperature of cycling (Electropaedia et al. 2014)

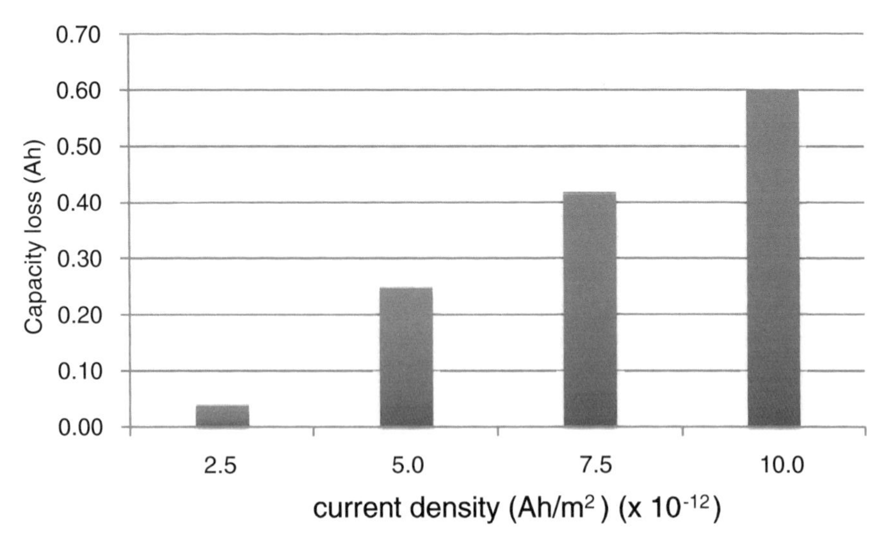

Fig. 14 Plot of capacity loss after 500 cycles with current density (i.e. charging rate) (adapted from Ning 2004)

Table 1 Derived capacity loss due to charging rate

Equivalent kW charge rate	Loss after 500 cycles	Loss per cycle	% loss per cycle
3	0.0125	0.000025	0.0025
7	0.03	0.00006	0.006
23	0.18	0.00036	0.036
50	0.43	0.00086	0.086

charging and discharging current. Charging at high SOC causes more damage to the battery than at lower values (Vetter et al. 2005). This is because the risk of stress cracking of the electrodes due to volume change and chemical breakdown of the battery's components is more likely at high SOC.

The average SOC for an EV battery depends on the SOC before and after charging and also before and after driving. It also depends on the time the battery spends in discharged state and that spent in charged state. In this research, the average SOC is calculated using the time of charging and the time when the car is charged ready for driving. The assumption is that the car is charged up ready for the next trip when it is connected, unless delayed charging or V2G is specified. The SOC whilst connected but not charging is then used to find the average SOC.

Figure 15 shows the results of testing $LiPF_6$ battery cells at different SOCs but at the same DOD, temperature and charge/discharge current rate. The results show that battery cycle number (capacity lifetime) reduces if the battery is cycled at high SOC. These results, and others obtained by the authors, demonstrate that battery life is prolonged by keeping the average SOC as low as possible. That is, using battery

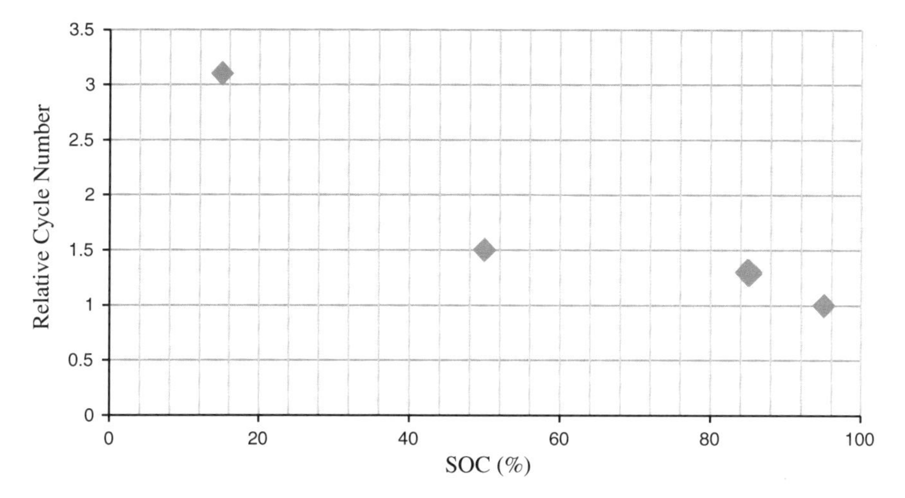

Fig. 15 Relative cycle number (capacity lifetime) for cells cycled at different SOC (adapted from Lunz et al. 2011a, b)

charging only when essential for the next trip, only charging what is required for the next trip and charging before driving, not immediately after.

iv. *Depth of Discharge*

Battery life is found by Peterson et al. (2010) and www.cars21.com (2010) to depend on the total charge throughput (Ah). Cyclic ageing is mainly due to mechanical stresses because of the volume change as the active material enters and leaves the electrode and is therefore dependent on the amount of charge transferred during charging and discharging. This can be isolated using the change in SOC, assuming a periodic charge/discharge cycle. Ignoring other ageing effects, the total energy throughput is fixed so that one cycle of 100 % change in SOC is roughly equivalent to 10 cycles at 10 % change in SOC and 100 cycles at 1 % change in SOC. The DOD is then the 100 % minus the minimum SOC in the cycle. Results of capacity loss with varying partial cycles, but grouped for average SOC, are shown in Fig. 16. The results back up what was stated earlier: that lower SOC means lower losses. The change in SOC (coloured bars) is not significant. Therefore, the DOD is not a factor, only the amount of charge transferred. If the DOD is defined as the change in SOC when the maximum is always 100 %, then the DOD is a factor insofar as it affects charge transfer.

In summary, experimental results show that the best temperature for battery cycling life is around 20 °C and that battery capacity loss increases with increasing current rate, SOC and number of charges transfer during charge and discharge. Thus the effect of each parameter in combination can be used to predict the battery degradation and useful life.

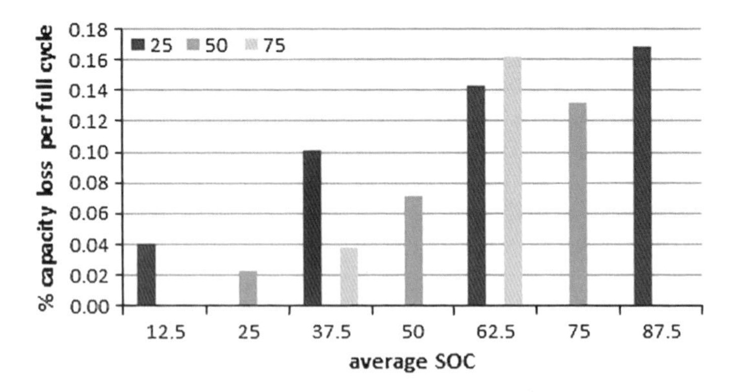

Fig. 16 Capacity loss for cells cycled at varying ΔSOC

6.4 The Effects of V2G and V2H Operation on Battery Lifetime

To evaluate the effects of cycling on the EV battery life time, experiments were conducted and results obtained in order to make a comparison between the effects of V2G and V2H (which would in themselves reduce battery life through the increased charge transferred) with the effects of uncontrolled charging (when the adverse effects of a higher average SOC would reduce battery life). The temperature and charge rate were kept constant.

Samples of Lithium Iron Phosphate ($LiFePO_4$) cells, which use similar chemistry to typical EVs currently on the market, were cycled at a rate equivalent to 3 and 7 kW, to represent the alternative rates for EV home charging. For each case, two cells were cycled with uncontrolled charging; where charging starts as soon as the EV would return home and is plugged in at the end of the day's work at 6.00 pm. Two other cells were controlled to charge later at night (to represent G2V) and two other cells were allowed to discharge to 10 % SOC during the evening (V2G) and then charged up at night.

Three possible scenarios were then analysed. The first in which the EV was charged (starting at 6.00 pm) to 70 % SOC and kept in this condition until it was required for driving at 8.00 am (uncontrolled charging). The second scenario is controlled (smart) charging, where the EV battery is kept at 30 % SOC until a later time when the charger brings the SOC to 70 % by 8.00 am (G2V). The other pathway involved carrying out V2G by discharging the EV battery down to 10 % SOC and then keeping the battery at this low SOC until the latest time during which the charger could bring the EV battery to 70 % SOC by 8.00 am. In all cases, the initial SOC at plug in is assumed to be 30 %.

The results obtained from these tests are shown in Fig. 17 (the results for G2V are not shown, as the capacity loss was negligible). As can be seen, under these conditions, the capacity loss is lower when using G2V and V2G, which is due to the

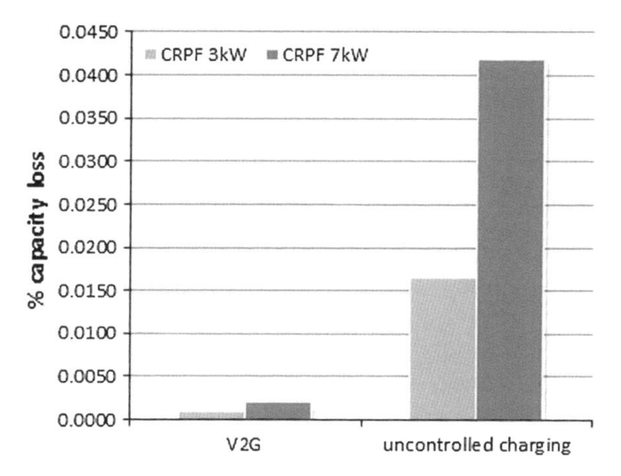

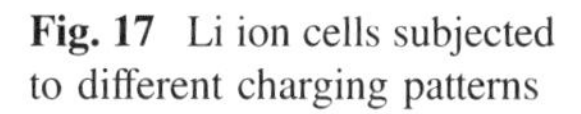
Fig. 17 Li ion cells subjected to different charging patterns

lower average SOC and hence less electric stress between the electrodes of the battery, as explained earlier. Further tests and analysis revealed that the effect on capacity loss for each pathway varies with the initial SOC upon connection of the EV. As the initial SOC at plug in increases, the battery lifetime seems to increase under conditions of uncontrolled charging. This is because of the effects of the extra degradation caused by the increased average SOC are offset to some degree by the reduced charge transfer experienced compared to that under the V2G regime. For the latter, as the initial SOC at plug in is increased, the percentage loss of capacity per cycle increases (leading to a shorter lifetime) due to increased degradation caused by the increase in charge transfer.

7 Conclusions

The work described in this paper shows that the most important feature of the EV, from the point of view of grid connection, is that the actual charging time and rate may be arranged to fit in with other demands upon the local distribution network. Controlled charging can help minimise the possibility of transformer overload and feeder voltage drop. In addition, the storage capacity of the EV battery may be used to intelligently reinforce the grid, by controlling the charging time and current rate to balance demand/supply and support the grid (G2V). Further, the battery may be used to supply power at times of scarcity, and absorbing it when in surplus, bringing supply and demand for power into equilibrium (V2G). However, concerns have been raised about the effects V2G may have on battery lifetime. Available literature shows that battery cycling reduces battery life due to increased charge transfer, as does faster charging.

The results of the research presented in this paper show that smart charging prolongs battery life as compared to what might be termed the 'standard' approach of uncontrolled charging at home. A combination of delayed controlled charging (G2V) and V2G results in comparable battery life, with the opportunity to earn revenue from carrying out grid support. This is an added value to the benefits smart charging brings to the grid by permitting a higher level of EV adoption and increased amounts of renewable energy penetration without the need for re-engineering the existing power grid.

The cost of the battery is the greatest single replacement cost associated with ownership of an EV. For example, in June 2014, the replacement cost of a Nissan Leaf battery is given at about $6500+tax (Ottaway 2014), compared to purchase prices for a new car ranging from $29,000–$35,000 (Car Ranking and Advice website 2014). Receiving payment from the grid operator for using the EV battery to provide balancing services to the grid (G2V or V2G) could be an attractive option for an EV owner concerned about the high capital outlay.

Acknowledgments The research work described in this paper has been supported by the Interreg NSR e-mobility project funded by the European Commission through the Interreg IVb programme. The authors would also like to thank Northumbria University in Newcastle for providing the resources to conduct this work.

References

Babrowski, S., Heinrichs, H., Jochem, P., & Fichtner, W. (2014). Load shift potential of electric vehicles in Europe. *Journal of Power Sources, 255*, 283–293.

Barbier, C., Maloyd, A., & Putrus, G. (2007). Embedded Controller for LV Network with Distributed Generation, DTI project, Contract Number: K/EI/00334/00/Rep, UK, May 2007.

Bates, J., & Leibling, D. (2012). Spaced Out. *Perspectives on parking policy, Royal Automobile Club Foundation*. Retrieved October 14, 2014, from http://www.racfoundation.org/assets/rac_foundation/content/downloadables/spaced_out-bates_leibling-jul12.pdf.

Car Ranking and Advice website. (2014). Retrieved October 14, 2014, from http://usnews.rankingsandreviews.com/cars-trucks/Nissan_Leaf/.

Department of Energy and Climate Change. (2010). *2050 pathways analysis, URN 10D/764, London, TSO*. https://www.gov.uk/2050-pathways-analysis.

Dubarry, M., Truchot, C., Liaw, B. Y., Gering, K., Sazhin, S., Jamison, D., & Michelbacher, C. (2011). Evaluation of commercial lithium-ion cells based on composite positive electrode for plug-in hybrid electric vehicle applications; part ii degradation mechanism under 2 C cycle aging. *Journal of Power Sources, 196*(23), 10336–10343.

Electropaedia. *Battery and Energy Technologies*. Retrieved October 17, 2014, from http://www.mpoweruk.com/lithium_failures.htm.

Haines, G., McGordon, A., Jennings, P., & Butcher, N. (2009). *The Simulation of Vehicle-to-Home Systems—Using Electric Vehicle Battery Storage to Smooth Domestic Electricity Demand, paper presented at EVRE Monaco*. Retrieved March 2009, from http://www.google.co.uk/url?sa=t&rct=j&q=&esrc=s&frm=1&source=web&cd=1&ved=0CCEQFjAA&url=http%3A%2F%2Fwww.researchgate.net%2Fpublication%2F228696724_The_Simulation_of_Vehicle-to-Home_SystemsUsing_Electric_Vehicle_Battery_Storage_to_Smooth_Domestic_Electricity_-Demand%2Flinks%2F0046352aac54cec4bf000000&ei=8y5JVPCQN4OQ7AbYpID4BA&usg=AFQjCNEx42C7mBoNtgenbpNu39PUTsu0Bg.

Harris, A. (2009). Charge of the Electric Car. *Engineering and Technology, 4*(10), 52–53 (Stevenage: IET).

Jiang, T., Putrus, G. A., Gao, Z., Conti, M., McDonald, S., & Lacey, G. (2014). Development of a decentralized smart-grid charge controller for electric vehicles. *Electrical Power and Energy Systems, 61*, 355–370.

Kampman, B., Leguijt, C., Bennink, D., Wielders, L., Rijkee, X., Buck, A., & Braat, W. (2010). Green power for electric cars. *Development of policy recommendations to harvest the potential of electric vehicles, Delft, CE Delft, report no 4.037.1*. Retrieved April 9, 2014, from http://www.transportenvironment.org/sites/te/files/media/green-power-for-electric-cars-report-08-02-10.pdf.

Kaneko, G., Inoue, S., Taniguchi, K., Hirota, T., Kamiya, Y., Daisho, Y., & Inami, S. (2013). *"Analysis of Degradation Mechanism of Lithium Iron Phosphate Battery", paper to EVS27 conference, Barcelona, Spain*, (pp. 1–7. Retrieved November 2013, from http://ieeexplore.ieee.org/stamp/stamp.jsp?arnumber=6914847.

Kempton, W., & Tomic, J. (2005). Vehicle-to-grid power implementation: from stabilizing the grid to supporting large-scale renewable energy. *Journal of Power Sources, 144*, 280.

Lacey, G., Putrus, G., Bentley, E., Johnston, D., Walker, S., & Jiang, T. (2013). *"A Modelling Tool to Investigate the Effect of Electric Vehicle Charging on Low Voltage Networks", EVS27 Barcelona, Spain*. Retrieved November 17–20, 2013, from http://www.google.co.uk/url?sa=t&rct=j&q=&esrc=s&frm=1&source=web&cd=2&ved=0CCoQFjAB&url=http%3A%2F%2Fieeexplore.ieee.org%2Fiel7%2F6902670%2F6914705%2F06914861.pdf%3Farnumber%3D6914861&ei=LTBJVMeRHc-u7AaYm4HYCQ&usg=AFQjCNEP3vOPc7XUnu25bosSKaKwcrr78A.

Lunz, B., Gerschler, J. B., & Sauer, D. (2011a). Influence of plug-in hybrid electric vehicle charging strategies on charging and battery degradation costs. *Energy Policy, 46*, 511–519.

Lunz, B., Walz, H., & Sauer, D. (2011b). Optimizing V2G Charging Strategies Using Genetic Algorithms Under the Consideration of Battery Aging. *Proceedings of EEE Vehicle Power and Propulsion Conference, Chicago*, (pp. 1–7). Retrieved September 2011, http://www.google.co.uk/url?sa=t&rct=j&q=&esrc=s&frm=1&source=web&cd=1&ved=0CCYQFjAA&url=http%3A%2F%2Fieeexplore.ieee.org%2Fiel5%2F6030111%2F6042961%2F06043021.pdf%3Farnumber%3D6043021&ei=YDBJVM7zJ-Gd7gbu9oD4Dw&usg=AFQjCNE1MQIPQ1Ps-sxpqVRmR9d9yORy1Q.

Ning, G. (2004). Cycle life modelling of Lithium ion Batteries. *Journal of the Electrochemical Society, 151*(10), A1584–A1591.

Office for Low Emission Vehicles. (2011). *Making the Connection; The Plug-In Vehicle Infrastructure Strategy*, London: Department for Transport, UK, June 2011.

Ottaway, L. (2014). *Replacement battery cost bodes well for used LEAF market*. Retrieved October 17, 2014, from http://www.torquenews.com/2250/replacement-battery-cost-bodes-well-used-leaf-market.

Peterson, S. B., Apt, J., & Whitacre, J. F. (2010). Lithium-ion battery cell degradation resulting from realistic vehicle and vehicle-to-grid utilization. *Journal of Power Sources, 195*(8), 2385–2392.

Putrus, G.A., Suwanapingkarl, P., Johnston, D., Bentley, E.C., & Narayana, M. (2009). Impacts of Electric Vehicles on Power Distribution Networks, *IEEE Vehicle Power and Propulsion Conference, Michigan*, (pp. 827–831). Retrieved September 2009, from http://ieeexplore.ieee.org/xpls/abs_all.jsp?arnumber=5289760.

Putrus, G. A., Bentley, E., Binns, R., Jiang, T., & Johnston, D. (2013). Smart grids: energising the future. *International Journal of Environmental Studies, 70*(5), 691–701.

Spotnitz, R. (2003). Simulation of capacity fade in lithium-ion batteries. *Journal of Power Sources, 113*(1), 72–80.

Sulligoi, G., & Chiandone, M. (2012). *Voltage Rise Mitigation in Distribution Networks using Generators Automatic Reactive Power Controls, paper to the Power and Energy Society General Meeting, IEEE July 2012 San Diego, CA*. Retrieved July 2012, from http://ieeexplore.ieee.org/stamp/stamp.jsp?arnumber=6345611.

Taylor, M., Pudney, P., Zito, R., Holyoak, N., Albrecht, A., & Raicu, R. (2009). Planning for Electric Vehicles in Australia—Can We Match Environmental Requirements, Technology and Travel Demand? *Proceedings of the 32nd Australasian Transport Research Forum, ATRF, 29 September-1, Auckland, New Zealand.* Retrieved October 2009, from http://ura.unisa.edu.au/

view/action/singleViewer.do?dvs=1414082933836 ~ 647&locale=en_GB&VIEWER_URL=/view/action/singleViewer.do?&DELIVERY_RULE_ID=10&adjacency=N&application=DIGITOOL-3&frameId=1&usePid1=true&usePid2=true.

Vetter, J., Noval, P., Wagner, M. E., Veit, C., Möller, K. C., Besenhard, J. O., et al. (2005). Ageing mechanisms in lithium-ion batteries. *Journal of Power Sources, 147*(1–2), 269–281.

www.cars21.com. (2010). *How to extend a Lithium EV battery life*, Retrieved August 14, 2014, from http://www.cars21.com/news/view/349.

Strategies to Increase the Profitability of Electric Vehicles in Urban Freight Transport

Tessa T. Taefi, Jochen Kreutzfeldt, Tobias Held and Andreas Fink

Abstract Electric vehicles (EVs) address the challenges global megatrends impose on freight transporting companies in urban areas. EVs decouple transport costs from depleting oil reserves and are free of tailpipe emissions. They are, technically, suitable for urban transport tasks which are often characterized by short, pre-planned tours and enable battery charging—or changing—at the depot. Despite these promising potentials, electric urban freight transport is still a niche market. The literature suggests the main obstacle for mass usage is the high purchase price, since profitability is considered the most important factor by nearly all companies. A descriptive statistical analysis of urban freight initiatives deploying EVs in the European North Sea Region identifies two current trends, and clusters profitability concepts of good practice examples in Europe. The study suggests that one trend is to deploy slow and light electric vehicles such as electric cargo bikes, scooters or heavy quadricycles, often combined with micro-consolidation hubs. In the second trend, medium heavy electric trucks substitute conventional vehicles in last mile logistics. Here, concepts that fully exploit the strengths of EVs to increase their productivity reach profitable operations. These include: (i) reducing the capital investment for EVs, (ii) increasing the kilometre range to benefit from low operational costs, (iii) capitalizing on the vehicles' sustainable image and (iv) exploiting of new business opportunities. The findings have implications for policy makers and companies, and they encourage the use of EVs in freight transport to abate freight transport-related emissions.

Keywords Urban freight · Electric vehicles · Profitability · Sustainable logistics

T.T. Taefi (✉) · J. Kreutzfeldt · T. Held
Department of Mechanical Engineering and Production Management,
Hamburg University of Applied Sciences, Berliner Tor 5, Hamburg 20099, Germany
e-mail: Tessa.Taefi@haw-hamburg.de

A. Fink
Institute of Computer Science, Faculty of Economics and Social Sciences,
Helmut-Schmidt-University, Holstenhofweg 85, Hamburg 22043, Germany

W. Leal Filho and R. Kotter (eds.), *E-Mobility in Europe*,
Green Energy and Technology, DOI 10.1007/978-3-319-13194-8_20

1 Introduction: Significance of Electric Vehicles in Urban Freight Transport

Passenger electric vehicles (EVs) are currently favourites with the media and shareholders of some companies. At the same time, commercial electric vehicles have reached, for instance in Germany, with much less media attention, a share four times higher within the licensed vehicle stock than passenger vehicles (KBA 2014). Deploying zero emission vehicles will become even more important in the future, especially in urban freight transport: The amount of inland freight ton kilometres grows (European Commission 2013), while the number of private car-owners is declining in urban centres (Kenworthy 2013). Accordingly, last mile distribution, which is carried out mainly on roads, increases (European Commission 2011). Utilizing freight EVs holds a potential for all relevant stakeholders—national and regional governments, municipalities representing the urban population, freight transporting companies, their EV drivers and customers:

- Reducing transport-related greenhouse gas emissions is of particular importance to European Member State and European Free Trade Area governments. Greenhouse gas emissions of the transport sector rose by 20 % between 1990 and 2010, while at the same time, emissions from all other sectors decreased (European Environment Agency 2012). Within the transport sector, road transport is responsible for 70 % of the greenhouse gas emissions, thus it is the main contributor by far (ibidem). This development counteracts the efforts of the EU Member States to reduce overall greenhouse gas emissions by at least 20 % by 2020, compared to the base year 1990 (European Parliament 2009). Moreover, the current Transport White Paper specifies the objective of a carbon-free city logistics in larger urban areas by 2030 (European Commission 2011). Hence, supporting electric mobility is an important measure for the Member State governments to adhere to the EU agreed and self-set climate goals.
- At a regional and local level, potential noise and air pollutant reductions of electric commercial vehicles are significant to municipalities. Especially in densely populated urban areas, fossil-fuel-based transportation-related air pollution is a serious concern for public health (Woodcock et al. 2007; World Health Organization 2013). Exemplary data from Hamburg shows that freight vehicles are emitting over 45 % of the city's traffic nitrogen dioxide (Böhm and Wahler 2012) and are the main contributors to noise (Ohm et al. 2012). While medium-sized commercial vehicles up to 12 tonnes emit around 100 times more nitrogen dioxide than a reference passenger car in Amsterdam (Verbeek et al. 2011), their electric counterparts operate locally tailpipe emissions free. Furthermore, they are more silent up to a speed of 50 km/h (Umweltbundesamt 2013), and this means nearly in the whole city area.

- Companies transporting freight in urban areas are increasingly facing the consequences of global megatrends. According to Altenkirch et al. (2011), urbanization and the demographic changes lead to a rising transport demand in e-commerce and growing home care and services. They suggest that at the same time, sustainability, another global megatrend, leads to a customer request for environmentally sound transportation. The necessity for companies to explore alternative urban transport solutions is enhanced by increasing fuel prices due to limited fossil resources, and through both the EU, national government and local public authorities tightening emission policies for vehicles and restricting the number of vehicles entering urban areas, the latter the prerogative of local public authorities.
- User acceptance of drivers and the attitude of shift managers towards freight EVs were "very positive over all", despite current limitations, such as the range and comparable high levels of investment, in a multi-agent empirical study of user needs by Ehrler and Hebes (2012). The positive feedback was complemented by customers and neighbours, who especially appreciated the reduction in noise and exhaust emissions. The city logistics study accessed user expectations and acceptance via interviews, questionnaires with shift managers, drivers, customers and neighbours in Berlin. The authors' recommendations include carrying out more quantitative studies of benefits as well as studies over a longer time span.

Raising freight EV numbers is significant for all stakeholder groups. Hence, projects at European, transnational, national and regional levels support battery electric freight vehicles.

Examples for European projects are the EGVI,[1] which researches alternative powertrains, including light and heavy commercial vehicles. The project BESTFACT[2] collects and provides the best practice examples for freight transport, with electric urban freight as one of three best practice categories. ELTIS[3] is a further website which facilitates the exchange of knowledge and contains European case studies on urban freight and city logistics, among other mobility topics. FREVUE[4] runs urban freight EV demonstration projects in major European cities. The fourth edition of the CIVITAS initiative (Citivas Plus II) improves urban mobility by integrating clean vehicles like EVs for personal, collective and freight transportation in the project 2MOVE2.[5] The ICT services developed in the

[1]EGVI "European Green Vehicles Initiative" (2014–2020) funded by the Horizon 2020 Framework Programme: www.egvi.eu/. The project followed on from the European Green Cars Initiative (2009–2013) funded by 7th Framework Programme.

[2]Bestfact (2012–2015) funded by the 7th Framework: http://www.bestfact.net/.

[3]ELTIS funded by the Intelligent Energy– Europe (IEE) programme: http://www.eltis.org.

[4]FREVUE "Freight Electric Vehicles in Urban Europe" (2013–2017) funded by the 7th Framework Programme: www.frevue.eu.

[5]2MOVE2 "New forms of sustainable urban transport and mobility" (2012–2016) funded by the CIVITAS PLUS II Programme: http://www.civitas.eu/content/2move2.

European EV pilot projects sm@rtCEM,[6] MOLECULES,[7] and ICT4EVEU[8] include services for vans or freight distribution. In the project ENCLOSE,[9] city logistics services for small and medium-size European Historic Towns are derived. The project CityLog[10] tested off-hours transportation of pre-packed mobile compartments ('BentoBox') by EVs and hybrids with partners in 6 European countries. The project SELECT[11] investigates the potential of electric mobility in the commercial transport. An electric MicroCarrier was tested in a city logistics concept in the EU project FiDEUS.[12] The EU funded project cyclelogistics[13] researches the potential to replace motorized freight transporting vehicles with cargo bikes and electrically supported cargo bikes.

Examples for transnational initiatives are the Interreg IVB projects E-Mobility NSR,[14] which integrates a freight dimension in promoting e-Mobility solutions *with a platform for knowledge exchange* (*CUFLOS*); and ENEVATE,[15] which includes research on light commercial vehicles in the European North West Region.

An overview of programmes and projects at national or regional level is provided in a project report of the "European Electro-mobility Observatory" (EEO 2013). However, the report states, it only provides a first overview; i.e. in Germany only the names of the most important national funding programmes are listed: the electric mobility pilot regions, the showcase programme and the clean energy initiative. Table 1 gives an overview of projects under the German national schemes, which fund electric freight mobility projects.

Despite the support in demonstration and research projects, and high potential for stakeholders, interested companies still often decide against deploying EVs for urban freight transport tasks. Only about one in a thousand commercial vehicles in Germany was an EV on 1 January 2014 (KBA 2014). The main obstacle for mass usage is the lack of profitability of utilizing the EVs, since profitability is considered

[6]sm@rtCEM (2012–2014) funded by the ICT Policy Support Programme: http://www.smartcem-project.eu.

[7]MOLECULES (2012–2014) funded by the ICT Policy Support Programme: http://www.molecules-project.eu.

[8]ICT4EVEU (2012–2014) funded by the ICT Policy Support Programme: http://www.ict4eveu.eu.

[9]ENCLOSE "ENergy efficiency in City LOgistics Services" (2012–2014) funded by the Intelligent Energy– Europe (IEE) programme: http://www.enclose.eu/content.php.

[10]CITYLOG (2010–2012) funded by 7th Framework Programme: http://www.city-log.eu/.

[11]SELECT (2013–2015) funded by 7th Framework Programme and national funding: http://www.select-project.eu/select.

[12]FiDEUS "Freight innovative Delivery of goods in European Urban Spaces" (2005–2008) funded by the 6th Framework Programme: http://www.2020-horizon.com/FIDEUS-FREIGHT-INNOVATIVE-DELIVERY-IN-EUROPEAN-URBAN-SPACE%28FIDEUS%29-s30333.html.

[13]Cyclelogistics "Moving Europe foreward" (2011–2014) funded by the Intelligent Energy Europe Programme: http://www.cyclelogistics.eu.

[14]E-Mobility NSR "North Sea Region Electric Mobility Network" (2011–2014) funded by the Interreg IVB North Sea Region Programme: www.e-mobility-nsr.eu.

[15]ENEVATE "European Network of Electric Vehicles and Transferring Expertise" (2010–2012) funded by the Interreg IVB North West Europe Programme: http://www.enevate.eu/.

Table 1 German national projects supporting battery electric freight transport

Region	Project name	Project runtime
Electric mobility pilot regions[a]		
Berlin/Potsdam	BeMobility 2.0	2012–2013
	E-City Logistik	2010–2011
Bremen/Oldenburg	UI ELMO	2012–2015
	PMC-Modul 3	2010–2011
Hamburg	hh = pure	2010–2011
	hh = wise	2010–2011
	Wirtschaft am Strom	2012–2015
Rhein-Main	E-LIFT	2012–2015
	EMIO	2012–2015
	Flottenversuch elektrisch betriebene Nutzfahrzeuge	2010–2011
Rhein-Ruhr	colognE-mobil	2009–2011
	colognE-mobil II	2012–2015
	E-Aix. Teilprojekte: Nutzfahrzeuge	2009–2011
	ELMO	2012–2015
	E-mobil NRW	2009–2011
Stuttgart	EleNa	2010–2011
	Elena II	2012–2013
	IKONE	2010–2011
Electric mobility showcase programme[b]		
Baden-Württemberg	e-fleet	2012–2016
	Get eReady	2013–2015
	Landesfuhrpark	2012–2015
Berlin-Brandenburg	DisLog	2013–2016
	KV-E-CHAIN	2013–2016
	NANU!	2013–2015
Bayern-Sachsen	Elektromobilität in Bereichen der Abfallwirtschaft der Landeshauptstadt Dresden	2013–2015
	E-Lieferungen im Allgäu	2012–2014
Others programmes		
Clean Energy Initiative	Ich ersetze ein Auto[c]	2012–2014
Erneuerbar Mobil	CO_2-neutrale Zustellung in Bonn[d]	2012–2016
Forschungsprogramme Stadtverkehr	komDRIVE[e]	2013–2015
	WIV-RAD[f]	2013–2015

[a]http://www.now-gmbh.de/de/projektfinder.html
[b]http://www.schaufenster-elektromobilitaet.org/programm/das-schaufensterprogramm-im-ueberblick
[c]http://www.ich-ersetze-ein-auto.de/
[d]http://www.erneuerbar-mobil.de/de/projekte/foerderung-von-vorhaben-im-bereich-der-elektromobilitaet-ab-2012/markteinfuehrung-mit-oekologischen-standards/erprobung-von-e-mobilitaet-im-flottenbetrieb-co2-freie-zustellung-in-bonn
[e]http://www.komdrive.de/
[f]http://www.dlr.de/vf/desktopdefault.aspx/tabid-2974/1445_read-39657/

the most important factor by companies (Fraunhofer IAO 2011). Since the capital investment for EVs is double to triple the price of conventional diesel vehicles (Taefi et al. 2015), but exemplary operational costs are about 50 % lower (Pommerenke 2014), a total costs of ownership calculation (TCO) is necessary to compare the costs. Hacker et al. (2011) have modelled a scheme to calculate the TCO for small passenger EVs and estimated the market potential of electric commercial vehicles, but they do not discuss the freight dimension. Element Energy (2012) outlines the total cost of ownership of low and ultra-low emission plug-in vans (fully electric, hybrid, hydrogen under 3.5 tonnes gross weight). Lee et al. (2013) compare the TCO of medium duty 7.49 tonnes electric and diesel trucks for different drive cycles and under various conditions. They exclude subsidies and regulatory effects and conclude that the TCO of the EV can be lower than that of a diesel truck, especially when EVs reach a high average daily mileage on an urban drive cycle with frequent stops and a low average speed. Feng and Figliozzi (2013) developed a deterministic TCO model and calculated break-even points for different scenarios for electric commercial vehicles. Davis and Figliozzi (2013) developed the model further and integrated internal and external factors, such as speed profiles, energy consumption, routing constrains and TCO into a competitiveness model for electrical trucks. Some web-based TCO calculators differentiate between private customers and companies.[16] Altenkirch et al. (2011) researched improved vehicle design and logistic processes for future strategies and services for electric commercial transportation. Suggestions in their report are based on possible technical innovations for future EVs, e.g. redesigned cargo compartments, but are not based on the vehicles available today. Kley et al. (2011) outline a methodology to derive business models for selling EVs or electric mobility services. The instrument can be utilized at a strategic level, but has no implications on the daily operation of EVs at freight transporting companies.

Some studies take a broader view of urban logistics and suggest further types of electric or electrically supported vehicles for road freight transport tasks: The baseline study of the EU funded cyclelogistics project calculated that 31 % of motorized professional urban goods transport in freight and service could be shifted to (electric) cargo cycles in European cities (Reiter and Wrighton 2013). Advantages of cargo cycles compared to motorized vehicles include the reduced space for kerbside loading space and overnight storage, an emission-free and very low noise operation, lower purchase and operational costs, and positive public perception (Leonardi et al. 2012). Results from a German project testing cargo cycles in courier services, in 7 of the 15 biggest German cities, show a substantial demand for transport services by electric cargo cycles (Gruber et al. 2014). The majority of interviewed car and bicycle messengers regarded the tested cargo cycles as "highly competitive for delivery tasks in their specific urban surroundings", but voiced concerns about the electric range and purchase price (ibidem). A simulation

[16]I.e. http://www.elbiler.nu/index.php/elbil-okonomi/oekonom-og-co2-beregner; http://www.gronnbil.no/calculator.

of energy, environmental and traffic impacts in the city of Porto (Portugal) showed that small electric vehicles, such as electric cargo cycles, improved traffic and environmental effects at a market penetration of 10 % and when utilized at street level (Melo et al. 2014). At the same time, the authors expressed reservations with regard to operational issues—such as the reduced travel distance and capacity to carry cargo weight and volume, which implies the use of an urban consolidation centre—and financial issues, such as higher vehicle purchase costs (ibidem). Two ex-ante studies of urban consolidation centres with electric delivery vans and tri-cycles in London and with electric quads with trailers (the Cargohopper) in Amsterdam proved successful with regard to the reduction of the total distance travelled (19 and 20 % reduction), CO_2-equivalents per parcel (at least 90 % reduction) and operational costs (Browne et al. 2011; Duin et al. 2013).

In summary, the current literature expresses the need for solutions to overcome the gap in the TCO, especially for electric vans and trucks. The available literature highlights single factors that positively influence the TCO calculation, but does not discuss a systematic approach to improve the profitability of freight EVs. Slow and light EVs, such as electric cargo cycles or electric quads with trailers are suggested —in some examples in conjunction with urban consolidation centres—as a viable alternative in certain transport cases, but their profitability remains disputed.

This leads to the research question explored in this paper: What measures are companies taking today, at an operational level, to increase the profitability of EVs in urban freight transport? To fill the perceived gap, this contribution discusses the current trends in deploying EVs for freight transport tasks. It builds on a comprehensive database and compilation report of 57 EV freight transport cases studies from the North Sea Region (E-Mobility-NSR 2013) and thus offers a transnational angle on the topic. Moreover, this article analyses successful concepts which the companies carry out to reach a profitable operation. In this way, the paper adds good practice examples of sound business models to the scarce literature. Finally, it includes guidance for policy makers, as well as highlighting gaps for further research and pilot projects. As a consequence, this paper provides insights for practitioners as well as the research community and policy makers at all levels and can serve as an inspiration to increase the share of EVs and reduce emissions in urban freight transport.

2 Methodology

This research study focuses on battery electric vehicles utilized in urban freight transport. Thus, hybrid vehicles or electric vehicles powered by fuel cells are excluded, as they do not fall under the definition of a battery electric vehicle. For better discussion of the current trends, the field of urban freight transport is divided into the following four segments:

- Commercial transport—The main business model of a freight forwarding company is to transport goods for another party. Examples are logistic providers such as mail-, courier-, express- and parcel transports or haulage.
- Transport on own-account—A company transports their own goods, within or between their own subsidiaries, with their own vehicles as a part of their value creation process. The transport is not for hire or rewarded (OECD 2002). One example is a bakery, which produces bread in a central bakery and transports it to various outlets for the product to be sold.
- Service—A company is providing a service at a customer's location which involves the transport or pick up of goods. Examples are craftsmen transporting tools and spare parts for repair, a fast food company delivering pizza, or a removal company picking up furniture.
- Municipal services—This includes fleets belonging to the municipality or private companies providing municipal services. Examples are municipal waste removal or foliage removal in public parks.

The explanatory research question demanded a qualitative, exploratory approach. A two-step research design was chosen (cf. Table 2):

Study A, see Sect. 3: Profitability concepts of 15 cases of urban freight with EVs in Germany were analysed in-depth. The cases where identified through a keyword search in the literature. The identified cases were only included if road legal EVs were tested or planned to be tested over a period longer than 1 year. This restriction was applied, since tests only provide realistic values on energy efficiency and thus profitability of EVs when deploying the vehicles for a full calendar year. The EVs' range is influenced by weather conditions: practical tests showed that at very low temperatures, the range of EVs can be halved (Taefi et al. 2015). This impacts the usability of EVs and decreases the savings from low operational costs per kilometre.

Table 2 Overview over the research process

	Study A: Profitability concepts	Study B: Trends and profitability measures
Region	Germany	European North Sea Region
# Cases	15 cases	72 cases
Research questions	1. Are freight EVs profitable?	1. Who uses freight EVs and how?
	2. How so, or why not?	2. How is EV productivity being improved?
Methodology	1. Literature review	Descriptive statistical analysis of freight EV state-of-the-art report
	2. Qualitative interviews	
Goals	Identify profitability strategy and concepts	1. Trends in freight EV usage
		2. Confirm profitability concepts and identify underlying measures
		3. Good practice portfolio

The data for each case were collected by semi-structured guideline-based expert interviews at company level in the period between May 2012 and April 2013. Questions included the company's objective on EV utilization, problems encountered, success factors, whether the EVs are profitable and how profitability is determined. Interviews were carried out face-to-face when possible or by telephone, due to a limited budget and time constraints. The negative impact of utilizing two different methodologies was reduced since the interviews were carried out by the same researcher. Furthermore, the succession and wording of the questions was kept, regardless of the interview method. Findings derived from a content analysis (Mayring 2010) were clustered in a matrix to detect a general profitability strategy and different concepts companies followed when deploying their EVs.

A deeper understanding of motivations and concepts can be gained through semi-structured interviews. However, the relatively small sample size of the resource intensive interviews is a potential weakness (Denscombe 2007). Thus, a second study (study B) verified profitability concepts and described single measures:

Study B, see Sect. 4: A descriptive statistical analysis examined a state-of-the-art database containing 57 case studies of urban road freight transport with electric vehicles.

The state-of-the-art case study was carried out within an EU funded project, *North Sea Region Electric Mobility Network*, during the year 2012. It included seven countries of the North Sea Region; Belgium, Denmark, the UK, Germany, The Netherlands, Norway and Sweden. Typical drawbacks of cross-country studies —such as different contexts and language barriers (Yin 2009)—were avoided, since the involved researchers utilized a common analytic framework and were based in the country of the respective case study (with one exception). The relatively small number of cases per country reflected the fact that utilizing EVs for urban freight transport tasks was still in its infancy. Case descriptions were detailed in the project report (E-Mobility-NSR 2013). The German companies questioned were a subset of companies in study A. However, the research focus of this set of interviews was different: Apart from case study descriptions, for each single case, the number of utilized freight EVs and enablers and barriers in deploying freight EVs were identified. Common enablers and barriers were identified and clustered (Taefi et al. 2015). The study confirmed profitability as being one of the most important barriers for companies in deploying freight EVs.

Through the descriptive statistical analysis, this paper adds a new perspective to the case studies by examining current trends in freight EV usage. Furthermore, this research confirms the suggested profitability concepts of study A. Through establishing measures increasing the profitability of freight EVs, good practice cases are identified and described. One possible limitation of the statistical analysis is the fact that the data were extracted from existing case studies; no new interviews were conducted. Since profitability measures for electric urban freight transport are newly defined in this article, the researchers conducting the interviews might have not have detected or recorded all relevant information. Future research might further enhance concepts and measures.

The four identified profitability concepts and their underlying measures serve three purposes:

1. To demonstrate how companies which consider deploying freight EVs can increase the vehicles profitability and thus reduce the perceived financial barrier.
2. To establish indicators for policy makers to support freight EVs.
3. To highlight gaps for future research projects.

3 Study A: Profitability Concepts of Freight EV Users in Germany

Even though the low quality of after-sales services was an obstacle, electric vehicles were often found, technically, suitable for urban freight transport tasks and welcomed by most stakeholders along the transport chain (Taefi et al. 2015). However, companies argue that vehicles are seldom profitable compared to diesel vehicles (ibidem). A study of 15 cases analyses the motivation of companies to deploy freight EVs and examines whether and how the EVs can be operated profitably. Table 3 gives an overview of the researched cases. An earlier publication provides detailed case descriptions (Taefi et al. 2013).

The 79 EVs in the study ranged from electric quad/trailer combinations and electric scooters, to light and heavy trucks up to 12 tonnes. The identified cases fell into three transport segments, commercial transport, transport on own-account and services. An emphasis on commercial transport was found. In Germany, the majority (62 %) of utilized EVs were vans between 3 and 3.49 tonnes, apparently due to a focus of local manufacturers on this weight class.

In nearly half the cases (7/15) the companies claimed or expected (if the test had only recently started) that usage of the EVs for freight transport is profitable. Each company factored in different variables to calculate the profitability of their EVs. Calculations ranged from a rough comparison of the higher capital investment against fuel savings in small companies to elaborate TCO calculations in larger companies, including deprecation time, residual value of the vehicle, costs for insurance, maintenance, taxes, price increases, etc. Though the methods of calculation were not comparable, in this study the relevant point is the fact that companies rated their EVs as profitable (or not) and, thus, might decide on this basis whether to invest in further EVs.

A cross-case synthesis of the case studies showed that profitability was not dependent on the geographical location, the type of vehicles used, the type of cargo or the transport segment. A correlation was found, however, between the profitability, the year the vehicles were acquired and the motivation to utilize EVs in urban freight transport. An exception is one case from 2000 which developed from an earlier city logistic project in the 1990s. In 2010 and 2011, practical freight EVs tests started in the German electric mobility pilot regions. Under this scheme, the

Table 3 Overview of profitability study cases and results

No.	City	Cargo	Year	Electric vehicles			Main purpose	Profitable	Profitability concepts			
				No.	Type	Gross weight			1	2	3	4
Commercial transport												
1	Stuttgart, etc.	Parcels	2011	7	Van	3 tonnes	Test EVs	No	x			
2	Nürnberg	Parcels	2000	2	Quad-trailer	3.5 tonnes	Cost reduction	Yes				x
3	Bonn, etc.	Parcels	2011	13	Van	3.5 tonnes	Test EVs	No	x			
4	Berlin, etc.	Parcels	2011	20	Van	3 tonnes	Test EVs	No	x			
5	Frankfurt, etc.	Parcels	2011	6	Truck	5.5 tonnes	Test EVs	No	x			
6	Karlsruhe	Parcels	2013	6	Truck	7.5 tonnes	Test concept	Yes	x			
7	Düsseldorf	Courier items	2011	1	Van	3 tonnes	Emission/Image	No	x	x	x	
8	Hamburg	Courier items	2010	1	Van	1.5 tonnes	Emission/Image	No	x		x	
			2012	6	Scooter	50 kg						
9	Berlin	Textiles	2013	2	Truck	12 tonnes	Test concept	Yes	x	x		x
Services												
10	Hamburg	Fast food	2012	6	Scooter	50 kg	Cost reduction	Yes	x	x		
Transport on-own-account												
11	Hamburg	Bakery products	2012	4	Van	3.5 tonnes	Emission/Image	Yes	x		x	
12	Hilden, etc.	Bakery products	2011	1	Van	3 tonnes	Emission/Image	No	x		x	
13	Hilden, etc.	Bakery products	2013	1	Van	3.5 tonnes	Emission/Image	Yes	x	x	x	
14	Solingen, etc.	Toiletries	2012	2	Truck	7.5 tonnes	Test EVs	No	x	x		x
15	Dortmund	Household articles	2013	1	Truck	12 tonnes	Test concept	Yes	x	x		

German Federal Government subsidized a total of 220 electric mobility projects in 8 pilot regions with a sum of 130 million Euros. In 2010 and 2011, the companies' main purpose of deploying freight EVs was to test their functionality and implementation into the daily routines. Reaching a profitable operation was not the main focus. Companies found that the vehicles were, technically, suitable for certain transport tasks, but also realized that a direct substitution of conventional vehicles did not fully exploit the strengths of EVs. Thus, in follow-up tests in 2012 and 2013, companies adapted and tested new delivery processes and concepts to capitalize on the advantages of EVs. When the main motivation of utilizing EVs was reducing emissions, process adaptations, similarly, led to a more profitable use of freight EVs in 2012/2013.

Aggregating measures that led to profitable operating of EVs revealed four main concepts:

1. Reduction of capital investment and operational costs.
2. Increase of vehicles mileage.
3. Capitalizing on green image.
4. Exploitation of new business opportunities.

The first two concepts aimed at reducing the total costs of ownership of electric vehicles: Costs of investment for an electric vehicle in the cases were twice to three times higher than the costs for a comparable diesel vehicle according to the companies. This gap grows with the weight of the vehicle, since the size of the cost-intensive batteries increases. In order to reduce the gap, companies relied on subsidies for the purchase price as offered in demonstration projects. Further costs saving measures were replacing passenger-sized cars for freight delivery with electric scooters, or converting depreciated diesel vehicles to EVs. Companies adapted their logistic processes to increase the vehicles mileage (concept two). Thus, they capitalized on the lower operational costs of the EVs to decrease the total costs of ownership of the vehicles. This was achieved by, e.g., recharging the vehicle while reloading cargo, quick charging, and deploying the EV on multi-shifts or training the driver in eco-drive strategies.

Through applying concepts three and four, the companies increased the profit generated through the EVs: Through communicating the benefits of electric mobility (concept three), some companies acquired new customers or justified a higher priced green product. The unique characteristics of the silent and locally emission free vehicle technology were exploited in concept four to generate new business opportunities. Examples are noiseless deliveries at night or deliveries in pedestrian zones limited for conventional vehicles by time windows.

The overall profitability strategy was, thus, to increase the productivity of the freight EVs by reducing TCO and increasing the profit.

4 Study B: Trends in Electric Urban Freight Transport in the North Sea Region

57 freight EVs case studies, in seven countries of the North Sea Region, were examined by a descriptive statistical analysis. The case studies were carried out in 2012; case study countries included Belgium, Denmark, the UK, Germany, the Netherlands, Norway and Sweden. This study analyses the data further to (i) detect common trends within the countries of the North Sea Region and (ii) to validate and enhance the findings of the study on profitability concepts and measures in Germany with a broader regional focus and more cases.

The study found a total of 57 cases comprising 5,239 electric vehicles. Case descriptions are detailed in the project report (E-Mobility-NSR 2013). A summary of the number of vehicles in the cases is depicted in Table 4. In order to identify trends, the data is analysed regarding vehicle size and transport segments.

4.1 Vehicle Sizes

In the above cases, two groups of EVs were most often recorded in urban freight transport: (i) slow and light vehicles; (ii) medium heavy trucks of 3–7.5 tonnes.

(i) Slow and light vehicles, including electric scooters, electric cargo cycles and heavy electric quadricycles, were deployed most often. Electric scooters were equipped with a carriage box to transport small freight up to 50 kg. Some scooters had a changeable battery to enhance their range. Swedish cases were dominated by a single company which deploys a large fleet of 4,500 small and light vehicles in commercial transport. In Germany, these vehicles were utilized in fast food delivery. Heavy electric quads had top speeds between 40 and 65 km/h with a payload ranging from 200 to 1,000 kg. They were utilized as an economic alternative to electric vans on slow and short routes. Examples include mail distribution or municipal services such as foliage removal in parks. Electric bicycles or tricycles had a payload of up to 300 kg. The identified models could electrically assist the driver for up to 100 km. Those electric vehicles were most often deployed in commercial inner city transport, such as courier services and parcel delivery. Slow and light vehicles proved advantageous in heavy traffic since they were often allowed on bicycle lanes, and also could be parked easily. Furthermore, electric cargo cycles could enter pedestrian zones anytime, while motorized delivery vehicles were restricted by time windows. To compensate for the lower cargo carrying capacity, slow, light vehicles were often combined with micro-consolidation hubs, enabling the driver to reloaded freight and continue delivery.

Table 4 Number of cases and vehicles identified in E-Mobility NSR state-of-the-art study 2012

		Total		Thereof: Vans			Thereof: Vans		Thereof: Slow and light vehicles			
		No. of Cases	No. of Electric vehicles	>12 tonnes	7.5–12 tonnes	3.5–7.5 tonnes	3–3.5 tonnes	<2.3 tonnes	Quads/ trailer	Heavy quad	Scooter	Cargo cycle
Total		57	5,239	3	8	158	140	70	2	995	1908	1910
Country	Belgium	11	134	3	1	2	46	25	0	48	0	9
	Denmark	10	19	0	1	8	6	4	0	0	0	0
	UK	11	133	0	0	111	0	6	0	5	0	11
	Germany	9	128	0	2	18	87	9	0	0	12	0
	Netherlands	11	26	0	4	18	1	0	2	1	0	0
	Norway	2	297	0	0	0	0	26	0	40	231	0
	Sweden	3	4,502	0	0	1	0	0	0	901	1,665	1,890
Segment	Commercial	27	4,958	1	6	34	86	35	2	943	1,902	1,904
	Own-account	9	22	0	1	2	8	6	0	5	0	0
	Service	12	178	0	1	95	35	26	0	9	6	6
	Municipal	9	81	2	0	27	11	3	0	38	0	0

(ii) Medium heavy trucks in the range between 3 and 7.5 tonnes substituted conventional trucks in all transport segments; commercial transport, transport on own-account, services and in municipal fleets and services. They constituted the majority of utilized EVs in Denmark, Germany, the UK and the Netherlands. These vehicles were all converted from mass produced diesel models.
(iii) The study found no vans between 2.3 and 3 tonnes and few light vans below 2.3 tonnes. The latter were utilized especially in services, courier and mail distribution, where lighter and fewer items were carried. Heavy trucks, above 7.5 tonnes, were even more rarely recorded. Heavy trucks were expensive compared to their diesel counterparts, due to their large, cost-intense batteries. A loss of payload because of the heavy batteries was found an additional barrier.
(iv) Different strategies of testing and utilizing freight EVs were found within the participating countries. The existence of local manufacturers or conversion companies proved an important bottleneck. As an example, in the UK a large number of heavy electric trucks from two British manufactures were used, while in Germany manufacturers and companies focused on light electric freight vans.

4.2 Transport Segments

Nearly half the cases, and the majority of vehicles, fell into the segment of commercial freight transport. An explanation is that EV characteristics were especially suitable in mail and parcel delivery. Here, EVs were often deployed on pre-planned routes and recharged at the companies' depots. The short urban delivery routes included many stops, which enabled energy recuperation through regenerative breaking. Furthermore, mail and parcel delivery companies often utilized large fleets of the same vehicle type. Thus, the companies were interested in testing EVs, which offered independence from fossil fuels and exemption from potential future emission charges. Cases were relatively evenly distributed between municipal services, transport on own-account and services. Municipal fleets were easily accessible to administrations which strove for environmentally friendly transport, and also served as role models in communicating the advantages of EVs.

4.3 Profitability Concepts

An analysis of the 57 state-of-the-art cases resulted in verification of the identified concepts and additions to the underlying measures. An overview of all identified measures is provided under the respective concepts in Table 5.

Table 5 Concepts and measures pursued to operate freight EVs profitably in the North Sea Region

Measures	Cases (n = 57)	
	Count	Share (%)
1. Reduction of capital investment and operational costs		
• Customizing freight EVs	5	9
• Purchase of discounted or second hand models	2	4
• Profit from purchase subsidies or EV project subsidies	32	56
• Use of slow and light EVs (scooters, cargo cycles, quads)	18	32
• Exemption from city toll	9	16
• Limit changes of business processes and daily routines	1	2
2. Increase of vehicles mileage		
• Intermediate charging, quick charging	10	18
• Battery swap	1	2
• Energy efficiency training of drivers (eco-driving)	5	9
• Seven days a week or multi-shift delivery	2	4
• Improvement of routing and scheduling of EVs (suggested or planned)	3	5
• External energy for heating, cooling, waste compacting	9	16
• Solar roof on EV to charge during operation	3	5
3. Capitalizing on green image		
• Enhancement of customers base with 'green' customers	13	23
• Environmental labels or awards	5	9
• Offering green products	18	32
• Marketing /Image building	35	61
4. Exploitation of new business opportunities		
• Night time delivery with noiseless EVs	2	4
• Access to zones with spatially or temporally limited access	9	16
• Freight bundling	10	18

Most companies relied on their environmentally friendly EVs for image building and subsidies when acquiring electric vehicles. However, the following good practice examples show that companies did apply measures other than just relying on financial subsidy of the purchase price. By exploiting the strengths of the EVs they increased their productivity and, thus, reached a profitable operation of their freight EVs.

4.4 Good Practice Examples

All of the following good practice examples feature a different combination of profitability measures (marked in italic) leading to a profitable operation.

(i) Inner City Last Mile Delivery
Cost reduction: customized EV; slow, light EV; subsidies.
Mileage increase: solar roof.
Image: new green customers; winning awards; green product; marketing.
New business opportunities: access to pedestrian zones; freight bundling.
A Dutch company combines a city logistic approach with a custom-made slow electric quad-trailer vehicle carrying boxes. The vehicle delivers pre-consolidated parcels and goods to shops in inner city areas and takes back recycling materials and returns on the way out. The interchangeable bodies are transported from the consolidation centre to the delivery area and loaded on the vehicle by forklift. The emission free vehicle is permitted to enter pedestrian zones outside of time windows and thus has an advantage in delivery times and costs. The existing customer base was enhanced by customers interested in green delivery. Solar panels, mounted on the roof of the vehicle, recharge the batteries via solar energy and extend the range during operation. Since the vehicle is very slow, young drivers are allowed to drive it, adding an educational perspective.

(ii) Courier Service
Cost reduction: small, light vehicles; exemption from city toll.
Mileage increase: intermediate charging.
Image: new green customers; winning awards; green product; marketing.
New business opportunities: freight bundling.
A London-based courier service, specializing in green urban freight delivery, utilizes electric cargo tricycles and, for bigger or heavier loads, electric quads. The cargo cycles are advantageous in congested traffic and are easy to park; the quads are exempt from city toll and road tax. The drivers take up cargo from micro-consolidation centres and change vehicles, once the battery is empty. Since the company operates an all-green fleet, the environmentally friendly delivery and gained sustainability awards are supporting the communication strategy.

(iii) Service
Cost reduction: exemption from city toll.
Mileage increase: intermediate charging; eco-driving strategies; multi-shift delivery.
Image: winning awards; marketing.
A company in London delivers groceries, ordered online, with electric trucks. In order to extend the range and benefit from the lower operational costs, the vehicles' batteries are partly recharged when taking up new freight, twice a day. To further increase the range, drivers are trained for eco-driving strategies, including regenerative breaking.

(iv) Haulage
Cost reduction: customized EVs; project subsidies.
Mileage increase: battery change, multi-shift delivery.
Image: Marketing.

New business opportunities: Night delivery.
A Berlin-based haulage company is testing a battery change concept for their heavy electric trucks in textile logistics. This will allow the EVs to be deployed on triple-shifts and thus amortize through low operational costs. Though night deliveries are not prohibited for conventional vehicles, a higher acceptance of night deliveries with EVs is expected.

(v) Parcel Delivery
Cost reduction: Customized vehicles; project subsidies; limited changes to daily routines.
A parcel delivery company utilizes a large fleet of similar 7.49 tonnes diesel vehicles for urban distribution. The combustion engines and powertrains of the trucks reach the end of their life after around 500,000 km, while the chassis is still sound. A conversion company fits a new electric motor, powertrain and batteries, at a cost similar to the purchase price of a new conventional diesel vehicle. Thus EVs having lower operational costs are profitable. By refurbishing old vehicles the company also reduces waste. An additional advantage is that the drivers are familiar with the vehicles. Operations stay the same (except charging/fuelling), which leads to reduced costs for maintenance and repair, compared to EVs of other manufacturers.

5 Conclusion

Charged by green energy, electric vehicles reduce greenhouse gas emissions as well as local air pollutants and noise. Many EVs in the case studies proved, technically, suitable for urban freight transportation tasks. So far, renowned car manufacturers offer only small electric vans below 2.3 tonnes. Despite the availability of those vehicles, the EVs most often deployed in urban freight transport were small and light vehicles, i.e. electric cargo bikes, electric quads and scooters; followed by converted medium heavy electric trucks between 3 and 7.49 tonnes.

Beyond reducing noise and emissions, small, light, electric vehicles reduced land use. For freight transporting companies in the case study they were advantageous in dense inner city traffic with limited parking spaces and when accessing pedestrian areas. In several cases deliveries were combined with micro-consolidation hubs close to the delivery area. These hubs enabled reloading of cargo to compensate for the limited mileage of the slow vehicles. In this way, the delivery efficiency was increased through the consolidation of goods. Another typical application was the delivery of mail and parcels.

Medium heavy electric trucks were mainly provided by local companies which convert existing diesel models. A common problem though, was the low quality of after-sales services. In the majority of cases medium heavy vehicles were used for commercial transport tasks. Here, transport was often characterized by pre-planned, reoccurring, short, urban tours with many stops, the possibility to recharge at the depot and large fleets of similar vehicles. These conditions were favourable for the current technology of freight EVs. Comparable cases were also found in service or transport on own-account under similar conditions. EVs in municipal fleets were an exception: they served as role models and were easily accessible for municipalities that wished to foster environmentally friendly transport.

As a general rule, the larger electric trucks become, the more heavy and expensive batteries are needed for propulsion. Thus, implementing profitability measures becomes increasingly important the heavier EVs get. A direct substitution of conventional commercial vehicles with EVs did not fully exploit their strengths, hence does not often lead to a profitable operation. One important reason was that the purchase price of EVs is two to three times higher, but their operational costs are about 50 % lower than of diesel vehicles. In financially successful cases, measures to increase the turnover generated with the EV were pursued, while at the same time the total costs of ownership were reduced. In this way the rentability of the commercial electric vehicles was increased.

Under the precondition that EVs were deployed on technically suitable urban tours, the vehicles were profitable in the perception of the companies, when the EVs strengths were exploited to increase their productivity. The profitability concepts included:

i. Reducing the capital investment for the EVs, for instance by saving costs for a city toll, customizing vehicles into freight EVs or, most importantly, benefiting from EV subsidies.
ii. Increasing the kilometre range to benefit from low operational costs. The most common measures to increase the range were to recharge the battery in-between tours and to utilize external energy for heating or cooling.
iii. Capitalizing on the vehicles' sustainable image by including the EVs in the company's marketing communication to gain new customers.
iv. Exploiting of new business opportunities, such as night time delivery with noiseless EVs or access to zones with spatially or temporally limited access for conventional vehicles.

5.1 Implications for Policy Makers

For policy makers who want to raise the share of environmentally friendly EVs, the following recommendations are proposed:

- Promote the usage of small, light EVs like electric cargo cycles, scooters or quads. If additional funds are available, infrastructure, such as bicycle lanes—wide enough for the fast and bulky cargo cycles—can be supported.
- Encourage local conversion companies to gain expertise with freight EVs and foster purchase of those EVs in R&D pilot projects. This ensures a supply of suitable vehicles, which are not yet provided by renowned manufacturers. At the

same time, reliable and fast after-sales and service structures are a key to ensure that delivery companies gain confidence in EVs. These structures need to be in place even if the supplying conversion company has to file for administration.
- Communicate that, through low operational costs, EVs can amortize. Eco-driving training and intermediate (quick) charging enhance the daily range and facilitate the amortization of the vehicles. Beyond this, regulations which encourage multi-shift delivery would enhance the profitability of EVs.
- Regulatory advantages, like the right to enter pedestrian zones throughout the day (in combination with a city logistics approach) or penalization of conventional vehicles, i.e. through a city toll, have proven to offer substantial advantages to companies deploying EVs.
- Finally, financial advantages for heavy freight EVs should be considered. These are currently applied at different levels in the various countries of the North Sea Region. They include subsidies of the purchase price, free recharging at public chargers, exemptions from parking fees or the city toll, or compensating for the disadvantages of heavy batteries by allowing drivers with a class B license to drive EVs up to 7.5 tonnes.

5.2 Limitations

The findings of this study have several limitations. The researched area covers the European North Sea Region. Thus, in other regions, different measures to reach profitable operations might be feasible. The paper focuses on a period between January 2010 and May 2013. Already, within this period, a change in motivation was observed, shifting from just testing vehicles to adapting logistics processes to benefit from the EVs' strength (compare Sect. 3 of this paper). Since electric mobility is evolving rapidly, new cases, concepts and measures might have been implemented in companies after mid-2013. These will have to be included in a future study.

5.3 Outlook and Implications for Research

Apart from the obvious need for further developments in the area of battery technology, the following further research is suggested: to identify and validate transport segments which are predestined for multi-shift delivery and to identity and rate appropriate policies supporting off-hour delivery; to understand eco-driving strategies for heavy (loaded) electric trucks; to support the implementation of slow, light EVs, such as electric cargo bikes, scooters or quads/trailer vehicles, in combination with micro-consolidation centres; to initiate a meta study, which collects further cases of urban electric freight transport, as i.e. provided by the BESTFACT project, and enhance the profitability concepts and measures suggested in Sect. 4 of this study.

Further development of electric urban freight transport holds a large potential for reducing dependency on fossil energy and enhancing the quality of life in growing urban areas. Although freight EVs might not become as popular in the media as some passenger EVs, they can become an important support for reducing freight transport-related emissions.

References

Altenkirch, C., Barth, G., Faul, F., Glatzel, G., Koch, J., Lienhop, M., Oltersdorf, K., Poser, A., Sauter-Servaes, T., Schuhmann, T., & Wiehle, M. (2011). *Konzipierung und Gestaltung elektromobiler Dienstleistungen im innerstädtischen Raum. Band 1 – Projektbericht*. HBK Braunschweig, Institut für Transportation Design, Braunschweig.

Böhm, J., & Wahler, G. (2012). *Luftreinhalteplan für Hamburg. 1. Fortschreibung*. Behörde für Stadtentwicklung und Umwelt, Hamburg.

Browne, M., Allen, J., & Leonardi, J. (2011). Evaluating the use of an urban consolidation centre and electric vehicles in central London. *IATSS Research, 35*, 1–6.

Commission, European. (2011). *White paper*. Brussels: Roadmap to a Single European Transport Area.

Davis, B., & Figliozzi, M. (2013). A methodology to evaluate the competitiveness of electric delivery trucks. *Transportation Research Part E, 49*, 8–23.

Denscombe, M. (2007). *The good research guide for small scale social research projects*. Berkshire, England: Open University Press.

Duin, J. H. R., Travasszy, L. A., & Quak, H. J. (2013). Towards E(lectric)-urban freight:first promising steps in the electric vehicle revolution. *European Transport*, Issue 54, paper # 9, ISSN 1825-3997.

EEO. (2013). *D1.1 Overview national and regional Electro-mobility pilot programs and projects*. European Electro-mobility Observatory. Retrieved June 08, 2014, from http://ev-observatory.eu/wp-content/uploads/2013/08/EEO_D1_1_Overview_national_and_regional_electromobility_programs.pdf.

Ehrler, V., & Herbes, P. (2012). Electromobility for city logistics – the solution to urban transport collapse? an analysis beyond theory. *Procedia—Social and Behavioral Sciences, 48*, 786–795.

E-Mobility-NSR. (2013). *Comparative analysis of European examples of schemes for Freight electric vehicles*. E-Mobility NSR. Retrieved June 08, 2014, from www.e-mobility-nsr.eu.

Energy, Element. (2012). *Ultra low emissions van study*. Cambridge: Final Report.

European Commission. (2013). *Eurostat freight transport statistics*. Retrieved April 10, 2014, from http://epp.eurostat.ec.europa.eu/statistics_explained/index.php/Freight_transport_statistics.

European Environment Agency. (2012). *EEA greenhouse gas—data viewer*. Retrieved April 10, 2014, from http://www.eea.europa.eu/data-and-maps/data/data-viewers/greenhouse-gases-viewer.

European Parliament. (2009). *Decision No 406/2009/EC*. Brussels.

Feng, W., & Figliozzi, M. (2013). An economic and technological analysis of the key factors affecting the competitiveness of electric commercial vehicles: a case study, from the USA market. *Transportation Research Part C, 26*, 135–145.

Fraunhofer IAO. (2011). *Schlussbericht IKONE Teilprojekt wissenschaftliche Begleitforschung*, Stuttgart.

Gruber, J., Kihm, A., & Lenz, B. (2014). A new vehicle for urban freight? An ex-ante evaluation of electric cargo bikes in courier services. *Research in Transportation Business & Management, 11*, 53–62. http://dx.doi.org/10.1016/j.rtbm.2014.03.004.

Hacker, F., Harthan, R., Hermann, H., Kasten, P., Loreck, C., Seebach, D., et al. (2011). *Betrachtung der Umweltentlastungspotenziale durch den verstärkten Einsatz von kleinen, batterieelektrischen Fahrzeugen*. Berlin: Öko-Institut e.V.

KBA. (2014). *Fahrzeugzulassungen (FZ) Bestand an Kraftfahrzeugen nach Umwelt-Merkmalen, 1. January 2014, FZ 13*. Kraftfahrt-Bundesamt, Flensburg.

Kenworthy, J. (2013). Decoupling urban car use and metropolitan GDP growth. *World Transport Policy and Practice, 19*(4), 8–23.

Kley, F., Lerch, C., & Dallinger, D. (2011). New business models for electric cars—a holistic approach. *Energy Policy, 39*, 3392–3403.

Lee, D., Thomas, V. M., & Brown, M. A. (2013). Electric urban delivery trucks: energy use. *Greenhouse Gas Emissions and Cost-Effectiveness, Environmental Science and Technology, 47*(14), 8022–8030.

Leonardi, J., Browne, M., & Allen, J. (2012). Before-after assessment of a logistics trial with clean urban freight vehicles: a case study in London. *Procedia—Social and Behavioral Sciences, 39*, 146–157.

Mayring, P. (2010). *Qualitative Inhaltsanalyse*. Grundlagen und Techniken: Belz Verlag, Weinheim and Basel.

Melo, S., Baptista, P., & Costa, A. (2014). Comparing the use of small sized electric vehicles with diesel vans on city logistics. *Procedia—Social and Behavioral Sciences, 111*, 350–359.

OECD. (2002). *Glossary for Transport Statistics*. Prepared by the Intersecretariat Working Group on Transport Statistics—Eurostat, European Conference of Ministers of Transport (ECMT), United Nations Economic Commission for Europe (UNECE). Retrived April 09, 2014, from http://stats.oecd.org/glossary/.

Ohm, D., Schüffler, M., & Thielemann, C. (2012). *Lärmaktionsplan Stufe 2 – 2012/2013*. Freie und Hansestadt Hamburg.

Pommerenke, K. (2014). *Dortmund—model region for E-mobility. Flagship Projects "ELMO – emobile urban freight services" & "metropole-E"*. Fuelling the Climate Conference 2014, Hamburg. Retrieved June 04, 2014, from http://www.haw-hamburg.de/fileadmin/user_upload/FakLS/07Forschung/FTZ-ALS/Veranstaltungen/Fuelling_the_Climate/05_Dortmund_HAW_FTC2014.pdf.

Reiter, K., & Wrighton, S. (2013). *Potential to shift goods transport from cars to bicycles in European cities*. Cyclelogistics—moving Europe forward. Retrieved June 12, 2014, from http://www.cyclelogistics.eu/docs/119/D7_1_Baseline_Cyclelogistics_final_15112012.pdf.

Taefi, T. T., Fink, A., Kreutzfeldt, J., & Held, T. (2013). On the profitability of electric vehicles in urban freight transport. In *Proceedings of the European Operations Management Association*, Dublin. Retrieved June 08, 2014, from http://www.publications.taefi.de/2013_Taefi_Proceedings_Euroma.pdf.

Taefi, T. T., Kreutzfeldt, J., Held, T., Konings, R., Kotter, R., Lilley, S., Baster, H., Green, N., Laugesen, M. S., Jacobsson, S., Borgqvist, M., & Nyquist, C. (2015). Comparative analysis of European examples of Freight electric vehicles schemes. *Proceedings of Dynamics in Logistics: Fourth International Conference*, LDIC 2014 Bremen, Germany. Berlin: Springer.

Umweltbundesamt. (2013). *Kurzfristig kaum Lärmminderung durch Elektroautos*. Dessau-Roßlau. Retrieved April 05, 2014, from http://www.umweltbundesamt.de/sites/default/files/medien/377/dokumente/position_kurzfristig_kaum_laermminderung_im_verkehr.pdf.

Verbeek, M., Lange, R., & de Bolech, M. (2011). *Actualisatie effecten van verkeersmaatregelen luchtkwaliteit voor de gemeente Amsterdam*. TNO-rapport MON-RPT-2010-03057, Delft.

Woodcock, J., Banister, D., Edwards, P., Prentice, A., & Roberts, I. (2007). Energy and transport. *The Lancet Series, 370*, 1078–1088.

World Health Organization. (2013). *IARC: outdoor air pollution a leading environmental cause of cancer deaths*. Press release #221. Retrieved June 04, 2014, from http://www.iarc.fr/en/media-centre/iarcnews/pdf/pr221_E.pdf.

Yin, R. K. (2009). *Case study research: design and methods* (4th ed.). Thousand Oaks, CA: Sage.

Erratum to: New Electric Mobility in Fleets in the Rural Area of Bremen/Oldenburg

Dirk Fornahl and Noreen Werner

Erratum to: Chapter 'New Electric Mobility in Fleets in the Rural Area of Bremen/Oldenburg' in: W. Leal Filho and R. Kotter (eds.), *E-Mobility in Europe*, Green Energy and Technology, DOI 10.1007/978-3-319-13194-8_13

The spelling of the "Noreen Wernern" name was incorrect. The correct name should be read as follows "Noreen Werner".

The online version of the original chapter can be found under
DOI 10.1007/978-3-319-13194-8_13

D. Fornahl (✉) · N. Werner
Centre for Regional and Innovation Economics (CRIE), Bremen University,
Wilhelm-Herbst-Str. 12, 28203 Bremen, Germany
e-mail: dirk.fornahl@uni-bremen.de

N. Werner
e-mail: n.werner@uni-bremen.de

W. Leal Filho and R. Kotter (eds.), *E-Mobility in Europe*,
Green Energy and Technology, DOI 10.1007/978-3-319-13194-8_21

Conclusions

The history of electric vehicles is long: a significant pre-history is recorded before the twentieth and twenty-first centuries in both the United States and Europe, as old as the petroleum-powered automotive vehicles (Mom 2012), and was outcompeted by then not necessarily on performance but based on social preferences and expectations, political choices and infrastructure, settlement and mobility pattern developments. Likewise, the latter half of the twentieth century also saw a false re-start in personal electric vehicles for a number of reasons.

In principle electric vehicles should now have a bright future, and overcome the challenges to widespread adoption (Fernandes Serra 2011). There are a range of challenges of integrated product-, process- and infrastructure development, the industrialisation of the product and the powertrain components which require innovation management and light construction, as well as infrastructure which is serviced and incorporated into spatial and urban planning, with a solid enough legal basis and appropriate business model developments (Kampker et al. 2013; Boesche et al. 2013). Electro-mobility is being advised as part of a toolbox of solutions for "smart cities" and "smart companies" (Stanek 2012), and a number cities and regions are piloting the coupling of Information and Communication Technologies and intelligent transport solutions with a range of forms of electro-mobility (http://www.smartcem-project.eu/).

There is still very considerable—and arguably intensified—investment and effort by the automotive industry to make the conventional, petroleum-based powertrains more fuel efficient and less-polluting (Liesenkotter and Schewe 2014), rather than fully focus on new alternative powertrains (Aswathanarayana 2010; Lee et al. 2007). This is in part because of the quite fundamentally revised value chain—and also employment and skills dimensions—that comes with an electrification of the powertrain (Kampker 2014b; Wallentowitz and Freialdenhoven 2011; Hans-Böckler-Stiftung et al. 2012; Waas 2012). The transition to electro-mobility requires sustained and coordinated innovation efforts at the interface of research in engineering and business (Proff 2013). For large automotive companies and their suppliers, this transition requires the pursuit of dynamic strategies to achieve competitiveness, reduced costs or product differentiation (Proff and Proff 2012). For the near future, there is still considerable uncertainty about developments,

W. Leal Filho and R. Kotter (eds.), *E-Mobility in Europe*,
Green Energy and Technology, DOI 10.1007/978-3-319-13194-8

which require decisions around business model evolutions, the development of the capacity of the necessary competences and economic (as well as environmental and social) evaluations of different options which could be pursued (Proff et al. 2014). Marketing strategies for electric vehicles are also to be considered (Rennhak 2013), with variable scenarios of market penetration of electric vehicles being projected (Schühle 2014), with some more conservative estimations based on contributing factors (Heymann et al. 2012). Evaluation studies of comparative assessments of electric and fossil fuel cars within context parameters contribute to an enhanced understanding of this (e.g. Bertram and Bongard 2014).

However, visions for a new automobile era, characterised by an electric-drive and wireless connections to become lighter, cleaner and "smarter" in movement, as well as being part of the "internet of things" linking also to mobility and social networks, whilst being recharged in a convenient and cost-effective way through three-way connections in a "smart grid" which makes increasing use of renewable energy sources, and being embedded in dynamically prized markets of electricity (Aichele and Doleski 2014), services and spaces for mobility and capacity of and within vehicles so as to optimise the management of urban and perhaps also increasingly rural mobility and energy systems have been outlined (Mitchell et al. 2010). In this sense, electric vehicles are seen by some as fundamentally (positive) 'cleandisruption' of energy and transportation (Seba 2014; Kampker 2014a; Korthauer 2013; Bauer 2013; Canzler and Knie 2011). Others' predictions and analysis, perhaps less radical in some ways but as focussed on radical transformations in other ways around the electricity industry and "smart power" connected by a smart grid and smart distribution and markets, are widely noted in industry, academia and policy-making (Fox-Penner 2014; Canzler and Knie 2013). Effects of the grid integration in countries and Europe are being studied also for potential capacity and disruption effects (Heinrichs 2014), as well as other advantages and potentially disruptive risks of up-scaled electro-mobility (De Haan and Zah 2012).

There are, in principle, still different possible routes to carbon-free vehicles, with electric vehicles and hydrogen vehicles arguably the key current contenders (the latter being currently behind in uptake and infrastructure in most countries, and certainly in Europe)—but with a potential to have hybrid electric vehicles based on hydrogen fuel cells rather than petroleum (Corbo et al. 2011).

As shown by the various chapters in this book, the development and testing of complex procedures and interfaces between different system components require the establishment of a complete information and infrastructure for electric vehicle management in relation to the electricity grid, which is tested in model projects and regions (Bucholz and Styczynski 2014, pp. 270–273). New mobility, infrastructure and energy concepts are emerging and being developed in the field of and relevant for electric mobility (Hülsmann and Fornahl 2013), and electric cars may make a major contribution in a number of ways to mobility concepts and practice (Keichel and Schwedes 2013), including through different and shared use in urban—but also rural—areas. Further research on, analysis of and policy advice for relevant industrial policy and the complex interplay of supply and demand, particularly during the wider market launch of electric mobility devices and infrastructure, will be welcome (Fornahl and Hülsmann 2015). A range of commissioned and independent research on concepts of e-mobility is available for policy-makers already (e.g. Buller and Hanselka 2013; Peters et al. 2013; Yay 2012; Acatech 2011)—and is reviewed from a range of perspectives in contributions in this book. As part two of this present book shows, there are both some convergences and differences in national but also regional and city policy field factors, choices taken and implementations pursued, which have a developmental history and dynamic (Koll 2013; Hickman and Banister 2014).

This is connected also to the field of innovation lobbying by the automotive industry to influence politically determined norms that set frameworks for commercial developments, investment and corporate operative behaviours (Langer 2013)—which also applies to the electricity/energy and information technologies industries—and where the options and limits of ecological innovation policy vis-à-vis personal vehicle transport are reviewed (e.g. Lehnert 2013). The search for and development of improved and viable business models based on and connected with electric vehicles will continue (Beeton and Meyer 2015). There is clearly much ongoing work in the field of bridging research and innovation on electric batteries (Korthauer 2013; Briec and Müller 2015), a key component in electric vehicles in terms of costs (production and purchase/leasing and warranty) which will also be influenced by the second life use, performance and environmental credentials in terms on their inputs—including the energy mix in the electricity used in their production (Ager-Wick Ellingsen et al. 2014). Future automotive technology, be it alternative range extenders in terms of different batteries or supercapacitors for electric cars or flywheel energy storage, mobility services that are more affordable, new powertrain solutions or vehicle concepts (e.g. Lienkamp 2012, 2013) will have a role to play to increase efficiency, performance and convenience whilst bringing costs down.

The range of psychological dimensions and impact on the sustained integration of electric vehicles into people's lives will also be relevant in shaping mobility, technology, design and commercial solutions relevant to e-mobility (Burgess et al. 2015). It shall be interesting to see what sociological dimensions the wider uptake of electric vehicles may have and result in (Meyer 2013). This is related also not just to technology acceptance but also different types of business models and access strategies to electric vehicles (purchase, lease, and car-sharing) (Fazel 2014), some of which are reflected in the chapters presented here.

It is hope that readers have obtained an overall view of some of the issues and challenges concerned with e-mobility in Europe today, including the integration of electric vehicles in the urban built environment. As to the future, organisational models of making such as carsharing (Barthel 2012), or the connections between electro-mobility and housing provision and markets (Clausnitzer et al. 2013) will become more important, and further investigations in these areas will be needed. The field of commercial transportation will have to receive sustained attention with regard to electro-mobility.

References

Acatech - Deutsche Akademie der Technikwissenschaften. (2011). *Wie Deutschland zum Leitanbieter für Elektromobilität werden kann*. Berlin: Springer

Aichele, C., & Doleski, O. D. (Eds.). (2014). *Smart Market: Vom Smart Grid zum intelligenten Energiemarkt*. Wiesbaden: Springer Vieweg.

Ager-Wick Ellingsen, L., Majeau-Bettez, G., Singh, B., Akhilesh, K. S., Valøen, L. O., & Strømman, A. H. (2014). Life cycle assessment of a lithium-ion battery vehicle pack. *Journal of Industrial Ecology, 18*(1), 113–124.

Aswathanarayana, U. (2010). Transport, Chap. 15. In U. Aswathanarayama, T. Harikhnan, & K. M. Thayybi Sahini (Eds.), *Green Energy* (pp. 183–199). Technology: Economics and Policy. CRC Press.

Bauer, A. (2013). *Elektromobilität - Realität und Chancen*. Windsor Verlag.

Barthel, S. (2012). *Elektromobilität im Carsharing: Chancen und Herausforderungen für ein nachhaltiges Mobilitätsmodell aus Sicht der Carsharingorganisationen*. Stuttgart: AV Akademikerverlag.

Beeton, D., & Meyer, G. (Eds.). (2015). *Electric vehicle business models: global perspectives*. Berlin: Springer (Forthcoming).

Bertram, M., & Bongard, S. (2014). *Elektromobilität im motorisierten Individualverkehr: Grundlagen*. Einflussfaktoren und Wirtschaftlichkeitsvergleich, Wiesbaden: Springer Vieweg.

Boesche, K. V., Franz, O., Fest, C. (Ed.), Gaul, A. J. (2013). *Berliner Handbuch zur Elektromobilität*. Műnchen: C.H. Beck.

Briec, E., & Müller, B. (Eds.). (2015). *Electric vehicle batteries: moving from research towards innovation: reports of the PPP European green vehicles initiative*. Berlin: Springer (Forthcoming).

Bucholz, B. M., & Styczynski, Z. (2014). *Smart grids—fundamentals and technologies in electricity networks*. Berlin: Springer.

Buller, U., & Hanselka, H. (Eds.). (2013). *Elektromobilität: Aspekte der Fraunhofer-Systemforschung*. Műnchen: Frauenhofer Verlag.

Burgess, M., Krems, J., & Keinath, A. (Eds.). (2015). *The psychology of the electric car: experiences of drivers from electric vehicle trials*. London: Psychology Press (Forthcoming).

Canzler, W., & Knie, A. (2013). *Schlaue Netze: Wie die Energie- und Verkehrswende gelingt*. Köln: Oekom Verlag Gmbh.

Canzler, W., & Knie, A. (2011). *Einfach aufladen: Mit Elektromobilität in eine saubere Zukunft*. Köln: Oekom Verlag Gmbh.

Clausnitzer, K.-D., Gabriel, J., & Buchmann, M. (2013). *Elektromobilität und Wohnungswirtschaft. Abschlussbericht*. Stuttgart: Fraunhofer Irb.

Corbo, P., Migliardini, F., & Veneri, O. (2011). *Hydrogen fuel cells for road vehicles*. Berlin: Springer.

De Haan, P., & Zah, R. (2012). *Chancen und Risiken der Elektromobilität*. Vdf Hochschulverlag.

Hülsmann, M., & Fornahl, D. (Eds.). (2013). *Evolutionary paths towards the mobility patterns of the future*. Berlin: Springer.

Fazel, L. (2014). *Akzeptanz von Elektromobilität: Entwicklung und Validierung eines Modells unter Berücksichtigung der Nutzungsform des Carsharing*. Wiesbaden: Springer.

Fornahl, D., & Hülsmann, M. (Eds.). (2015). *Electric mobility evolution: theoretical, empirical and political aspects*. Berlin: Springer (Forthcoming).

Fernandes Serra, J. V. (2011). *Electric vehicles: technology, policy and commercial development*. London: Routledge.

Fox-Penner, P. (2014). *Smart power. climate change, the smart grid & the future of electric utilities* (Anniversary ed.). Washington: Island Press.

Hans-Böckler-Stiftung et al. (2012). *ELAB: Wirkungsanalyse alternativer Antriebskonzepte am Beispiel einer idealtypischen Antriebsstrangproduktion*. Dűsseldorf: HBS. http://www.boeckler.de/pdf/pub_ELAB_2012.pdf

Heinrichs, H. U. (2014). *Analyse der langfristigen Auswirkungen von Elektromobilität auf das deutsche Energiesystem im europäiischen Energieverbund*. Karlsruhe: KIT Scientific Publishing.

Heymann, E., Koppel, O., & Puls, T. (2012). *Evolution statt Revolution - die Zukunft der Elektromobilität*. IW-Analysen 84. Köln: Institut der Deutschen Wirtschaft.

Hickman, R., & Banister, D. (2014). *Transport, climate change and the city*. London: Routledge.

Kampker, A. (2014a). *Herausforderungen disruptiver Innovationen am Beispiel der Elektromobilität*. Apprimus Wissenschaftsverlag.

Kampker, A. (2014b). *Elektromobilproduktion*. Wiesbaden: Springer Vieweg.

Kampker, A., Vallée, D., & Schnettler, A. (Eds.). (2013). *Elektromobilität: Grundlagen einer Zukunftstechnologie*. Berlin: Springer.

Keichel, M., & Schwedes, O. (Eds.). (2013). *Das Elektroauto: Mobilität im Umbruch*. Wiesbaden: Springer Vieweg.

Koll, F. (2013). *Regionale Felder Der Elektromobilitaet: Entstehungsbedingungen und Formationsprozesse*. Frankfurt a. M.: Peter Lang.

Korthauer, R. (2013). *Handbuch Elektromobilität 2013: Grenzenlos mobil mit Batterie und Brennstoffzellen*. Stuttgart: Ew Medien und Kongresse.

Korthauer, R. (Ed.). (2013). *Handbuch Lithium-Ionen-Batterien*. Wiesbaden: Springer Vieweg.

Langer, R. (2013). *Innovationslobbying. Eine Analyse am Beispiel der Elektromobilität*. Wiesbaden: Springer VS.

Lehnert, M. N. (2013). *Möglichkeiten und Grenzen der ökologischen Innovationspolitik im PKW-Verkehr*. Frankfurt a.M.: Peter Lang.

Lee, S., Loyalka, S. K., & Speight, J. G. (2007). *Handbook of alternative fuel technologies*. Boca Rotan: CRC Press.

Lienkamp, M. (Ed.). (2013). *Conference on future automotive technology. Focus electro mobility*. Wiesbaden: Springer VS.

Lienkamp, M. (2012). *Elektromobilität: Hype oder Revolution?*. Műnchen: VDI Buch.

Liesenkotter, B., & Schewe, G. (2014). *E-Mobility: Zum Sailing-Ship-Effect in Der Automobilindustrie*. Wiesbaden: Springer Gabler.

Meyer, T. (2013). *Elektromobilität: Soziologische Perspektiven einer Automobilen (R)evolution*. Frankfurt a. M.: Peter Lang.

Mitchell, W. J., Borroni-bird, C., & Burns, L. (2010). *Reinventing the Automobile* (New ed.). Cambridge, MA: MIT Press.

Mom, G. (2012). *The electric vehicle: technology and expectations in the automobile age* (Reprint ed.). New York: John Hopkins University Press.

Peters, A., Doll, C., Plötz, P., Sauer, A., Schade, W., Thielmann Wietschel, M., & Zanker, C. (2013). *Konzepte der Elektromobilität. Ihre Bedeutung für Wirtschaft, Gesellschaft und Umwelt*. Berlin: Edition Sigma.

Proff, H., Proff, H., Fojcik, T. M., & Sandau, J. (2014). *Management des Übergangs in die Elektromobilität: Radikales Umdenken bei tiefgreifenden technologischen Veränderungen*. Wiesbaden: Springer Gabler.

Proff, H. (Ed.). (2013). *Radikale Innovationen in der Mobilität: Technische und betriebswirtschaftliche Aspekte*. Wiesbaden: Springer Gabler.

Proff, H., & Proff, H. (2012). *Dynamisches Automobilmanagement: Strategien für international tätige Automobilunternehmen im Übergang in die Elektromobilität* (2nd ed.). Wiesbaden: Gabler Verlag.

Rennhak, C. (Ed.). (2013). *Zukunftsfeld Elektromobilität: Herausforderungen und Strategien für Die Automobil- und Zulieferindustrie*. Reutlingen: Ibidem.

Salminen, J., Kallio, T., Omar, N., Van den Bossche, P., Van Mierlo, J., & Gualous, H. (2014). Transport energy—lithium-ion batteries. In: Letcher, T. M. (ed) *Future energy. improved, sustainable and clean options for our planet*. Amsterdam: Elsevier.

Schühle, F. (2014). *Die Marktdurchdringung der Elektromobilität in Deutschland: Eine Akzeptanz-und Absatzprognose*. Essen: Reiner Hampp Verlag.

Seba, T. (2014). *Clean disruption of energy and transportation*. Silicon Valley, CA: Clean Planet Ventures.

Stanek, G. (2012). *Elektromobilität 2012: Erfolgreiche Lösungen für Smart Companies und Smart Cities*. Echomedia Buchverlag.

Waas, A. (2012). *Dynamic capabilities, die Resourcesbasis und die Veränderungen in Unternehmen. Auswirkungen der Elektromobilität auf die deutschen Automobilindustrie bis ins Jahr 2020*. Berlin: Logos Verlag.

Wallentowitz, H., & Freialdenhoven, A. (2011). *Strategien zur Elektrifizierung des Antriebsstranges: Technologien, Märkte und Implikationen* (2nd ed.). Wiesbaden: Vieweg +Teubner Verlag.

Yay, M. (2012). *Elektromobilität* (2nd ed.). Frankfurt a.M.: Peter Lang.